Innovation Governance in an Open Economy

Paradoxically, in an increasingly globalized world, regional innovation clusters have moved to the forefront of attention as a strategy for economic and social development. Going beyond international success cases, like Silicon Valley and Route 128, as sources of lessons, successful high tech clusters in niche areas have had a significant impact on peripheral regions. Are these successful innovation clusters born or made? If they are subject to planning and direction, what shape does this take: top down, bottom up or lateral?

This book shows how regional economic and social development issues have become increasingly amenable to policy influence. Governance is seen as a broader phenomenon than governmental actions – the result of confluence of the collaboration of various actors, including, but not limited to, government, business and academic organizations and individuals. This phenomenon is discussed in a variety of case studies, primarily in Sweden, a country that may be seen as an "early adapter" of regional innovation policy. The introduction and conclusion infer general factors that may be applicable in similar situations in other parts of the world.

Governance of regional innovation clusters is too important to be left to government alone. Governance is best accomplished laterally through the development of consensus among key actors in knowledge-based innovation such as university, industry and government (the triple helix). The case studies and analysis of peripheral regions in this volume provide exemplary models for transference to similar situations elsewhere. This work will be invaluable for students and researchers of regional innovation policy and regional development.

Annika Rickne is Professor of Entrepreneurship and Innovation at the University of Gothenburg, Sweden.

Staffan Laestadius is Professor of Industrial Dynamics at the Swedish Royal Institute of Technology.

Henry Etzkowitz is Senior Researcher in the Triple Helix Group, H-STAR at Stanford University, USA.

Regional Studies Association

THE INTERNATIONAL FORUM FOR REGIONAL DEVELOPMENT POLICY AND RESEARCH

The Association welcomes as members all individuals and organisations who are interested in the study and understanding of regions and regionalism

Regions and Cities

Series editors: Ron Martin, University of Cambridge, UK; Gernot Grabher, University of Bonn, Germany; Maryann Feldman, University of Georgia, USA; Gillian Bristow, University of Cardiff, UK.

Regions and Cities is an international, interdisciplinary series that provides authoritative analyses of the new significance of regions and cities for economic, social and cultural development, and public policy experimentation. The series seeks to combine theoretical and empirical insights with constructive policy debate and critically engages with formative processes and policies in regional and urban studies.

1. **Beyond Green Belts**
   Managing urban growth in the 21st century
   Edited by John Herrington
2. **Retreat from the Regions**
   Corporate change and the closure of factories
   Stephen Fothergill and Nigel Guy
3. **Regional Development in the 1990s**
   The British Isles in transition
   Edited by Ron Martin and Peter Townroe
4. **Spatial Policy in a Divided Nation**
   Edited by Richard T. Harrison and Mark Hart
5. **An Enlarged Europe**
   Regions in competition?
   Edited by Louis Albrechts, Sally Hardy, Mark Hart and Anastasios Katos
6. **The Regional Imperative**
   Regional planning and governance in Britain, Europe and the United States
   Urlan A. Wannop
7. **The Determinants of Small Firm Growth**
   An inter-regional study in the United Kingdom, 1986–90
   Richard Barkham, Graham Gudgin, Mark Hart and Eric Hanvey
8. **The Regional Dimension of Transformation in Central Europe**
   Gorzelak Grzegorz
9. **Union Retreat and the Regions**
   The shrinking landscape of organised labour
   Ron Martin, Peter Sunley and Jane Wills

10. **Regional Development Strategies**
A European perspective
Edited by Jeremy Alden and Philip Boland

11. **British Regionalism and Devolution**
The challenges of state reform and European integration
Edited by Jonathan Bradbury and John Mawson

12. **Innovation Networks and Learning Regions?**
James Simmie

13. **Regional Policy in Europe**
S. S. Artobolevskiy

14. **New Institutional Spaces**
TECs and the remaking of economic governance
Edited by Martin Jones and Jamie Peck

15. **The Coherence of EU Regional Policy**
Contrasting perspectives on the structural funds
Edited by John Bachtler and Ivan Turok

16. **Multinationals and European Integration**
Trade, investment and regional development
Edited by Nicholas A. Phelps

17. **Unemployment and Social Exclusion**
Landscapes of labour inequality and social exclusion
Edited by Sally Hardy, Paul Lawless and Ron Martin

18. **Metropolitan Planning in Britain**
A comparative study
Edited by Peter Roberts, Kevin Thomas and Gwyndaf Williams

19. **Social Exclusion in European Cities**
Processes, experiences and responses
Edited by Judith Allen, Goran Cars and Ali Madanipour

20. **Regional Development Agencies in Europe**
Edited by Charlotte Damborg, Mike Danson and Henrik Halkier

21. **Community Economic Development**
Edited by Graham Haughton

22. **Foreign Direct Investment and the Global Economy**
Corporate and institutional dynamics of global-localisation
Edited by Jeremy Alden and Nicholas F. Phelps

23. **Restructuring Industry and Territory**
The experience of Europe's regions
Edited by Anna Giunta, Arnoud Lagendijk and Andy Pike

24. **Out of the Ashes?**
The social impact of industrial contraction and regeneration on Britain's mining communities
Chas Critcher, Bella Dicks, David Parry and David Waddington

25. **Regional Innovation Strategies**
The challenge for less-favoured regions
Edited by Kevin Morgan and Claire Nauwelaers

26. **Geographies of Labour Market Inequality**
Edited by Ron Martin and Philip S. Morrison

27. **Sustainable Cities**
Graham Haughton and Colin Hunter

28. **Regions, Spatial Strategies and Sustainable Development**
David Counsell and Graham Haughton

29. **Clusters and Regional Development**
Critical reflections and explorations
Edited by Asheim Bjorn, Philip Cooke and Ron Martin

30. **Regional Competitiveness**
Edited by Ron Martin, Michael Kitson and Peter Tyler

31. **Regional Development in the Knowledge Economy**
Edited by Philip Cooke and Andrea Piccaluga

32. **The Rise of the English Regions?**
Edited by Irene Hardill, Paul Benneworth, Mark Baker and Leslie Budd

33. **Geographies of the New Economy**
Critical reflections
Edited by Peter W. Daniels, Andrew Leyshon, Michael J. Bradshaw and Jonathan Beaverstock

34. **European Cohesion Policy**
Willem Molle

35. **Creative Regions**
Technology, culture and knowledge entrepreneurship
Edited by Philip Cooke and Dafna Schwartz

36. **Devolution, Regionalism and Regional Development**
The UK experience
Edited by Jonathan Bradbury

37. **Intelligent Cities and Globalisation of Innovation Networks**
Nicos Komninos

38. **Whither regional studies?**
Edited by Andy Pike

39. **Business Networks in Clusters and Industrial Districts**
The governance of the global value chain
Edited by Fiorenza Belussi and Alessia Sammarra

40. **China and Europe**
The implications of the rise of China as a global economic power for Europe
Edited by Klaus Kunzmann, Willy A Schmid and Martina Koll-Schretzenmayr

41. **Globalizing Regional Development in East Asia**
Production networks, clusters, and entrepreneurship
Edited by Henry Wai-chung Yeung

42. **Manufacturing in the New Urban Economy**
Willem van Winden, Leo van den Berg, Luis Carvalho and Erwin van Tuiji

43. **The Impacts of Automotive Plant Closures**
A tale of two cities
Edited by Andrew Beer and Holli Evans

44. **The Futures of the City Region**
Edited by Michael Neuman and Angela Hull

45. **Migration in the 21st Century: Rights, Outcomes, and Policy**
Kim Korinek and Thomas Maloney

46. **Leadership and Place**
Edited by Chris Collinge, John Gibney and Chris Mabey

47. **Beyond Territory**
Edited by Harald Bathelt, Maryann Feldman and Dieter F. Kogler

48. **The Recession and Beyond**
Local and regional responses to the downturn
Edited by David Bailey and Caroline Chapain

49. **Cultural Political Economy of Small Cities**
Edited by Anne Lorentzen and Bas van Heur

50. **Just Growth: Inclusion and Prosperity in America's Metropolitan Regions**
Chris Benner and Manuel Pastor

51. **Industrial Policy Beyond the Crisis**
Regional, national and international perspectives
Edited by David Bailey, Helena Lenihan and Josep-Maria Arauzo-Carod

52. **Promoting Silicon Valleys in Latin America**
Luciano Ciravegna

53. **Regional Development in Northern Europe**
Peripherality, marginality and border issues
Edited by Mike Danson and Peter De Souza

54. **Creating Knowledge Locations in Cities**
Innovation and integration challenges
Willem van Winden, Luis de Carvalho, Erwin van Tujil, Jeroen van Haaren and Leo van den Berg

55. **Complex Adaptive Innovation Systems**
Relatedness and transversality in the evolving region
Philip Cooke

56. **Innovation Governance in an Open Economy**
Shaping regional nodes in a globalized world
Edited by Annika Rickne, Staffan Laestadius and Henry Etzkowitz

Innovation Governance in an Open Economy

Shaping regional nodes in a globalized world

Edited by Annika Rickne, Staffan Laestadius and Henry Etzkowitz

Routledge
Taylor & Francis Group
LONDON AND NEW YORK

First published 2012 by Routledge
2 Park Square, Milton Park, Abingdon, Oxon, OX14 4RN

Simultaneously published in the USA and Canada
by Routledge
711 Third Avenue, New York, NY 10017

Routledge is an imprint of the Taylor & Francis Group, an informa business

British Library Cataloguing in Publication Data
A catalogue record for this book is available
from the British Library

Library of Congress Cataloging-in-Publication Data
Innovation governance in an open economy : shaping regional nodes in a globalized world / edited by Annika Rickne, Staffan Laestadius, and Henry Etzkowitz.
p. cm.
1. Regional planning—Sweden. 2. Sweden—Economic policy—Case studies. 3. Technological innovations—Economic aspects—Sweden. 4. Diffusion of innovations—Sweden. 5. Business and education—Sweden. 6. Organizational learning—Case studies. I. Rickne, Annika, 1966– II. Læstadius, Staffan. III. Etzkowitz, Henry, 1940-
HT395.S94I56 2012
338.9485—dc23
2011042745

ISBN: 978–0–415–50493–5 (hbk)
ISBN: 978–0–203–12130–6 (ebk)

Typeset in Times New Roman
by Florence Production Ltd, Devon

Contents

Illustrations

Figures

Tables

Contributors

The editors

Annika Rickne is Professor of Entrepreneurship and Innovation at the University of Gothenburg. Her interest is economic growth initiated by new knowledge, reshaping existing knowledge fields and industries or giving rise to the evolution of new ones. This involves issues of public policies as well as firm and university strategies and behavior, of commercialization in the form of academic entrepreneurship, new firms or the diversification of established ones, and of globalization versus the role of regional arenas.

Staffan Laestadius is Professor of Industrial Dynamics – and vice Dean – at the School of Industrial Technology and Management at Royal Institute of Technology (KTH) in Stockholm. Basically interested in the mechanisms of industrial and technical change, he has been engaged in several research projects on non-science-based innovations and on low-tech industries, on knowledge formation and on innovation systems.

Henry Etzkowitz is Faculty Research Fellow at Stanford University and Visiting Professor at Edinburgh University Business School. He is President of the Triple Helix Association and General Advisor to the International Triple Helix Institute, Madrid. He is originator of the "entrepreneurial university" and "triple helix" concepts that link university with industry and government at national and regional levels.

The authors

Martin Andersson is Professor of Innovation Studies at CIRCLE, Lund University and is also affiliated to Blekinge Institute of Technology. He is an economist by training and focuses on innovation, trade and location and their interplay, where local and international entrepreneurship is an important aspect. His research also attempts to account for non-linear relations and dynamics in models and empirical applications, in particular non-linear features in innovation and learning processes.

Bo Carlsson is Professor of Economics and former Director of the Doctor of Management Program and Associate Dean for Research and Graduate Programs at the Weatherhead School of Management at Case Western Reserve University in Cleveland, Ohio. His research focuses on Industrial Dynamics, especially the nature and role of innovation systems and entrepreneurship in economic growth, and the formation of industry clusters, especially the role of knowledge creation and knowledge flows.

Linda Gustavsson is a researcher at the Royal Institute of Technology, Stockholm. Her research interest is in industrial development, and particularly the organization of innovation and knowledge formation – from a policy as well as from a corporate management perspective. She is currently involved in an international research program on the dynamics of natural resource industries.

Börje Johansson is Professor in Economics at Jönköping International Business School and Adjunct Professor at the Royal Institute of Technology. He is director of the Centre of Excellence for Science and Innovation Studies. He specializes in transaction costs analysis, industrial dynamics and game theory, innovation and technology change, structural and location dynamics, as well as specialization and trade.

Johan Johansson is a member of the Swedish Parliament and the committee on industry and trade. He was previously Assistant Professor in entrepreneurship and industrial organization at Luleå University of Technology. His research interest involves collaborative processes and entrepreneurial orientation, especially among constellations including small- and medium-sized enterprises, universities and policy actors.

Jens Laage-Hellman is Associate Professor at Chalmers. He is also affiliated with the Institute for Management of Innovation and Technology (IMIT) and works as a consultant. Laage-Hellman's main research field is technological innovation, with special focus on networks and collaboration. Much of his research deals with life sciences, including studies of the pharmaceutical, biotechnology and medical device industries.

Fredrik Lavén is a researcher and teacher at University of Gothenburg. His main interests are organizing, innovation and strategy, and particularly the relationship between plans and action. He is also a management consultant at the research-based consultancy firm NormannPartners, focusing on scenario planning, strategy and business innovation.

Malin Lindberg is a political scientist and researcher at Luleå University of Technology. Her research interest concerns innovation policy discourses and effects, gendered aspects of innovation and policy, as well as the quadruple helix concept and its organizational features.

Cali Nuur is Associate Professor at the Royal Institute of Technology. His research interest relates to industrial dynamics and industrial development. He is currently involved in an international research program on the dynamics of natural resource industries.

Bo Persson is Associate Professor in Political Science at Linköping University. His research interests lie in the area of comparative public policy, especially research and innovation policy. The focus is on formulation and implementation of research and innovation policy in Sweden, the role of research funding agencies, and the function of ideas in policy-making.

Marina Ranga is a Research Fellow at Stanford University, Visiting Assistant Professor at Stony Brook University and Visiting Fellow at Sussex University. Her research includes national and regional innovation systems, integration of research and innovation capabilities, triple helix interactions, the evolution of the entrepreneurial university, and the gender dimension in science, technology and innovation.

Björn Remneland-Wikhamn is a researcher at the School of Business, Economics and Law at the University of Gothenburg. His main research interests are the organizing aspects of innovation, entrepreneurship and networks.

Emelie Stenborg is a PhD candidate at Research Policy Institute, Lund University. Her dissertation project investigates how chemical risks in consumer goods are communicated to and by the general audience. Her broader research interests cover aspects of organizational structures, communication and management in different types of organizations in society.

Torbjörn Stjernberg is Professor Emeritus at the School of Business, Economics and Law at the University of Gothenburg. He is interested in development processes in organizational networks, in particular the creation, distribution and protection of values created in networks. He has a long-term interest in viability and diffusion of organizational innovations, especially in innovations aimed at improvements of quality of work life.

Jukka Teräs is Senior Researcher of Economic Geography at Turku University, Finland. His interest is regional development catalyzed by new knowledge, including evolution and internationalization of non-metropolitan, research-based industrial concentrations. Teräs has been involved in research and consultancy assignments ranging from international co-operation programs to regional and company-specific issues.

Håkan Ylinenpää is Professor of Entrepreneurship and Innovation and research director for the Centre for Inter-organisational Innovation Research at Luleå University of Technology and Umeå University. His research is focused on entrepreneurial innovation activities and strategies, innovation system research, and knowledge management, and preferably involves research in close collaboration with practice and/or policy-makers in regional, national or international programs and projects.

1 Regional governance in global innovation processes

Annika Rickne, Staffan Laestadius and Henry Etzkowitz

Introduction

Regional governance has moved to the forefront of the innovation agenda in recent years in many countries, and this is a book about governance of regional innovation in a global world. Our interest is to move beyond the success cases of Silicon Valley and its like, and focus on the multitude of regions that, while having fewer resources and less of a leading role, still strive to make their way to global competitiveness. In fact, governments around the globe name their specific regions as contestants in the race to be the most influential hub in a certain sector. Even so, regional innovation governance has yet to come up with solid and all-encompassing tools for stable innovation and growth. Instead, perhaps too much effort has been placed on repeating the patterns found useful in some successful regions without coming to grips with the underlying mechanisms. The literature lacks in systematic examples of innovation governance of "aspiring" regions. Therefore, we will discuss if and how small or peripheral regions can introduce and realize innovative strategies, and what role innovation governance has in such processes.

Our aim is to discuss this theme by setting national innovation policy in relation to specific initiatives to spur regional innovation hubs, thereby understanding such initiatives and their development in the specific governance context. In particular, we want to illustrate how modern innovation governance is a balance between benefitting from regional hubs and taking advantage of globalization processes. Empirically, this is well illustrated in small, open economies dependent on other countries for knowledge generation as well as production.

With this in mind, our focus in this book is on Sweden, where a strong interest in the regional sources of innovation and growth has emerged in recent years.[1] Sweden initially began to pay attention to regional issues in the 1960s, in response to depopulation of its already sparsely populated Northern regions, where national government offered incentives for firms to locate manufacturing facilities and create jobs. By the 1990s, all types of regions were targeted, with significant authority and funding devolved from the national level to invigorate the regions, shifting "from a distribution policy to a growth enhancing one" (Nilsson, 2010: 9). According to the Regional Innovation Scoreboard, the eight Swedish regions rank high to medium high on enablers, such as tertiary education, lifelong learning, R&D expenditures and broadband access (Hollanders *et al.*, 2009).[2]

Regions were traditionally based on a natural feature such as New York harbor – the starting point for a global business and financial hub (Sassen, 2001) – or the Tennessee Valley watershed that became the setting for a hydraulic electrification and economic development scheme.[3] A region may also emerge as a cluster of traditional firms – e.g. an Italian or Danish industrial district in shoes or food products – or from new business concepts such as New York City's Silicon Alley multimedia firms (Heydebrand, 1999). An innovative region may be built around universities or other research organizations, or be initiated by political and business coalitions (as in North Carolina's Research Triangle) and academic and political collaborations (as with Sophia Antipolis). Globally, the concept of a region is undergoing a fundamental transformation from a geographical, cultural and industrial area to an innovation entity. A region is traditionally an area of imprecise dimension that does not necessarily constitute a political entity but may have political institutions of various orders, including multi-functional governments (such as the German länder and US states), special purpose districts for transportation (e.g. pollution control or business improvement), or quasi-political entities (e.g. hi-tech councils) that lack official status.

Independent of starting point or jurisdiction, the ultimate goal is self-sustaining innovation and growth dynamics. Instead of resting on a single base that is inevitably subject to "creative destruction," an innovating region, with multiple knowledge bases, has the ability to move across technological trajectories and renew itself (Etzkowitz and Klofsten, 2005). Regional development is an evolutionary process that takes place non-linearly, with different possible starting points and configurations of actors.

The conditions for hi-tech economic growth can be identified and put in place by explicit measures (Nelson, 2002). As regions formulate knowledge-based innovation strategies, the constellations of actors and their relative importance in the local political economy is transformed. With knowledge assuming increased significance as a factor of production, in both high technology and older manufacturing industries, the traditional elements of land, labor and capital reduce in importance with various consequences, including the inclusion of universities in regional growth coalitions. The enhanced involvement of the university in knowledge and technology transfer and in the formulation and implementation of strategies for regional development, in collaboration with industrial and government partners, are key elements of this strategy. The role of government, and the nature and level of its contribution, is also changing.

Indeed, our specific research focus for this book on governance of *regional* innovation and growth opens up for the debate on how the global, national and regional levels are balanced as illustrated in the Swedish case, where the government has taken a strong role in governance and where theoretical and policy debates have intertwined and pollinated one another. Most small economies – especially if export oriented – by necessity become highly globalized. This is also true for Sweden, which has for a long period maintained a "top" position as regards globalization.[4] Sweden combined this feature with an economic growth slightly above (recently, much above) EU and OECD average, and with innovation-related activities above most other countries. As regards R&D, firms located in Sweden spend (BERD) 4.5 percent of VA on research, second only to Israel (OECD, 2009). Patenting activities are high per capita; rank three after Japan and Switzerland (OECD, 2010). Although small as regards population – and thus in total GDP – Sweden is large enough to show clear regional differences: physical and cultural distances are important also in this "homogenous" country. This means that the issue of the regional in connection to the global becomes specifically pertinent. In short, Sweden is not only an advanced industrial nation with relatively high innovation input and output, but is also small and open.

We do not argue that Sweden is "typical." In fact, there is probably only one country similar to Sweden in the dimensions we analyze: Finland has a size as well as an economic and industrial structure comparable to Sweden. Most other European economies of similar size in population are smaller geographically, and those of similar geographical magnitude are larger in population and production. Although Sweden – except regarding geography – can be compared to Canada in its economic structure, comparisons with the US should be done with care. Some US states – notably Ohio, Massachusetts and, to some extent, Washington and Michigan – display more relevant associations in the economic structure. Not the least do such comparisons have some relevance because the internal US open-state economies may be looked upon as similar to what, for small European economies, is experienced as "global." Thus, Sweden is not "typical," in that its size or structure is to be found elsewhere. Instead, Sweden may serve as a case to illustrate the mechanisms involved in the puzzle of regional and global scales of innovation processes. Below, we will introduce the themes in the book and sketch the content of each of the chapters, giving an outline of the arguments of the book as a whole.

Innovation governance and the role of theoretical models

Processes to spur development and diffusion of innovations are most often in need of some form of governance. In our context, three issues are important to highlight: the meaning of governance, governing actors, and the theoretical foundation of governance in any specific country.

First, the governance concept has a wide range of meaning, not only referring to a top-down government or policy influence, but to something more nuanced. Indeed, it indicates that there is a multitude of actors that may be involved in the intricate actions and procedures that together lead to innovation and growth. These may be firms, universities, financial organizations, lobbying groups, users, individuals, or governmental bodies with different focus of interests and geographical scope. Governance should be seen as the sum of various, intentional measures that together direct an area or phenomenon in a certain direction. Measures may include governmentally induced laws and regulations, but also firms' strategic moves, customer preferences made known and thereby influencing innovative behavior, etc. It is, therefore, an empirical matter to what extent the governance of a specific area – e.g. a technology or a sector – is consciously orchestrated by some leading organizations, or if the governance can be described as distributed and uncoordinated, being the sum of a wide spectrum of actions. An example can be found in the case of fuel cell and hydrogen technology on the European level, where some subareas were orchestrated and others evolved in an uncoordinated fashion (Nygaard, 2008).

Second, as regards to governing actors, some contend that there has, during recent decades, been a move from traditional top-down or orchestrated governance modes to more distributed forms. The argument is that "hard" governance arrangements such as regulation or taxes have given way to "softer" modes such as consensus building, and various forms of public-private partnership. Authors mean that, in this way, the use of instruments have varied over time, and where governance earlier was seen as almost synonymous with government actions, today government often takes a less profound role or collaborates with private actors. The working of government bodies may instead be to facilitate network formation, to encourage voluntary collaboration modes, etc. Some writers go as far as to say that such governance – with a more withdrawn government role – is to be preferred and to be labeled as more "modern" governance. However, others take a different stance and mean that the mix of governance arrangements and the role of government as related to other public and private organizations will inevitably shift over time and institutional context, and that hybrid variants mixing hard and soft measures are common. There is even evidence pointing to the pendulum already swinging back, making top-down forms such as regulation and taxation more frequent in some areas such as sustainability (Hillman *et al.*, 2009).[5]

Third, in any specific country or region, the mode of innovation governance rests on a number of grounds: theoretical understanding and interpretations, political choices, earlier experiences, and global trends being some of them. In our context of understanding the workings of regional innovation initiatives and the role of governance, we will therefore – as a first theme of the book – dwell on the theoretical contexts and choices in the Swedish setting.

It is a fact that theory informs policy not only in a top-down mode, but involves loops and learning processes. Likewise, policy influences theory formation. Not least has that been the case with the theoretical approaches in focus for this book. Therefore, we start off our book with an exposé of theoretical concepts that have been strongly related to the governance approach taken in Sweden. Sweden is not unique in this area: in many countries, innovation and technology policy increasingly feeds on – and provides influence to – the theoretical and conceptual approaches provided by the academic community.

Accordingly, in Chapter 2, Laestadius and Rickne bring us an understanding of the underlying theoretical concepts of innovation systems and the triple helix approach, the very basis on which much of innovation policy rests today. The aim of the chapter is to especially provide a comprehensive analysis on how dynamics of innovation may be traced and how governance issues may be treated through theoretical concepts, and how policy, practice and theory interact. The Swedish benefits and challenges from leaning on these theoretical concepts in its formation of innovation policy are specifically highlighted. It is clear that modern innovation theory may be seen as a reaction to the equilibrium focus of conventional economic theory. As an alternative approach, the evolutionary view of innovation acknowledges learning processes and path dependencies, and thereby centers on variety creation, selection and retention mechanisms. Based on such assumptions, the innovation system concept has emerged – in a number of complementary variants – with strong focus on institutional determinants and knowledge flows. Moreover, the triple helix concept is another way of addressing much of the same issues, albeit emerging from different grounds.

On the same note, Chapter 3 by Etzkowitz and Ranga leads us further along this road, and expands the dynamics and the policy uses of the theoretical tripe helix concept. Universally applicable measures for regional development are difficult to construct, given divergent conditions such as emerging versus declining industrial regions, urban versus rural areas, etc. The chapter presents a policy model that captures the initiation and evolution of knowledge-based regional economic development. The model helps policy actors on both national and regional levels to conceptualize the early phases of industrial development and cluster formation within a region. In this way, the authors address the issue of how regional innovation can be governed, and argue that the triple helix model of the interaction among the university, industry and government spheres takes local conditions into account and provides a strategy for building on strengths and filling gaps. The model has been synthesized from international success cases, such as Boston and Silicon Valley, into a strategy to enhance the success of knowledge-based regional development projects.

Regional innovation processes in a globalized world

As stated above, the concept of *innovation governance* – in this book – refers to the processes where a set of different actors in various ways – together or separately – implement arrangements that aim to influence the innovation input, process and outcome (in general or for a specific area). The governance dilemma partly relates to the fact that modern innovation processes, to a large extent, are geographically distributed, where knowledge flows often are global, and activities such as research and development (R&D) and production can take place at, and be coordinated between, nodes around the world. Researchers argue that geographical distances will – and already have – become less significant in today's society, as knowledge generated in one part of the world readily may be transferred to, as well as utilized, developed and exploited, in another. At the same time, there are strong spatial dimensions in processes of knowledge generation and innovation. Therefore, in this time of globalization, the region as a sub-national concept is still – and perhaps even increasingly so – seen as one important level of analysis to understand entrepreneurship, innovation and growth processes. Once summarized as the existence of "sticky places in a slippery space" (Markusen, 1996), this is found in numerous studies. For example, Sonn and Storper (2008) showed that, in generating knowledge, inventors ever more turn to national, regional and local knowledge sources – as indicated by the citation of local patents. It may, in fact, be so that globalization presupposes regional knowledge formation and innovation processes, where strong and concentrated hubs form the backbone. Specialization on the local, regional or national scene may in itself be necessary for more far-reaching knowledge flows (Sölvell and Birkinshaw, 2000).

There are indeed a multitude of reasons why geographical proximity matters in such processes. In essence, geographical and physical proximity is beneficial for some facets of innovation due to localized learning processes (Asheim and Isaksen, 2002; Gertler, 2004). Exchange of tacit knowledge, which is difficult to express as blueprints, texts or formulas, is less pricey when face-to-face interaction is possible (Nonaka, 1994; Nonaka and Takeuchi, 1995). Transaction costs are abridged, and valuable information and knowledge can be expressed through the display of practical skills and through observation of an actor's behavior. Also, social and cultural proximity – and thereby a short cognitive distance and a common language – that can be built up within a delimited geographical space as a region or country may favor trust building (Malmberg and Maskell, 2002). Moreover, local "buzz" on specific technological solutions, market development, etc. may result in making a regional setting precious for actors within – e.g. a certain sector (Bathelt *et al.*, 2004; Storper and Venables, 2004). The co-location of related actors within a field to a specific geographical site may then result in agglomerations of human resources, specialized component suppliers and other joint resources, further enhancing its attractiveness. In fact, even small regional advantages may count when most other major differences are leveled out in the globalized economy.

This influence of regional hubs and clusters has been acknowledged widely. The literature is filled with examples of regions that have or have not been able to make an impact, and of how firms and other organizations may benefit from being located within knowledge-intensive clusters. The question we are faced with relates to what role the regional scene may play in the processes of global knowledge production and innovation.

First, we may ask to what extent the mechanisms whereby knowledge flows comes about and lead on to innovation are the same or differ depending on the spatial level (Lagendijk and Oinas, 2005; Lorenzen, 2005). Some studies have spoken about "local buzz" as key, when combined with "global pipelines," with such mechanisms as discrete types of knowledge creation (Bathelt *et al.*, 2004; Moodysson, 2008). Other studies argue that the types of knowledge mechanisms do not differ between spatial levels, but instead between the nature of innovation, where more radical innovations require a larger variety of sources and diffusion mechanisms (Trippl *et al.*, 2009). This highlights that we need a deeper understanding of the workings of knowledge production, and whether the regional level poses specific challenges.

Second, while clustering of actors and activities is an undisputed phenomenon all over the globe, it is less clear what makes such co-location lead to knowledge sharing, innovation and/or growth. The fact that activities co-locate on a specific geographical spot does not, in itself, lead to true knowledge or resource sharing or development, as has been shown over and over again. It is when actors connect and people and knowledge move around that new ideas can be spurred and contribute to dynamic processes. Thereby, we need to ask ourselves what it is that makes each specific regional innovation system function well.

Third, it is clear that some actors have the ability to successfully draw on – and contribute to – regional hot spots, and simultaneously proficiently act on the global arena. Examples are multinational companies that may be deeply connected into several regional settings, using the knowledge generated there to further their international operations (see e.g. Laage-Hellman *et al.*, 2009). An important issue is thereby to distinguish what exactly these organizations want to source from the region, and for what resources they turn to for more distant actors and events.

The challenge for governance of innovation relates to these three questions above, indicating that it is by no means enough to lend efforts to impelling co-locations of actors within various sectoral areas. Instead, full-fledged regional innovation systems imply a thorough understanding of when and how regional dimensions are of importance for innovation, and how such effects are achieved.

To come to grips with these questions, the second theme of the book concerns how – in a global and yet regionalized world with both international and localized knowledge formation – regional innovation processes are poised towards globalization. In promoting regional growth, there is a major challenge of balancing regional knowledge and resource formation processes with global ones, and the fact that markets are often highly international even for young or small companies. The three chapters in this section of the book base their analysis on detailed analysis of selected regional development processes in Sweden and Finland.

Teräs and Ylinenpää stress that to develop cluster-based regional advantage involves the challenge to form local/regional critical mass of actors and resources, global partnerships, and a more global mindset (Chapter 4). They develop their argument as based on a longitudinal analysis of two knowledge-based clusters. One is based in Sweden and the other in Finland, and what makes the comparison particularly relevant is that both are non-metropolitan regions that, despite a similar history and seemingly similar development conditions, reveal quite different development trajectories. The dynamic development, in a striking way, reveals the growing importance and effects of globalization on regional development.

Laage-Hellman, Rickne and Stenborg (Chapter 5) assume a firm perspective and ask what role regional knowledge nodes may have for individual companies, and how policy actors could approach regional cluster formation. The empirical data based on life science firms in Western Sweden shows that, while knowledge interaction sometimes takes place regionally, working with national or international partners predominates, especially for the large and well-established companies. What is interesting for the regional dynamics is their conclusion on for what purposes the firms do feel that regional partners are crucial, and what benefits they perceive the building of a critical mass of regional life science-based actors add. The discourse highlights that, in a small country such as Sweden, policy-makers may sometimes miss the mark by focusing on a too small unit of analysis – the region – where the firms in many instances would benefit more from a national competence focus.

Digging deeper into the requirements and needs of firms as related to regional knowledge nodes, Gustavsson, Nuur and Laestadius contrast the characteristics of corporate knowledge formation with the priorities of a regional development initiative (Chapter 6). Building on case data from a Swedish regional initiative on robotics and a sectorally related MNC located within the region, the authors refer to the specific Swedish situation, with a limited domestic market and a high percentage of the workforce employed in MNCs. The findings point to a paradoxical situation between globalization and regionalization as regards knowledge formation processes in industry and technology. Whereas the policy initiative reveals an approach to knowledge formation as an essentially geographically sticky process, the MNC is not necessarily dependent on geographically proximate relations and interactions. Regional interactions and networks can be of great importance in certain cases, whereas non-regional connections may be more important in others, and one aspect ought not to outshine the other.

Governance of regional innovation

A third theme in our discourse concerns governance aspects of innovation processes, and to what extent innovation and regional growth can be constructed and orchestrated by policy intervention. For the Swedish case, Andersson and Johansson (Chapter 7) give an empirical take on the issue of governance of regional innovation. Their study focuses on challenges and potentials for policy in the presence of fundamental change processes that influence the long-term evolution of regions. The perspective in the chapter implies that policy can be viewed as "management of change." The authors introduce a paradigm model for understanding the nature of fundamental change processes. This model emphasizes slowly changing regional characteristics and invariant self-organized response mechanisms. It is supported by empirical examples of the invariant character of regional development and innovation phenomena, such as long-term population growth, export dynamics and persistence in new firm formation across regions in Sweden. Policy conclusions are derived from analyses of the behavior of dynamic systems designed in concordance with the paradigm model, and examples of how policy may support new trajectories are provided.

The contribution by Persson (Chapter 8) takes its offspring in the fact that, in recent decades, the traditional role of the state in science and technology policy has been challenged. In this "innovation paradigm," the role of the state is to promote innovation, integrate or coordinate complementary policy areas and change the institutional context, instead of – in a linear manner – supporting only a limited set of actors (such as universities or research institutes performing R&D). This means that government preferably should engage in, for example, facilitating collaborations and entrepreneurship. Such a role places new expectations and competence needs on governmental agencies. The challenges include a move from a reliance on instruments such as research councils or mission-oriented agencies to other more flexible instruments. The new role also implies the need for building legitimacy among new stakeholders, and to prove competence as an interactive partner. Persson analyses how government adapts to and builds competencies and legitimacy for this new role in innovation governance. In the Swedish case, the novel responsibility was addressed by the establishment of a new governmental agency for innovation systems – VINNOVA. The new idea paradigm impacted the functioning of this agency, in particular as regards the need to maintain backing from a wide array of actors. Also, where the previous model of research councils stipulated high-quality research as an output, the mission of an innovation agency involves tangible effects on innovation and growth.

In the chapter by Lavén, Remneland-Wikhamn and Stjernberg, the workings of the Swedish Governmental Agency for Innovation Systems, VINNOVA, is scrutinized in greater detail (Chapter 9). The authors pose the question of how policies crafted on the theories of innovation systems and triple helix interaction (as discussed in Chapters 2 and 3) are interpreted into practice on the regional level. The authors compare how the policy guidelines have been translated into practical organization in two regional initiatives. The paper argues that policy scripts may both enhance and restrain regional innovation practice, and must therefore be enacted with care, taking the local interests and environment into consideration.

Addressing the question of whether regional innovation systems can be "constructed," Johansson and Ylinenpää focus on a main issue of governance (Chapter 10). The authors use a regional initiative related to the process industry in Northern Sweden to show that, as the development of a regional innovation system is rarely a linear process, it might be risky to spur the development solely through a top-down approach. Instead, the complex system develops through interaction and learning loops, and a bottom-up tactic appears more appropriate. As the endeavor to develop a regional innovation system is characterized by the importance of know-who and path dependency, orchestrated planning is not always meaningful. Therefore, policy actors may find it useful in their managerial practices to find approaches that lean on more entrepreneurially and effectuation-oriented systems.

Lindberg introduces the issue of gender into the governance equation (Chapter 11). In the literature, the concept of "doing gender" refers to the fact that gender aspects are constantly constructed trough human behavioral patterns and choices. This is true also for processes of innovation and growth, and therefore a pertinent issue for innovation policy. The issue for the chapter is how governance brings on modification in gendered practices. She sketches three types of tentative measures for gender mainstreaming, relating to impact of combined strategies, influence of government prescriptions and importance of critical examinations.

Sweden in a global context

Having focused on the experience of Sweden – a small, open and highly globalized economy where innovation governance has been both theory and government driven – we, in our concluding section of the book, set our understanding in a global context. Thus, in order to fully comprehend and make use of the Swedish experience, the international empirical context is painted with a broad brush. Much theoretical and policy understanding of the emergence and governance of regional innovation systems or clusters builds on the experience of regions that differ from the Swedish reality. Such regions may be much larger than the Swedish regions, have an internationally excellent position (such as Silicon Valley) – whereas Sweden as a country seems to have a stronger position than its regions – and/or may have come to a mature phase of development. To enrich the literature with awareness of mechanisms in smaller and less mature innovation regions, it is important to set Sweden in relation to the empirical context of mechanisms in some leading regions.

Therefore, providing an international outlook and setting the example of Sweden in perspective, we offer two chapters based on complementing empirical examples. First, the contribution by Carlsson examines the role of entrepreneurship and public policy in some emerging technology-based industrial clusters (Chapter 12). Based on observed empirical regularities in the form of path dependence, knowledge spillovers from previous or related activities, the necessity of a triggering event, and an entrepreneurial/organizational response, Carlsson proposes a few necessary but not sufficient functions of public policy in cluster formation. It is clear that the policy requirements in each phase vary a great deal both over time and between clusters, and no single policy is universally applicable – even sustained and substantial policy efforts may fail to yield the desired results. In the end, success or failure depends on the creativity and persistence of the entrepreneur – with an element of luck, as well.

A second international outlook is provided by Etzkowitz, analyzing regional innovation policy and practice in the US (Chapter 13). The path-dependent elements in US regional innovation policy include an enhanced role for universities and an ambivalent role for national government. Nevertheless, knowledge-based regional development has been shaped by the unintended consequences of federal policies, as well as measures targeted at encouraging universities to participate in innovation. Federal research funds, largely confined to support of agricultural research before the Second World War, became available for a variety of civilian and military purposes, on an ongoing basis, after the war. An assisted linear model of coordinated innovation mechanisms has been constructed on this base to translate inventions into economic activity via firm-formation.

The concluding chapter extracts crucial lessons across sectors and regions, and discusses implications for the intertwining of theory and practice. In particular, we dwell on the complexity of proximity, the size and structure of a region, path-dependent processes, and implementation of a concept-driven policy.

Empirical setting

To address these themes, several of the chapters in this book discuss and analyze the experience of regional innovation and development initiated by – or further developed with the help of – long-term, government-financed programs. Such policy initiatives have addressed both already established regional competence areas or sectors, and ones that have a substantial research-related and/or industrial potential but have not yet formed into a regional area of strength. The rhetoric of this government approach to such regional innovation processes has been widely influenced by the theoretical concepts of innovation systems and triple helix. The idea is not only that a shared vision should be formed between various actors, but that strong relations will form between them, ensuring a long-term regional commitment to the specific growth field. In fact, the importance of a "regional leadership" is high on the policy agenda, not limiting this leadership to (local/regional) policy actors, but assuming a common platform where firms, research organizations, etc. participate. The joint efforts from these regional actors should ideally lead to coordination of existing goals, competences and resources, an accumulation of new resources, new innovative opportunities, knowledge sharing and – in the end – to innovation and economic growth. Exactly how the triple helix actors ought to collaborate and the division of labor between the local/regional is not specified by the national agency, but rather leaves the issue open for regional solutions.

The national agency responsible for this policy is the Swedish Governmental Agency for Innovation Systems (VINNOVA), which is the main government "tool" for innovation policy in Sweden, aiming to "promote sustainable growth by developing effective innovation systems and funding problem-oriented research" (VINNOVA, 2006). In particular, the focus is to spur what they call "need driven" research and development, and thereby the Swedish industry's and general society's situation is what directs their investments. VINNOVA works both by financing short- and long-term projects and by supporting, for example, competence building, network formation and learning between actors.

Several of the chapters in this book discuss and analyze the experience of regional innovation and development as spurred by a specific policy program initiated by VINNOVA. This program, called "Regional Growth through Dynamic Innovation Systems" (in Swedish, the program is abbreviated as VINNVÄXT), was started in 2001. The aim was to "promote sustainable growth by developing internationally competitive research and innovation environments in specific growth fields" (VINNOVA, 2008: 1). Both already established regional competence areas or sectors and ones that had a substantial research-related and industrial potential but had not yet formed into a regional area of strength were encouraged to submit their applications. The VINNVÄXT call looked for sectors that, on a regional basis, could become internationally competitive in their respective fields within a 10-year period. The program aimed to form sustainable regional innovation systems, i.e.

> [a] system of organizations, individuals and rules and regulations within which the creation, dissemination and innovative exploitation of technology and other branches of knowledge takes place. When the interaction between the different players works well, new, valuable knowledge is generated which is quickly put to practical use. This creates the foundations for innovations and attracts investment.
>
> (VINNOVA, 2008: 2)

One point of departure behind the VINNVÄXT program is that a dynamic triple helix constellation is of essence for the regional growth process. This means that firms and other private sector actors are expected to contribute, as are public research organizations and policy actors. Behind the triple helix idea was the common view – in the literature, as well as in industrial and policy circles – that geographical closeness can lead to many positive effects if handled correctly.

> Geographical proximity between many different players creates competitive advantages in terms of co-operation, learning, access to know-how and expertise and development and business collaboration. Regions that realize this and that have a capacity for renewal can radically develop their competitive advantages. However, this also requires that companies, researchers, and the political/public sector collaborate and work actively towards realizing a joint strategic idea.
>
> (VINNOVA, 2008: 2)

The idea of VINNVÄXT is not only that a shared vision should form between actors, but also that strong relations should ensure a long-term regional commitment to the specific growth field. In fact, the importance of a "regional leadership" is high on VINNOVA's agenda, not limiting this leadership to policy actors, but assuming a common platform where firms, research organizations, etc. participate. The joint efforts from these regional actors should ideally lead to coordination of existing goals, competences and resources, an accumulation of new resources, new innovative opportunities, knowledge sharing, and – in the end – to innovation and economic growth. Exactly how the triple helix actors ought to collaborate and what organizations should be responsible for the various parts of the growth agenda is not something that the national agency specifies, but rather leaves the issue open for regional solutions.

As a basis for the selection criteria was the foundation in both a well-established research base and an industrial cluster, or potential for such a cluster. Also, established networks to key actors both nationally and internationally were highly valued. While the initiatives were to be based in a functionally delimited region, nothing in the call hindered that their area of operations could span over the entire country, aspiring to become a national node within their specific sector.[6]

The calls for participation in the VINNVÄXT program in 2003 and 2004 led to, in total, around 70 applications from regions throughout all of Sweden, and 8 of these were awarded with funding (Laestadius *et al.*, 2007). In the third round, in 2005, the criteria were different, encouraging younger initiatives and sectors to apply, not requiring the same industrial history. This call for emerging regional innovation systems attracted 86 initial applications, where some were given seed financing, and the 26 final applications resulted in 4 granted initiatives. Thus, in the end, 12 regional initiatives have been awarded between 6 and 10 million SEK yearly for 10 years each. The financing was awarded to the following 12 initiatives:

1. Biomedical Development in Western Sweden, Gothenburg region: supports long-term growth in the biomedical field.
2. Biorefinery of the Future, Örnsköldsvik/Umeå: supports development of biorefineries based on forest raw materials and energy crops.
3. Fiber Optic Valley, Hudiksvall: coordinates development and testing of products and services based on fiber optics.
4. Food Innovation at Interfaces, Skåne: supports the food industry, especially prioritizing functional food and nutritious food in schools and hospitals.
5. New Tools for Health, Östergötland: spurs individually adapted solutions for home-based care and health.
6. Peak of Tech Adventure, Åre-Östersund: supports the establishment of an internationally leading environment for research and business development in winter sports and tourism.

7. Printed Electronics Arena, Norrköping/Linköping: coordinates commercialization and exploitation of printed electronics.
8. ProcessIT Innovations, Luleå/Umeå: supports ICT-related services and products in mining, steel, paper and pulp and manufacturing industries.
9. Robot Valley, Mälardalen: supports research, development and manufacturing of robotics-related products.
10. Smart Textiles, Sjuhärad: focuses on design, development and production of next-generation textile products.
11. Triple Steelix, Bergslagen: focuses on manufacturing, processing and knowledge-based services in the steel industry.
12. Uppsala BIO, Uppsala: promotes growth of the biotechnology sector.

It is indeed a major investment from the Swedish government, involving financing of in total 760 million SEK over a rather extended time period, spanning from the first design of the program in 2001 until the completion of the program in 2013, and the measuring of effects thereafter (VINNOVA, 2009). What is particularly striking with the program is, however, not the volume of financing. In fact, one may argue that 760 million SEK, spread out over a 10-year period and to many regional actors, is not a large investment at all in an international comparison. Instead, the extraordinary features of the policy program – in a Swedish context, as well as internationally – involves its systematic and long-term approach. In truth, in addition to this funding, on the national level, a comprehensive support structure has also been provided to the initiatives, including, for example, management education and several meetings and seminars yearly to ensure learning between and within the initiatives. Moreover, an interesting and somewhat innovative way of supporting the initiatives is the interactive research projects set up on a long-term basis, and the consultancy backing offered to the regional management groups (more on this below). VINNOVA also engages in ongoing evaluation of the initiatives through, for example, collecting and analyzing effect indicators yearly, through yearly meetings with the regional management teams, as well as by the means of an internationally composed evaluation team every third year. On the regional level, all of these initiatives also have matching financing from regional policy units, firms, universities and other actors. For the year of 2009, the total co-financing amounted to approximately 88 million SEK, matching VINNOVA's contribution (VINNOVA, 2010). In addition, several initiatives have secured financing from the EU Structural Funds, gender-related research schemes, EU programs, etc.

As mentioned above, an integral part of the VINNVÄXT concept is long-term interactive research or consultancy support to the initiatives, as well as to VINNOVA. From the start of the program, VINNOVA had the idea to initiate learning processes where the initiatives' leadership teams got the opportunity to reflect on their organization, the development process and the results of their activities. The interactive researchers/consultants would provide independent analysis of the regional development initiatives, and to set this in relation to their accumulated knowledge about regional innovation process to give guidance to the initiatives' leadership teams. The exact issue focused was to be decided individually by each regional initiative, as was the competence profile of the external researcher/consultant. In reality, VINNOVA's only requirement was that each initiative did earmark resources for such externally guided analysis, evaluation and reflection, that they took a long-term perspective and engaged in longitudinal studies and evaluation exercises. The result today is an array of different issues addressed by the various initiatives. For example, the initiatives themselves, and VINNOVA, have invested in analyses of how regional actors form joint visions, how they organize and manage regional processes, how policy is interpreted by regional actors, the mechanisms of interaction among firms and with universities, the role of the region in an international context, etc. In addition, the empirical reality of the VINNVÄXT program has spurred the interest of other independent researchers, resulting in a number of interesting studies.

In essence, this book contains the outcome of a number of research-related projects mentioned above. Earlier results have been presented in, for example, Laestadius *et al.* (2007), as well as in some journal papers. It is our hope that the result from these studies will bring guidance to not only regional practitioners, but also highlight some issues relevant to policy-makers on regional, national or supra-national levels. Moreover, we aim to address some key issues in today's social science debates pertaining to governance and globalization of knowledge.

Notes

1. We are grateful to the Swedish Governmental Agency for Innovation Systems (VINNOVA) for funding of substantial parts of the book project, as well as to the Institute for Innovation and Entrepreneurship at Gothenburg University.
2. Following the National Units for Territorial Statistics in Eurostat.
3. Regions shaped by an ethnic or cultural identity (within or across national boundaries such as the Southern US states or Kurdistan across Turkey and Iraq) are beyond the scope of our discussion, except insofar as proto-national identity provides a motivation for economic development, as for example, in Catalonia and Quebec (cf. Greenfeld, 2001).
4. In the Ernst and Young 2010 Globalization Index (Ernst and Young, 2010), Sweden scores No 7 (of 60). In the KOF Index, developed by Dreher (2006), Sweden scores No 4 (of 208). In the Foreign Policy Index (Kearney, 2007), Sweden scores No 10 (of 72).
5. It may be argued that mitigating, and adapting to, climate change may necessitate much more of classical top-down regulation, thus favoring their return. Those governance problems fall outside the scope of this book, however.
6. Given the commitment and financing from regional actors, and thereby their wish to keep innovation and growth within the region, this national reach has, in practice, been more difficult to accomplish.

References

Asheim, B.T. and Isaksen, A. (2002) "Regional Innovation Systems: The Integration of Local 'Sticky' and Global 'Ubiquitous' Knowledge." *The Journal of Technology Transfer* 27 (1): 77–86.

Bathelt, H., Malmberg, A. and Maskell, P. (2004) "Clusters and Knowledge: Local Buzz, Global Pipelines and the Process of Knowledge Creation." *Progress in Human Geography* 28 (1): 31–56.

Dreher, A. (2006) "Does Globalization Affect Growth? Evidence from a New Index of Globalization." *Applied Economics* 38: 1091–1110. Recent data on http://globalization.kof.ethz.ch.

Ernst and Young (2010) *Winning in a Polycentric World: Globalization and the Changing World of Business*. Available at: www.ey.com/GL/en/Issues/Business-environment/Winning-in-a-polycentric-world-globalization-and-the-changing-world-of-business-The-Globalization-Index-2010 (checked 16/1/2012).

Etzkowitz, H. and Klofsten, M. (2005) "The Innovating Region: Towards a Theory of Knowledge Based Regional Development." *Research Management* 35 (3): 243–255.

Gertler, M. (2004) *Manufacturing Culture: The Institutional Geography of Industrial Practice*. Oxford: Oxford University Press.

Greenfeld, L. (2001) *The Spirit of Capitalism*. Cambridge, MA: Harvard University Press.

Heydebrand, W. (1999) "Multimedia Networks, Globalization and Strategies of Innovation: The Case of Silicon Alley." In H.-J. Brazcyck, G. Fuchs and H.-G. Wolf (eds), *Multimedia and Regional Economic Restructuring*. London: Routledge.

Hillman, K., Nilsson, M., Rickne, A. and Magnusson, T. (2011) "Fostering Sustainable Technologies: A Framework for Analysing the Governance of Innovation Systems." *Science and Public Policy* 38 (5): 403–415.

Hollanders, H., Tarantolo, S. and Loschky, A. (2009) *Regional Innovation Scoreboard 2009*. Maastricht: PRO INNO Europe.

Kearney, A.T. (2007) "The Globalization Index 2007." *Foreign Policy*.

Laage-Hellman, J., Okamoto, Y. and Rickne, A. (2009) "Effects of Regional Multilateral R&D Collaboration: A Case Study of a Stem Cell Project in Western Sweden." Report to GöteborgBio and RIDE. Working paper 84426-031, Chalmers University.

Laestadius, S., Nuur, C. and Ylinenpää, H. (eds) (2007) *Regional växtkraft i ewn global ekonomi*. Stockholm: Santérus Academic Press.

Lagendijk, A. and Oinas, P. (2005) "Proximity, External Relations, and Local Economic Development." In A. Lagendijk and P. Oinas (eds), *Proximity, Distance and Diversity: Issues on Economic Interaction and Local Development*. Aldershot: Ashgate, 3–22.

Lorenzen, M. (2005) "Knowledge and Geography (Introduction to Special Issue on Knowledge Geographies)." *Industry and Innovation* 12 (4): 399–407.

Malmberg, A. and Maskell, P. (2002) "The Elusive Concept of Localization Economies: Towards a Knowledge-Based Theory of Spatial Clustering." *Environment and Planning A* 34 (3): 429–449.

Markusen, A. (1996) "Sticky Places in a Slippery Space: A Typology of Industrial Districts." *Economic Geography* 72: 293–313.

Moodysson, J. (2008). "Principles and Practices of Knowledge Creation: On the Organization of 'Buzz' and 'Pipelines' in Life Science Communities." *Economic Geography* 84: 449–469.

Nelson, R. (2002) "The Problem of Market Bias in Modern Capitalist Economies." *Industrial and Corporate Change* 11 (2): 207–244.

Nilsson, J.-E. (2010) "Expert Evaluation Network Delivering Policy Analysis on the Performance of Cohesion Policy 2007–2013: Sweden, 2010." A report to the European Commission Directorate-General Regional Policy.

Nonaka, I. (1994) "A Dynamic Theory of Organizational Knowledge Creation." *Organization Science* 5 (1): 14–37.

Nonaka, I. and Takeuchi, H. (1995). *The Knowledge Creating Company*. Oxford/New York: Oxford University Press.

Nygaard, S. (2008) *Co-Evolution of Technology, Markets and Institutions: The Case of Fuel Cells and Hydrogen Technology in Europe*. Ph.D. thesis. CIRCLE, Lund University.

OECD (2009) *OECD Science, Technology and Industry Scoreboard 2009*. Paris: OECD.

OECD (2010) *OECD Science, Technology and Industry Outlook 2010*. Paris: OECD.

Sassen, S. (2001) *The Global City: New York, London and Tokyo*. Princeton, NJ: Princeton University Press.

Sölvell, O. and Birkinshaw, J. (2000) "Leading Edge Multinationals and Leading Edge Clusters." *International Studies of Management and Organization* Summer, 30 (2): 3–9.

Sonn, J. W. and Storper, M. (2008) "The Increasing Importance of Geographical Proximity in Knowledge Production: An Analysis of US Patent Citations, 1975–1997." *Environment and Planning A* 40: 1020–1039.

Storper, M. and Venables, A.J. (2004) "Buzz: Face-to-Face Contact and the Urban Economy." *Journal of Economic Geography* 4 (4): 351–370.

Trippl, M., Tödtling, F. and Lengauer, L. (2009) "Knowledge Sourcing Beyond Buzz and Pipelines: Evidence from the Vienna Software Sector." *Economic Geography* 85 (4): 443–462.

VINNOVA (2006) "In Search of Innovation Systems." VINNOVA Policy VP 2006:02, Swedish Governmental Agency for Innovation Systems, Stockholm.

VINNOVA (2008) "VINNVÄXT: A Programme to get Sweden Moving! Regional Growth through Dynamic Innovation Systems." VINNOVA Information 2008:22, Swedish Governmental Agency for Innovation Systems, Stockholm.

VINNOVA (2009) "Årsredovisning 2008." VINNOVA Information 2009:06, Swedish Governmental Agency for Innovation Systems, Stockholm.

VINNOVA (2010) "Årsredovisning 2009." VINNOVA Information 2010:04, Swedish Governmental Agency for Innovation Systems, Stockholm.

2 The theoretical foundation for Swedish innovation policy

Staffan Laestadius and Annika Rickne

Introduction

Swedish industrial and technology policy has become more concepts-driven than ever before. This has especially been the case within the Governmental Agency for Innovation Systems (VINNOVA), and in particular with the "VINNVÄXT program," a program that, in short, may be understood as a means to make policy based on the theories of innovation systems (IS), and not least on a variety of that discourse labeled triple helix (TH). As the VINNVÄXT program has been one of the core programs in Swedish industrial and technology policy during the first decade of the new millennium, it is – when analyzing this policy – important to have a reasonably good insight in the theoretical foundations for this policy.

The aim with this chapter is to contribute to an understanding of the theoretical foundations of innovation systems for Swedish innovation policy in general, and the policy behind the VINNVÄXT program in particular. The ambition is pedagogical and synthetic – to provide a review of state of the art – rather than being fundamentally critical or to develop theory itself. In extension, the aim, of course, is to sharpen the intellectual tools, to make them still better and capable to analyze – and handle – the great industrial transformations we have reasons to expect for decades ahead.

Let us, for the sake of convenience, at this stage provide a broad definition: an innovation system (IS) is, in this chapter, defined as an interrelated structure of institutional and actor-based condensations in an economic space that has specific industrial and innovative properties compared to its environment/context with which it has exchange. Commonly, the components of an IS are defined as all the various types of actors within the system that are relevant for innovation processes to come about, the networks and knowledge flows between them, the knowledge spaces, technologies and artifacts involved in the specific system, and the institutional set-up guiding the behavior of the components, as well as of the system, as such. The approach of innovation systems – be they sectorally, technologically, nationally or regionally defined – thus focuses on the fact that emergence, specialization and competitiveness of sectors/technological areas/countries/regions relate to the institutional conditions, the set-up of actors and their specific, path-dependent knowledge base and how this is shared.

The chapter is structured in the following way. We start with laying the foundation on which these system concepts rests: an evolutionary view of innovation. Indeed, we see modern innovation theory as a reaction to the often static views of economic theory, with its focus on equilibrium rather than dynamic changes, selection mechanisms, learning processes and path dependencies. In the following section, we return to the innovation system concept, presenting its basic characteristics, paying particular attention to the importance of institutions and the problems of openness and intentions in systems, as well as the identification of systems borders.

Paving the way for a deeper understanding of the policy uses of the IS approach, we then make an in depth analysis of the various IS-related concepts. This section is primarily of a typological character, with the intention to give an overview of the conceptual jungle of partly overlapping concepts. As this is particularly relevant for VINNOVA, we also, in the following section, turn to the triple helix concept.[1]

Before the conclusion we focus on two problems for which the IS approach has been criticized. First, one set of critiques says that, in spite of claiming to address evolutionary aspects of innovation, there is a risk of rather static and structural analyses. Recently, several authors have addressed this critique by developing a functional approach to the analysis. Second, critiques assert that, while the institutional set-up is used to delimit the systems analyzed, institutional changes and influences, as well as governance aspects, have more seldom been captured by IS analyses.

An evolutionary economic view of innovation

Our analysis focuses on the innovative properties of innovation systems (IS). We can, in the definition of IS above, find an implicit criticism of orthodox economic theory, with its focus on equilibrium. Innovation theory, with a cultural heritage dating back to the Austrian economist Joseph Schumpeter, centers on the breaking up from equilibrium, on the creative destruction when new products and processes substitute for old ones, and on the transformation and growth taking place as a result (Schumpeter, 1934/1962).

In innovation literature, innovations and entrepreneurship are two sides of the same coin. The entrepreneur is the actor who introduces the innovations on the market; be they science-based, technological, service-related, organizational or market-related innovations. An entrepreneur without innovations is not perceived.[2] The strong interrelation between innovation and entrepreneur is, in fact, an important part of the original theory pointing towards the action-oriented element in the innovative economy, not only the exploring (focusing on inventions and R&D).

Basically, innovation theory is related to what may be labeled *evolutionary economics*. The logic is simple: the essentials of the Schumpeterian message

100 years ago was that innovations continuously destroyed the equilibrium process and forced the actors of the economy to a behavior that ended in industrial transformation and change. In evolutionary theory, equilibrium is thus a special case, maybe an attractor on systems level, but never realized, as the actors in the system always strive to avoid "the circular flow of economic life," as Schumpeter formulated it (Schumpeter, 1934/1962: Chapter II). Competition in this world is not primarily a price competition between homogenous products among independent and fully informed actors, but a continuous struggle to use different solutions (e.g. advanced design) to create variety that attracts certain consumers. As an outcome, this creates temporary monopolies in all areas, thus providing a short period of profitability before the competitors catch up with new solutions; more or less influenced by or deviating from the original ones (Schumpeter, 1942/2000: Chapter VII).

The attempts to combine the strengths of evolutionary theory with equilibrium theory have a long tradition among economists and, in fact, anticipate the contributions of Schumpeter. Most interesting, and still relevant, are probably the serious attempts made by Alfred Marshall in the late nineteenth century (Marshall, 1890/1990; Laestadius, 1992; 1999). Marshall did not solve the problem of combining evolutionary theory with equilibrium; during the process of trying, however, he, introduced the concept of "externalities," which became the cornerstone in his analysis of learning effects, knowledge spillover and cultural influences, transmitted outside the market but over the borders between firms as "were they in the air." For Marshall, this was a core concept in his analysis of the economic and industrial dynamics in industrial districts. Although he did not use the concept "innovation," what he analyzed was, in fact, the conditions for innovative processes.

This takes us back to the IS concept. Originating in the analysis of Marshall, there is an insight that externalities is one of the core concepts for understanding the learning and knowledge-formation process characterizing dynamic systems – be they industrial districts or innovation systems of "higher" order. On all levels, actors – professional craftsmen and engineers, as well as firms and R&D units – learn and develop new knowledge more or less influenced by each other, contributing to development of the actors themselves, as well as of the whole system. This analytical approach is an essential part of understanding the processes that occur in industrial districts (and the innovation system) and in the economy. Learning, however, is not the same as innovation, as we will see below.

Knowledge formation and learning are thus core processes in the transformation of economies. Not least, the American economists Nelson and Winter (1982), who probably have written the most important book on *Evolutionary Theory and Economic Change*, have a strong focus on knowledge, skills and competences on an individual and organizational level. They were among the first scholars who focused on the role of non-codified or *tacit knowledge* – a concept originally developed by Polanyi (1967) – in the learning processes of the economy.

The evolutionary perspective of Nelson and Winter focuses on the tension or balance between the *creation of variety* (the invention process), on the one hand; on the *selection process*, on the other, and, on the third, on the *retention* created in the system by the *development of routines* to keep the system (and production) going. In particular, it is important not to ignore the selection processes, something that is often done by innovation researchers, as well as policy units. Most innovations fail, and have to fail, before, during or after their introduction to the market. Nelson and Winter argue that we here find one of the classical failures among innovation researchers: the neglect of the demand side.

Actors (networks, systems) who do not – through learning – have the capability to adapt their routines, products and processes to compete with innovations in their fields of activities (markets) have to close down or restructure their activities. There is thus a connection between learning systems and innovative ones (Lundvall, 1994), but they are not identical. Although innovations, following Schumpeter, on the one hand, may be characterized as new creative combinations of knowledge that may be recently acquired as well as known for a long time – and thus have to be (re)learned – they, on the other, must be looked upon as "break aways" from ingrained and learned paths – i.e. processes of *unlearning*. Innovations destroy existing structures and create something new in the ruins of the old. This destruction may be more or less radical; nevertheless, it represents an *act of creation* that is more than just adapting to old routines.

Marshall realized that intra-district (local) learning and knowledge transfer from one generation of craftsmen to another, under certain circumstances, could contribute to cementation of old competencies and to an incapacity to innovate new products or processes. The diligent and quick-to-learn pupil is far from always the most innovative, and the innovative violinist breaks at least some of the routines learned from the master. Schumpeter related the innovative process to "creative destruction" (Schumpeter, 1942/2000: Chapter VII). Using modern evolutionary theory, we may understand that as "creation of variety," on the one hand, and "selection," on the other. In this context, "learning" should not be identified as identical with or enough for the variety – selection – retention processes.

This view on innovation as a creative process combining learning and unlearning also challenges the conventional view – still common among policy-makers and some parts of academia – on the innovation process as fundamentally linear. Strongly simplified, *the linear model of innovation* displays an understanding of innovation activities, as did they start with basic knowledge (e.g. basic science), followed by more applied knowledge (e.g. applied science), product or service development (e.g. by engineering) and later close-to-market activities. As argued for a long time by many innovation researchers, this model – strongly connected with the US post-World War II report *Science the Endless Frontier* (Bush, 1945) – ignores the fact that innovation processes take place over and between different systems levels and domains of knowledge. Innovation is, in essence, a highly interactive and iterative process, where various knowledge components (e.g. market knowledge) repeatedly come onto the scene, interact, change and shape new understanding (Rickne, 2000). For example, the *design* elements can be, and often are, substantial, and (recent) scientific knowledge may be either present or absent in this creative process (Utterback *et al.*, 2006). Although there are many academic contributions to this discourse (e.g. Faulkner, 1994), probably the most well known challenge of the linear model has been written by Kline and Rosenberg (1986). Of central importance for their analysis is the stock of knowledge to which all activities – also R&D – contributes and may be utilized in all phases of the innovation process, which in its turn is characterized by significant feedback loops and engineering and design activities.

Related to the move away from a linear model is the realization that the bulk of scientific research does not (within a reasonably short time) result in new products or processes and the main part of innovations is based on knowledge that, for a long time, has been known outside research laboratories, although not necessarily spread over the world (Freeman, 1979; Kline and Rosenberg, 1986; Faulkner, 1994). In addition to the often too strong science focus view on innovation, it may be argued that the distinction between basic research and applied research is far from easy to uphold. A large part of what sometimes is labeled engineering science is also of a clear design character. This has, for a long time, been argued by Herbert Simon (1969/1996), who, in his *Sciences of the Artificial*, focused on the differences between natural scientists, who explain nature, and engineering scientists, interested in the artificial world. While there are many examples of Swedish innovations where new science and technology have played a crucial role, it is also true that several of the most innovative enterprises during recent decades – of which IKEA, H&M, the successful music industry and the innovative Swedish *Absolut Vodka* concept are perhaps the most internationally well known – the role of (basic) science was very limited. Instead, their success was based on creative combinations of logistics and design in interaction with the transformation of market demand.

The general conclusion is that a pure linear model is relevant only for a minority of those creative activities that contribute to innovation, growth and transformation. Moreover, only a subset of innovations relies on basic research or technology push. The importance of this subset *is* an empirical question, and may differ between sectors and countries. The question of how important it *can, and should, be* is another and in itself also important policy issue.

In order to handle the different modes of knowledge mobilized in innovation processes, it could be useful to depart from a distinction of the character of the knowledge involved rather than its assumed relation to science. One means to do that is to introduce a conceptual distinction between three main modes of knowledge: synthetic, analytic and symbolic – the SAS typology (Laestadius, 1998; 2000; Asheim and Gertler, 2005; Asheim *et al.*, 2007).

Synthetic knowledge is, in short, related to combinatorial capability. We may here include different forms of entrepreneurial skills (creative combinations) as well as polytechnical abilities, i.e. engineering competence to combine various knowledge elements, technologies and systems to higher and still more complex levels. Analytic, on the other hand, is the kind of knowledge that dominates within natural science, e.g. within biology and chemistry, being to a large extent codified and non-tacit. Symbolic knowledge is the notion used for that kind of knowledge related to esthetical attributes, to design work and creation of cultural artifacts (including texts). The knowledge developed among marketing people as well as among designers may be included in this category. Science is thus not based on analytic knowledge only. Neither is engineering and design based only on synthetic knowledge. Rather, we may assume that the balance between analytic and synthetic knowledge differs between different sciences and human activities. Industries may thus have different knowledge profiles and still be innovative.

In addition, it may be argued that the distinction between synthetic and symbolic knowledge is not obvious. Design activities may, for example, fit in both classes. Partly, this is a question of classification, i.e. what activities we include in the design category: the design of a production system is not necessarily an activity close to design of wine glasses. Partly, however this is also an epistemological question: the creation of artifacts – irrespective of whether it is a new advanced airplane or a piece of IKEA furniture – is a creative and optimizing activity where form and function has to fit (Alexander, 1964). Although the boundary conditions are set by nature (and have to be understood more or less in detail depending on the task at hand), the degrees of freedom may differ between the design of the airplane and the wine glass; and thus also the kind of knowledge mobilized in the process.

Of importance in this case is that learning processes basically differ, that the means to develop individual excellence are different and, consequently, the means to develop industrial capabilities and competitiveness. Not only does the balance between synthetic and analytic knowledge differ between different domains of knowledge; synthetic and symbolic knowledge is, in addition, more related to the process of creation ("the science of the artificial") rather than understanding, as suggested by Simon (1969/1996).

The SAS typology fits well into the discourse once introduced by Gibbons *et al.* (1994) on the new mode of integrated knowledge (mode 2) assumed to be part of the advanced industrial society and assumed to replace the disciplinary-based knowledge as developed and taught from traditional universities. The mode 2 knowledge concept developed by Gibbons *et al.* (1994) has an obvious family resemblance with what we here label synthetic knowledge, and partly also with symbolic knowledge. Also, mode 1 knowledge has family resemblance with analytic knowledge. The difference between our position and Gibbons *et al.* is that we argue that synthetic (mode 2) knowledge has been there all the time, but not necessarily explicitly identified; in fact, it has always been an important element in engineering science and polytechnic education, and a cornerstone in the creation of successful industrial enterprises.

With the approaches discussed above, we do not have to assume that some forms of knowledge (read "scientific") are more advanced than others, more difficult to master or more important for innovation activities and industrial development. Neither do we have to assume that university-based knowledge is the most important for industrial competitiveness. Synthetic knowledge is always needed in industry; analytical knowledge (e.g. research competence) is not enough to build up industrial competitiveness. And, as noted above, it is far from always the case that those knowledge and competence areas that are largely dependent on analytical knowledge are also the most complex, and thus most difficult to imitate for competitors.

Nevertheless, the linear model – with its strong focus on the analytically biased basic sciences – is still alive as a blueprint for significant parts of the innovation and technology policy discourse. To some extent, this is probably because parts of the academic community has knowledge interests in it; partly also because the linear model is easy to communicate; and finally because it is attractive for politicians and policy-makers who can argue that money to basic science contributes to the long-run development of industry.

The body of IS literature that more or less uncritically adopts and/or ends up in the linear model is significant, although, as argued here, this mode is just a subset of all possible forms of innovation (Kline and Rosenberg, 1986). It may, however, also be argued that the IS approach may be used as a framework for identifying the complexity and network character of different innovation processes.

The innovation system concept defined

Let us return to the IS definition given in the introduction: an interrelated structure of institutional – including cultural – and actor-based condensations in an economic space that has specific industrial and innovative properties compared to its environment with which it has exchange.

The "innovation system framework" may, in fact, be looked upon as a set of concepts with family resemblance,[3] containing terms with more or less similar meaning.[4] Members of the core family are notions such as "national innovation system" (NIS), "regional innovation system" (RIS), "technological (innovation) system" (TS),[5] and "sectorial innovation system" (SIS). We may also include notions such as "development block" (DB), "competence block" (CB), "industrial districts" (ID) and, not least, the "diamond" and "cluster" terms once introduced by Michael Porter (1990).[6] Worth mentioning – but not part of the IS family as such – are the multi-level perspective (Geels, 2002; 2004), and sociotechnical system concepts such as those presented by Bijker (1995) or Hughes (1992), even though they will not be explicitly discussed in this chapter.

As the IS concept embraces such a range of related terms, we will here point to what we see as common denominators. A first core feature is the explicit identification of the importance of institutions in conditioning economic processes. Second, we discuss the general understanding of the system concept, particularly focusing on openness and intentions. Third, we scrutinize the identification of systems borders that may be related to geography – but also to other forms of proximity (knowledge/technology).

The role of institutions

In the definition of innovation systems we used above, there is an implicit although fundamental criticism of the core message in standard economics: the notion of pure atomistic equilibrium markets – i.e. characterized of all the requirements for full competition – is not a good approximation of real economies, which are always embedded in an institutional and cultural web conditioning activities in the long run, as well as in the short. Our focus on the role of institutions – contributing to the creation of the system, as well as being created by it – follows a long tradition among economists and industrial analysts (North, 1990; Hodgson, 1998). The IS approach may be looked upon as a holistic attempt to understand the institutional web of which the classical economic agents "firms" and "labor" are just two elements of many.

The role of institutions has been a core research topic for economists for at least a century. Among the most well known international classics are Thorstein Weblen, Douglas North, Ronald Coase and Gunnar Myrdal. Among Swedish economists, not so well known in this area, we also find Johan Åkerman and Ingvar Svennilsson (Hodgson, 1998). Among some social scientists outside the economics discipline – e.g. sociologists and political scientists – the important role institutions take is, by definition, obvious. What is the book by Putnam on the *Working Democracy*, if not an institutional analysis of the North and South Italian "innovation systems" (Putnam, 1996)?

Understanding innovation systems as an institutional and relational condensation is thus essential. That leads us to the meaning of the "institution" concept. Here, we follow conventional social science theory using a very broad definition for the concept where it is seen as "sets of common habits, routines, established practices or rules which regulate the relations and interactions between individuals and groups" (Edquist and Johnson, 1995: 9).[7] According to this view, an institution is every social construct – be it intentional or not – regulating social interaction. Thus, an institution is not only a government authority; neither is it primarily a hospital, a home for the elderly or an opera – although these three categories may all be included in the institution concept. Institutions may be acting on the market – e.g. patent authorities charge applicants – but their importance is primarily not in selling and buying, but conditioning and influencing the behavior of societies.

Accepted business behavior, the working moral, patent systems, property laws and registration systems, tax laws, school system, entrepreneurial spirit and prevailing views on gender, nature, equality and religion all belong to the set of institutions that contributes to the embeddedness that the IS approach intends to capture. Obviously, there is variety as regards institutional embeddedness, and thereby there is also variety between IS as regards the impact certain institutions may have on the behavior of actors. IS researchers typically have the ambition to examine differences and similarities between different innovation systems, maybe also to point to their perceived or revealed weaknesses or strengths analyzed from a certain perspective. While such analyses could theoretically contain a very large number of institutions, they are, in practice, often focused on a rather limited set. For example, one may scrutinize what role different countries' institutions for property rights have for inventiveness. However, while the analysis of cultural norms and behavior may be just as relevant, such institutions have more seldom been centered in the IS literature. In part, this may be due to the knowledge background of the community of IS scholars, and also on the theories and methodologies developed to measure such phenomena. For example, do we really have the tools to evaluate whether the welfare system, the role of ethnical minorities, the gender attitudes or the R&D policy is most important for the innovation capacity of a certain region?

The systems concept

As we are dealing with a "system," we need to first discuss what this implies. In general, systems are made up of components, attributes and relationships. The components are institutions, actors and artifacts, and attributes are the properties of the components. Relationships are the links between these components: market, as well as non-market, links. Feedback loops and interaction provide the dynamics of the system. As Carlsson *et al.* (2002: 234) put it: "One result of interaction (feedback) among actors is that capabilities shift and grow over time, and therefore, the system configuration also changes." Also, the system's propensity to be both robust and flexible, and its ability to induce and handle change are crucial. Another way to define a system is to see it as "a group of components (devices, objects or agents) serving a common purpose, i.e. working towards a common objective or overall function" (Bergek *et al.*, 2008: 408). In this context, many researchers agree that an innovation system has the overall function of developing, diffusing and utilizing innovations.

Classical systems theory struggled with how to handle the transformation of the systems approach from the natural world (e.g. physics or biology) to the social (Emery, 1969; Ingelstam, 2002). An important step – especially interesting due to the evolutionary foundation of innovation theory – taken by von Bertalanffy (1950) is his distinction between the equilibrium (closed) systems of physics and the steady state, on the one hand, and open biological systems, on the other. Similar to biological systems, social systems are open, which makes it possible for them to develop and to maintain, or even change, their systems structure while transforming their parts.[8] Clearly, systems can be more or less open (closed) towards their context. An extreme interpretation with a "totally open" system renders the concept without meaning, as one cannot discriminate it from its context. One way to come to grips with the degree of openness is to specify the *types of elements* that are to be included in the system. In the case of innovation systems literature, one often includes the relevant institutions, knowledge areas, artifacts, actors, networks and the associated knowledge flows. We will come back to this issue of delineating the system and deciding upon relevant elements in a more detailed way below.

Another important distinction of the system concept – relating to how the IS concept is used by various theorists and practitioners – is whether or not the system is seen as an analytical construct or as a system that exists in reality. Phrased differently, one may use the IS concept to describe an empirically observed set of interconnected actors, networks and institutions. Another way to use the IS framework is to see it as an analytical construct, meaning that it is a model for us to better describe and explain system structure and dynamics. With this use of the IS concept, one may sketch a potential system that could emerge but that does not exist in practice, or where links or functions are not in place (Bergek *et al.*, 2008).

Innovation systems as self-organized or constructed is yet another crucial distinction between approaches. In this context, the role of intentions in systems may be analyzed. As innovation systems are social constructs, they contain human actors and organizations that have intentions – some of which may be more successful in fulfilling them. Actors may have intentions for their activities (e.g. firms are often assumed to maximize profit and individuals may have certain goals for action), and these intentions may include that some functions of the system need to be strengthened. In this way, actors such as firms may actively work towards supporting (their view of) a well functioning system (Rickne *et al.*, 2009). Rarely, however, firms take an overall responsibility for all functions of a system, but are more likely to focus on particular sub-functions. Moreover, policy or bridging organizations may have the intention to construct a system with certain properties. In this quest, such organizations may strive to create a system from scratch or to strengthen an existing system. In order to achieve this, they may be engaged in the construction and transformation of institutions that are parts of the system. Having said this about actors' intentions, our broad definition of innovation systems does not, as a prerequisite, assume that innovation systems are the result of intentional systems building. In fact, while in some cases collective and coordinated action may be the case, it seems to be more common that innovation systems emerge through the aggregated effect of uncoordinated intentional actions over long time periods. From an analytical point of view, such systems may be looked upon as self-organizing, where intentions, entrepreneurial culture, etc. may be more or less formed endogenously. As Bergek *et al.* put it:

> Actors do not necessarily share the same goal, and even if they do, they do not have to be working together consciously towards it (although some may be). Indeed, conflicts and tensions are part and parcel of the dynamics of innovation systems.
>
> (Bergek *et al.*, 2008: 408)

When scrutinizing innovation systems, the borders may, of course, be defined narrowly enough to allow for intentionally constructed systems. Not the least has this "voluntaristic" approach been popular among policy-makers with intentions to create innovation systems. This phenomenon – whether (local/regional) innovation systems can be "constructed" – is analyzed by Johansson and Ylinenpää in this volume (Chapter 10).

System borders

In our definition of innovation systems above, we have situated the IS to an *economic space*, i.e. the domain where all activities of an economy take place. Inspired by the French economist François Perroux (1950), who originally introduced the concept of economic space, we choose not to restrict our definition to a territorial – *geonomic* – space. The possible territorial anchoring of an innovation system thus becomes an empirical question, unless we do not explicitly restrict ourselves to geographical proximity. Consequently, geographically defined systems – like regional ones – thus become a subset of the large family of IS. Similarly, national innovation systems become one of many possible interpretations of an IS, although – as we will come back to below – the NIS concept is the original (Lundvall, 1992; Nelson, 1993).

One may argue that geographical borders become increasingly meaningless when speaking about innovation processes. In this epoch of globalization, what is left of territorially defined innovation systems when not only industrial production but also knowledge formation takes place on a global scale? Is the world, in fact, flat, as formulated by the American journalist Thomas Friedman (2005), i.e. can everything more or less take place everywhere with zero friction and instant connection? And do product and service innovations require ever more knowledge domains to be integrated, rendering the analysis of specific technologies or sectors less meaningful?

While there may be more than a grain of truth in such statements, innovation processes are nevertheless characterized by space, as well as by the specificity and logics of knowledge domains, and the IS concept tries to capture such features not only statically, but also over time. This means that each IS analysis needs to define the borders of the system in question. By default, each such attempt to delineate a system must carry its inaccuracies: there are no absolute borders to an open system, but only the more or less arbitrary ones set for the purpose of analysis. In general, there are three dimensions that are used to delineate any specific innovation system: geography, knowledge domain and time. And for each of these dimensions, issues of level of analysis become important. Depending on which of these starting points the analyst chooses – geography, knowledge domain and/or time – he or she will see different systems. The same set of components (institutions, actors, artifacts) or relations will not be included in the structural description of the IS, nor will necessarily identical attributes of the components come to focus. This implies that the choice of perspective from which to look at the innovation processes will be crucial for the findings. For example, being interested in the emergence of environmentally friendly alternatives to the combustion engine, one may partly stress dissimilar aspects, having chosen different geographical scopes and levels of analysis (e.g. the RIS of Western Sweden, the NIS of Sweden, or the IS of EU). Clearly, the question at hand is what will guide the choice of borders for the system (Carlsson *et al.*, 2002).

What may be seen as a weakness of the IS framework – that there are no clear guidelines as how to discriminate the system and its parts from its environment – may also be considered the strength of the framework, giving it its flexibility. As we do not a priori (i.e. by definition) know what phenomena or what components to include in the various innovation systems, the choice of delineation will be essential and has to be well informed. With this choice comes a set-up of components, attributes and relationships to be included in the IS. Each of these approaches will be exemplified in our overview of the IS family members below.

While these issues of border setting are decisive – and far from easy – they will also lead to further questions relating to the level of analysis and measurement matters. For example, should we only count condensations of (which?) firms, or should we include (which?) individuals (and in what of their roles)? What level of analysis should one chose for the demarcation of a technological innovation system? How do we count regional industrial service offices and technical universities? How do we include change and transformation in the systems analysis? These questions have been dealt with in several methodological texts related to the IS concept (Carlsson *et al.*, 2002; Bergek *et al.*, 2008), but more is yet to be done in this vein of research.

The IS family members

Let us, after this general introduction to the foundations of the IS approach, turn to a more detailed presentation and analysis of the different members of the IS family. *National innovation system* (NIS) was the original concept introduced around 1986/87 first, probably, by the British economist Christopher Freeman (who, in fact, had used the concept already in 1983, cf. Freeman, 1987; 1983/2004) in his analyses of the Japanese economy. The NIS concept was also – in connection to Freeman – used by the Swedish-Danish economist Bengt Åke Lundvall in his analyses of the competitiveness of small European economies (Freeman and Lundvall, 1988). Around 1990, many texts were published where the NIS concept was used (Dosi *et al.*, 1988; Lundvall, 1992; Nelson, 1993)

It is easy to recognize that the 1980s was an ideal and logical foundation for these kinds of theories. By then, the post-World War II catching-up processes from the first wave of Asian Tigers was maturing: not only Japan, but also South Korea, Taiwan, Hong Kong and Singapore challenged the incumbent industrial nations. Several of the newly industrialized countries were not neo-liberal, unfettered market economies, but were rather characterized by strong institutions with high ambitions to favor industrial dynamics. The success stories from these catching-up countries are in sharp contrast to the stagnation and lack of policy that was revealed in the incumbent world (Freeman, 1987; Nelson, 1993; Dertouzos *et al.*, 1988).

There is also a connection between the NIS concept, formulated in the late part of the 1980s, and the heritage from the German economist Friedrich List. In the first known text on NISs, there are significant references to the *national production system* concept once developed by List in his analyses of the industrial competition between the incumbent and free trade oriented UK – benefitting from economies of scale and being far ahead on the learning curve – and the emerging German and US catching-up economies (Freeman, 1983/2004).

The 1980s was, in addition, a period characterized by globalization as well as emerging pervasive technologies, such as ICT, which in extension were assumed to threaten small countries' possibilities to stay competitive (Laestadius, 1980; Sharp, 1983; Freeman and Lundvall, 1988). From a Northern European perspective, there was also the problematique whether small countries, such as Norway, Denmark, Sweden and Finland, could maintain competitiveness on country level within an integrated Europe. All in all, this contributed to give the national innovation systems discourse a "small country bias" (Freeman and Lundvall, 1988; Lundvall, 1992).

It may be somewhat simplified to argue that this is also mirrored in the two dominating NIS texts from the early 1990s: one European and one American. The comparative approach in the American anthology (Nelson, 1993) is more focused on the analysis of the *existence* of innovation systems than the European anthology, edited in Aalborg, Denmark (Lundvall, 1992), where most papers focus on the *character* of the systems. Several of the papers in the American anthology argue for caution as regards the potential to understand, still less to manage, innovation systems (Mowery and Rosenberg, 1993).

The national innovation system obviously fulfilled a need (for academics) in analytically knitting evolutionary innovation theory with institutional theory, as well as to connect industrial transformation/dynamics with policy, here illustrated with an influential text by Stan Metcalfe:

> A national system of innovation is that set of distinct institutions which jointly and individually contribute to the development and diffusion of new technologies and which provides the framework within which governments form and implement policies to influence the innovation process. As such it is a system of interconnected institutions to create,

store and transfer the knowledge, skills and artefacts which define new technologies. The element of nationality follows not only from the domain of technology policy but from elements of shared language and culture which bind the system together, and from the national focus of other policies, laws and regulations which condition the innovative environment. In the operation of national systems, governments play an important part in their support of science generally and in their procurement of technologies to meet the needs of the executive. To define such a system empirically one must locate the boundaries, its component institutions and the ways in which they are linked together.

Metcalfe (1995: 462–463)

The NIS concept developed into an organizing metaphor that successively colonized more and more space in international policy documents (OECD, 1997; 1999). Not least, in the OECD (1999) document – *Managing Innovation Systems* – national innovation systems appear to be a question primarily of *governance*. By then – the late 1990s – the original NIS concept had developed into a general concept, but also bifurcated into a family of varieties (Edquist, 1997). In a world where knowledge formation/learning as well as industrial activity takes place on a global scale, it is far from obvious that the essential and systemic condensations are national or even territorial in any real sense. With introducing the concept *sectorial innovation system*, the Italian economists Breschi and Malerba intend to focus on the industrial rather than the territorial dimension (Breschi and Malerba, 1997; Malerba, 2004). Analogously, although with focus on technology rather than industrial sectors, there also emerges a research tradition on *technological systems* (Carlsson and Stankiewicz, 1991; 1995). The difference, in short, between these approaches can be explained by the fact that the logic of technological change does not necessarily follow the logic of the industry. ICT, for example, may penetrate and influence various industries in different ways: some advanced process industries may, to a certain extent be more influenced by ICT than the IT industry itself. Also, biotechnology is often realized in unexpected industries: functional food is, for example, the result of biotechnology that has colonized a small niche within a traditional low-tech industry. Technological cognition processes, in addition, do not necessarily take place within "communities" or thought worlds that follow the industry classification or national borders. This is, in fact, one aspect of the mode 2 discourse mentioned above. This distinction is far from unimportant and has policy implications: if *industries* are targets for industrial policy, we have reason to expect other comparable results if policy focus is on *technologies*. Not only will there be different receivers of the resources connected to the policy, but it will also develop needs for other policy tools and different routines.

The concept *development bloc* (DB), once formulated by the Swedish economist Erik Dahmén (1950), deserves to be mentioned in this context. It may be looked upon as an important analytical contribution to the family of theories we are analyzing here. The DB concept contributed to the development of the NIS concept among the Danish economists. Still more has it influenced the technological systems approach (Carlsson, 2000). The development bloc – which, theoretically, may be argued to slide between a sectorial and a technological logic – is based on the notion of *structural tensions*, i.e. imbalances occurring due to the very character of the innovation process. There are always – in every period – *necessities* and problems that have to be handled, as well as *opportunities* that may be utilized. These imbalances may develop on the level of technology in the form of bottlenecks or local breakthroughs as analyzed by Tomas Hughes, with the concepts *salients*, *reverse salients* and *critical problems* (Hughes, 1992). But they may also occur on industry/sector level: the rapidly increasing electricity output due to large scale construction of hydro power plants in Sweden contributed to the establishment of energy consuming TMP pulp and paper plants. The DB concept – which never explicitly was formulated in territorial terms – may be looked upon as a means of making Schumpeterian dynamics concrete (Laestadius, 2005). It provides a tool to probe the dynamics of structural tensions in industry and technology. Dahmén is also clear that development blocs may be used for *ex post* analyses by academics, as well as intentionally by industrialists (or policy-makers) to create and/or exploit systems imbalances.

If sectorial/technological approaches may be argued to focus on the non- or a-territorial dimensions of innovation systems, the *regional innovation system* (RIS) approach represents the opposite. The RIS concept, however, illustrates the difficulties in distinguishing a distinct level of territorial analysis in a globalized economy. It is a concept with a clear inspiration from the NIS concept and is, to large extent, based on a similar analytical approach (Asheim and Gertler, 2005).

It may, however, also be argued that the RIS concept has entered the innovation system family from aside, that the regional dynamics dimension has been there all the time, although branded differently. Specifically, Alfred Marshall's concept *industrial district* is highly relevant in this context (Marshall, 1890/1990). The Marshallian district was, for example, an important tool for the understanding of the industrial dynamics of Northern Italy (Brusco, 1982; Amin and Thrift, 1994).

Also, the rapid industrial development in Silicon Valley – which, from many aspects, differed from the Italian – has contributed to the growing interest for the local and for the territory, as well as for analyzing the socio-cultural networks and processes that contribute to an innovative atmosphere – or to the absence of such climates (Saxenian, 1994). Silicon Valley – the dynamics of which, to a large extent, was the carrier of the ICT boom during the 1970s and 1980s – has been the victim of often too far reaching generalizations based on one extreme case: not least, the interpretation of Silicon Valley as a "science park" has contributed to exaggerated expectations – often connected to policy failures – on how to use policy to create industrial dynamics based on local universities

The differences between, among other things, the low-tech industrial creativity in some regions, on the one hand, and the significant clustering of R&D labs in some localities, on the other, has contributed to a family of RIS concepts. One way to capture these differences is to make a distinction between *institutional regional innovation systems* (IRIS) and *entrepreneurial* ones (ERIS) (Asheim and Gertler, 2005). Although this distinction is not necessarily exhaustive – both systems may, in fact, have a low-tech/hi-tech dimension – it serves the purpose to reveal the fact that government, academia, firms and other institutions play different roles in different forms of regional innovation systems. This is, of course, highly relevant for policy institutions such as VINNOVA: the variety of regional innovation systems may respond differently to various types of policies – and countries such as Sweden show such a variety.

Globalization, it may be argued, paradoxically has increased the interest for the regional dimension in industrial dynamics. The existence of "sticky places in slippery space," as formulated by Markusen (1996) more than a decade ago, brings up the question on the role of regional mechanisms – if any – in creating industrial creativity in a totally open economy.

Just to assume that innovativeness grows in a region as a consequence of territorial co-location of firms does not mirror reality in the globalization process of our time (Cooke, 2005). Although the presence of social interaction, trust and local institutions is essential for the development of clusters, this does not exclude that significant non-local knowledge may also be necessary to maintain or develop competitiveness (Maskell *et al.*, 1998; Cooke *et al.*, 2000; Asheim and Herstad, 2003; Isaksen, 2005). Regional clusters must develop interfaces with other parts of the world. Interactivity, as regards learning and knowledge formation, thus must not be restricted to geographic (territorial) proximity only (Torre and Gilly, 2000; Coenen *et al.*, 2004; Boschma, 2005; Malmberg and Power, 2005; Gustavsson, 2009).

The complex interrelationship in the regional system is characterized by interdependence on several levels, of which the territorial level is just one of many possible (Howells, 1999). Within the framework of these interrelationships – which are more or less open towards the rest of the world – certain activities tend to locate to specific regions (Cooke *et al.*, 2000; Asheim and Coenen, 2006). A regional innovation system is thus not a national innovation system writ small. Even if there are similarities between the NIS and RIS level of analysis – e.g. in the institutional structure – we argue that a significant part of the difference is in the "embeddedness" (Lundvall and Maskell, 2000). In short, the national innovation system is primarily institutional from a strict point of view: politics, laws, authorities, school system, culture politics, incentive structure in the economy, etc. The regional system, on the other hand, is primarily social and cultural, and contains cooperation, exchange of experiences (transfer of knowledge). The regional level – it is argued – is more based on social relations than to institutions (Granovetter, 1985). But this is far from always obvious: as will become clear below, the distinction between

RIS and NIS becomes blurred when comparing countries of very different sizes, such as China and Sweden.

Even if the world has become "slippery," human activities have to be grounded somewhere on the globe, and in a "globalized" world – with global access to presumably zero costs – all kinds of even marginal competitive advantages in a certain location may serve as a mechanism for competitive advantage, and maybe initiate a path dependent process. From these mechanisms, tensions are created between the local and the global. It is, thus, no paradox that the interest for territorial analyses increases parallel to the globalization process.

The competitive analyses by Michael Porter also belong to this research territory; his analyses, in fact, span between the national and the regional (Porter, 1990). Although the concept of "innovation system" is missing in his magnum opus, his analyses show family resemblance with the IS approach. The Porter *diamond* may be looked upon as a pedagogically attractive formulation of a (national) innovation system, although the "government" in the original diamond (Porter, 1990: Chapter 3) is given a much more limited role than what European analysts probably would do. In the more empirical-based part of the book, the concept *cluster* – which Porter may be given the credit for having introduced into the analyses of industrial dynamics – has a strong family resemblance with the Marshallian industrial district and the regional innovation system. The clusters of Porter, which are analyzed through the diamond approach, are characterized by competition and rivalry, as well as cooperation and networking. As was the case with the NIS concept, the analysis by Porter naturally received its inspiration from the territorially located industrial dynamics, which was part of the boom during the 1980s and 1990s, as well as the development, of the IC technologies and globalization. Over time, government has taken a more prominent position in Porter's research. This is especially the case in the paper "Green and Competitive," where the importance of a high-end domestic demand, created by an advanced environmental policy, for the creation of green industrial clusters and competitiveness is analyzed (Porter and van der Linde, 1998). It may be argued that part of their argument misses the point: the transformation pressure discussed did not have its origin in government, but in the environmental movement (Cerin, 2004). Institutions are more than governments.

Common for these research approaches is the ambition to identify the combination of locally unique and relatively sticky mechanisms and cultures favoring innovativeness and creativity, i.e. the origin of competitiveness: to have the capability to develop something that does not immediately slip over to other actors, networks or regions.

More sophisticated discourses on the contextual ground for knowledge formation (and learning) do, however, take us away from a narrow interpretation of the territory. Introducing the concept of *spaces of knowledge* – reminding us of the concept of *economic spaces* introduced by Perroux (1950) already – the geography of knowledge is no longer fettered to the territory

(Amin and Cohendet, 2004: Chapter 5). This also takes us to the approaches common in modern sociology and management theory. Specifically, a significant part of the knowledge management theory is focused on how to develop and maintain learning and knowledge creation on corporate level in global firms, and thus basically independent from territorial aspects (von Krogh *et al.*, 2000).

Creative and innovative processes can nowadays be organized on the global level, as well as on the regional. Multinationals that traditionally have had their R&D units in Europe or the US (Pavitt and Patel, 1991) now relocate their R&D activities. Within some scientific and technological areas, there are also research communities strongly connected in networks with low physical proximity but maintaining and developing professional relations on distance. Phenomena such as these do change over time: mechanisms that contributed to proximity some decades ago – e.g. in the industrial districts of Alfred Marshall – may be less relevant today. And information technology development may create potentials for communication that, at least partly, may substitute for direct human interaction. How far this globalization of knowledge formation has gone and whether there are limits to this process is the topic for much present-day research. It may be assumed that innovation systems are changing in character due to internationalization (Carlsson, 2006).

What constitutes a region or a regional system is, thus, far from evident. We can imagine a set of relations that all have different geographical anchoring – and distributed over different geographical territories – but still being important for the system in question. The condensations of the different relations may be interpreted as layers of different systems added to each other; this set of layers may reveal a certain territorial structure in the form of a set of nodes (hubs) connected through a set of network relations, together constituting the economic space. The non-territorial systems we discussed above may thus be included in this model.

One way to conceptually handle this variety of relations is – like VINNOVA, and other Swedish government authorities – to introduce the concept of *functional regions* (Hallin, 2005). Theoretically, we thus get a set of regions, each of which is defined from a certain identified function and not necessarily identical with a classical geographical or administrative structure. The classical Swedish definition of a functional region has been related to revealed labor market behavior. In the VINNOVA case, the concept is – potentially, at least – different. A functional region from an innovation point of view must not necessarily coincide with the regional labor market.

It may be argued whether functional regions – strictly speaking – are territorial or not; that is of no importance for our analysis here. Of importance, however, is that (innovation) system borders no longer – if they ever could – can be assumed to coincide with those administrative borders created by and conditioning the activities of policy makers.

Enlarging the conceptual ambition to the global level the innovation systems approach faces a lot of problems. The meaning of concepts such as global or *continental innovation systems* – as introduced by Freeman (2002) – is far from obvious. This problem was revealed in the early analysis by Mowery and Rosenberg (1993), relating to the US innovation system. Still more do we face the problem of analyzing the industrial transformation in China, which has – with 1,400 million inhabitants – during three decades (since 1978–79) shown a unique industrial transformation; longer and more rapid than any other country in history. But hitherto, most of this dynamic has taken place in a segment of Chinese society covering approximately 300 million people, of which the majority live in three large regions (Pearl River Delta, Yangtse River Region and Bo Hai Rim) and in another half a dozen big urban areas, with populations up to 30 million (e.g. the Chonquing area). But in these fast-transforming regions there are large segments of the population that are outside, or even losers, in this dynamic. It is difficult to imagine that the NIS and RIS concepts, as used in the European way, could have a similar meaning when applied to countries such as China. Every dynamic "region" in China is larger than the biggest European countries, and the big industrial cities have a larger population than a typical small- or medium-sized European country. In addition, it should be observed that much of this transformation, or catching-up process, has the character of classic growth – i.e. more roads, canals, railways, cars and houses – rather than innovations as usually defined (Laestadius *et al.*, 2008). In short, we are here probably approaching the limits of the innovation system concept.

The triple helix concept and its relation to IS

In this section, we turn shortly, and with specific reference, to Swedish innovation policy as it plays out in practice, to the triple helix concept. This approach has, during a decade, played the role of a toolbox more instrumental than the conceptual world of innovation systems. In fact, even though it is not formally included in the IS family, we mean that it provides a practice-oriented amalgamation of the IS approach and the mode 2 approach, as will be discussed further below. The triple helix (TH) concept provides a simplified – and therefore appealing and useful – metaphor for an innovation system. While the TH approach partly rests on other theoretical grounds, and has different aims (more practice/policy-oriented), somewhat dissimilar issues are therefore addressed as compared to the IS approaches discussed above.

The TH concept is presented as a model based on three interrelated helices that together illustrate the three sets of interactions and relations assumed to constitute the systemic character of innovation processes. Rhetorically, this model is much more convincing than the rather vague innovation system concept: it focuses on three actors – or, strictly speaking, three families of actors – the interaction of which is assumed to determine the dynamics of the system. The three (families of) actors are "government," "industry" and "academia." All the others – which contribute to the vagueness of the IS concept – are excluded from the triple helix rhetoric, thus contributing to its sharp policy implications. Learning, industrial relations, entrepreneurial climate, cooperation habits, etc. are located to an implicit or secondary level. They may be assumed to be part of the main helices, and they can be targets for policy, but disappear from the direct analysis. In the TH case, simplicity contributes to rhetoric elegance.

The triple helix approach, introduced by Henry Etzkowitz and Loet Leydesdorff in the mid-1990s, contributes – we argue in this chapter – to a synthesis between the IS approach and the mode 2 approach (Etzkowitz and Leydesdorff, 1997; Etzkowitz and Leydesdorff, 2000). They describe the growth of TH systems based on growing interaction of the three main actors: academia, universities and industry (Etzkowitz *et al.*, 1998). This resembles the theory of co-evolution developed by innovation systems researchers, i.e. how interacting systems, components and actors co-develop and become more and more interdependent as time goes by (Nelson, 1994). The triple helix of innovation is described as an endless transition on the one hand based on the co-evolution (and self-organization) between the three identified helices, on the other dependent on the internal transformation within these helices. "When two helices are increasingly shaping each other mutually, coevolution may lead to a stabilization along a trajectory" (Etzkowitz and Leydesdorff, 2000: 113f).

Assuming that the dynamics is located to the interplay between the three main (families of) actors – government, academia and industry – it is, of course, possible to imagine a set of potential models for interaction. Etzkowitz and Leydesdorff mention three variants (we may label them *ideal types*): one embedded/socialist; one extremely liberal/atomistic; and one trilateral, where actors are integrated in each other's areas of competence and become knitted together. The latter is also the model they prefer and for which they argue. Strictly, they do not argue that universities should develop into firms, that government should engage in business and that firms should develop into universities, but the vision is that all three groups of actors should adopt significant parts of competencies of the others. Universities should, thus, be entrepreneurial enough to become penetrated by an entrepreneurial logic in research and teaching. Teachers can, for example, become employed under conditions assuming them to be innovative one day a week to contribute to the contact between academia and industry (Etzkowitz, 2005). Universities do nowadays establish science and technology parks, as well as incubators within their normal activity and, in doing so, they take an entrepreneurial role to which government contributes, e.g. with risk capital through focused and strategic sectorial policy measures.

The traditional institutional roles for universities, government and industry change and develop when these actors, in addition to their traditional roles and functions, start to perform also parts of functions traditionally performed by the others (Leydesdorff and Etzkowitz, 1998). It may, somewhat paradoxically, be argued that the linear model dissolves parallel to a process where the role of universities becomes more important than ever. This takes place first through the fact that all actors – all helices – become equally important in the interrelated triple helix model: universities have no a priori special advantage or position and are not identified as the mother of all knowledge. Second – and this is an important part of the triple helix approach – the character of universities is transforming: their activities become more and more directly growth focused. In addition to teaching and basic research they adopt to the changing needs of society. And this is not what they were assumed to be in the linear model put forward by Vannevar Bush (1945).

Although the Bush (1945) model was well in line with established European and US traditions, and related to what Gibbons *et al.* (1994) label mode 1 knowledge, Etzkowitz and Leydesdorff (2000) argue that it is a parenthesis in academic history. Mode 1 is a historical construct, they argue: university-based knowledge formation was, until the late nineteenth century, closely related to industrial demand and societal problems. And so it evolves once again.

The role of universities in the knowledge-based society is, in fact, a core question for the TH approach – as it is in *The New Production of Knowledge* (Gibbons *et al.*, 1994). It is not only that universities are assumed to play an important role in basic science, but also that they are given a fundamental and direct task in the industrially relevant innovation process together, and in cooperation with, government and industry.

The core message in that latter study is the proposition that it, parallel to the old disciplinary based science (mode 1), has developed a new knowledge tradition, i.e. both new means of knowledge formation and new forms of knowledge. This double phenomenon, labelled mode 2, is characterized as follows:

- knowledge is produced close to its application;
- knowledge is cross disciplinary;
- knowledge is characterized by heterogeneity and organizational pluralism;
- knowledge is social and reflexive; and
- this knowledge necessitates new forms of quality evaluation.

This Gibbons *et al.* (1994) proposition on mode 2 knowledge is – as discussed on page 23 – not totally new. Nevertheless, this discourse provides a source for critical reflection. The ongoing process of strengthening the disciplinary character of academic research – especially driven by the internal quality criteria within academia – opens a territory, or maybe even creates a necessity, for more applied and integrating knowledge processes. A significant part of R&D in our time is, in addition, performed by industrial and applied units/firms, where the task ahead is of a cross-disciplinary character. These applied R&D efforts – which, in Sweden, constitute 3 percent of GDP, or 75 percent of all Swedish R&D – are not only more driven by results that "work" than by intra-disciplinary criteria; they are also dependent on the potential to commercialize the results and of appropriability conditions (Teece, 1986).

The TH approach is less positive and more normative than "classical" IS theory: the TH model is explicitly introduced to increase growth and innovation capacity in countries and regions. It may be argued that the authors – Etzkowitz and Leyesdorff – in addition to analyzing knowledge formation processes and social phenomena, "are sometimes involved in stimulating entrepreneurs, university administrators and public figures to rethink policy and conduct in response to changing cognitive, technical, economic and international trends" (Shinn, 2002: 605). This sometimes also has the effect that they disregard whether the processes they analyze and comment on are truly global or embedded in various national contexts (Shinn, 2002).

It may be argued that the Swedish Parliament's decision some years ago that all Swedish universities have to focus on the third mission was directly in line with the triple helix ideas. Parallel to their traditional tasks – academic teaching and research – universities and colleges are assumed to cooperate with firms and organizations in their neighborhood – in fact, with the surrounding society as a whole (Benner and Sandström, 2000; Nilsson and Uhlin, 2002). In particular, this was clear in the 1996 government bill on R&D (Prop. 1996/1997:5). Similar processes take place – not least on a regional level – all over the world (Brulin, 1998).

Even broadly accepting the premises of the TH approach, it is, however, far from obvious what policy conclusions that are the most natural, given the transformation of knowledge production as assumed in the triple helix (or mode 2) models. It may also be argued that the TH approach – although the helix metaphor is elegant – lacks a clear model for describing and analyzing how interactions between the helices take place, as well as on what they should look like (Benner and Sandström, 2000).

The fact that functions, as well as structures, are far from given in the interrelation between the TH actors may be illustrated from the area of intellectual property rights in relation to research and innovations. How should universities react on the achievements from their employees? First of all, in a strict linear model – if it ever existed – the IPR problem is non-existent: basic science results, by definition, fall outside what should be patented. Universities all over the world, however, perform research that is obvious candidate for patenting. Given that it may be argued that "applied" research should be allocated to independent research institutes outside universities proper, that is also the case in several countries. It may also – on good grounds – be argued both that IPRs should belong to the researchers themselves and that the IPRs should be the property of the universities, due to the fact that researchers create their innovations on working time in the premises of the universities. Both models exist among OECD countries (Sellenthin, 2006).

Moving from the specific IPR problem to a more general one, it may be argued that it is far from obvious that an "entrepreneurial university" – as discussed in the TH literature – is the best way to favor an "entrepreneurial society." It is thus far from evident that – even if there is a growing self-organizing entrepreneurial transformation of academia – this process should be enforced and supported. Maybe – just because there are strong utility demands on universities – it is extra important to guarantee the relative independence of academically based knowledge formation and of knowledge that is not primarily validated based on its usefulness. Maybe – and just because there is a growing focus on patenting also on academic knowledge – it is important to guarantee the non-patented academic knowledge. In fact, highly innovative and entrepreneurial societies, characterized by rapid and creative enterprise activities and exploiting still unexploited niches, may be preferred in the long term by the existence of solid and non-opportunistic knowledge institutions – in Sweden, responsible for approximately one quarter of the 4 percent of GDP allocated to R&D – searching for solutions and knowledge beyond the immediately profitable.

The optimal division of labor in the triple helix world of government, academia and industry is thus far from obvious, even if there is agreement on the necessity to favor industrial transformation, growth and technical change. The problem becomes still more complex if we consider potential differences between various systems and systems levels: what seems to be a reasonable model on regional level in one country is not always obvious on the national level in another country. And what is useful for some industries and technologies may be less useful for others.

And, as is increasingly the case, when academic networks are global and firms – like the Swedish ones – are highly internationalized and establish their R&D processes on a global scale, this makes the co-evolution process assumed to take place in the TH world problematic. Still more troublesome are the policy conclusions to be drawn.

Critiques and responses: Dynamics and governance

A common set of critique towards the IS approach says that, in spite of claiming to evolutionary aspects of innovation, there is a risk of rather static and structural analyses. There are several reasons for that critique. One is the relative lack of tools to capture dynamics in structure, as well as in the function of a system. That problem has, for some time, been in focus among systems analysts (cf. Emery, 1969). In short, the problematique may be described as the relation between *structure* and *function* in defining the system. As regards the structure – functional dimension in social systems – this was identified early by, for example, Merton (1949/1969).[9] Is the system basically defined from its structure (its components) and, if so, has the system changed (or is it another system?) if components disappear and/or transform? Or, is the system basically defined from a set of essential functions or purposes, with the implication that the system is basically the same, irrespective of structural change, as long as the basic functions are there? And, more fundamentally, what are the functions of the system, and how can these be confined? This is, of course, also relevant for the analysis of innovation systems.

Another problem for (IS) systems researchers is how to measure the performance of a specific system, as well as the relative importance of its specific attributes. A third and related problem is how to obtain relevant data for comparative analyses between systems that largely deviate from another.

In social sciences, this balance between structure and function of systems – and the epistemological consequences of that – has, for some time, been in focus for many disciplines: anthropologists (Radcliffe-Brown, 1952), sociologists (Merton, 1949/1969; Parsons and Smelser, 1956) and political scientists (Almond, 1960). The general problem is well illustrated with a case from political science: in his classical paper from 1960, Almond identifies the political systems more or less completely from its functions, of which he identifies four essential input, and three essential output, functions.

During the last decade, several authors have addressed this critique by developing a functional approach to IS dynamics (Galli and Teubal, 1997; Johnson, 1998; 2001; Rickne, 2000; Johnson and Jacobsson, 2001; Liu and White, 2001; Bergek, 2002; Bergek and Jacobsson, 2003; Carlsson and Jacobsson, 2004; Hekkert *et al.*, 2007; Bergek *et al.*, 2008). In essence, these researchers argue that each IS – basically, their focus is on technological innovation systems (TIS) – can be portrayed by a number of functions describing what actually "happens" in the system. The set of functions identified is, from the beginning, empirically derived. Complemented by findings from, for example, sociology, organizational theory and political science, the functions arrived at have a broad resonance in the literature. The approach describes how to assess the "functional pattern" of the IS: this implies to determine how the key processes currently work in any specific IS. Bergek *et al.* (2008: 414) stress that the "functional pattern of a TIS is likely to differ from that of other TISs and is also likely to change over time. Thus, the concept should not be interpreted as implying that the pattern is either repeated or optimal."

The Swedish "TIS functionalists" suggest that the systems are based on the following functions (Bergek *et al.*, 2008):

- knowledge development and diffusion;
- influence on the direction of research;
- entrepreneurial experimentation;
- market formation;
- legitimation;
- resource mobilization; and
- development of positive externalities.

This list can be argued and the definitions questioned: do the functions constituting TIS differ from those of the NIS, the SIS or the RIS? Or, do the set of functions differ over countries or time? While the functional approach by no means solves the entire dilemma of measuring dynamics or performance, it is nevertheless a valuable step along that road. The epistemological question on how to identify the functions of a system is indeed not trivial. It may be argued – and has been – that such functionalistic approaches run the risk that the list of functions gets an ad hoc character, that it can be extended or reorganized with no obvious theoretical ground. How do we validate that the 4 + 3 Almond functions make the toolbox complete? Or, in general terms: is my list of essential functions more reliable than yours?[10] Nevertheless, this functional approach opens for research activities to get answers on these questions.

Moreover, the functional approach among IS analysts may be looked upon as one answer (among others) to the "governance dilemma." This impasse relates to the fact that there are many actors who, individually or jointly, endeavor to put various schemes into practice (let us call these governance arrangements), with the aim to influence innovation input, its processes and its outcome. We call this process innovation governance. The dilemma, as such, lies in the difficulties in guiding or controlling such complex and ever-changing innovation processes, where so many actors and knowledge areas are involved. The governance concept is wider than public policy: we include all actions and actors involved in guiding or directing the process at hand. This implies that public policy actors may or may not be crucial actors: there is nothing in the governance approach admitting a special role for public policy. Instead, various types of actors may take on different responsibilities depending on, for example, technological or sectorial area, geographical conditions or current trends (cf. this volume, Chapter 1).

Not least, a functional approach may contribute in relation to the globalization problematique: sorting out the various functions to be filled within an IS and making them comparable over space and time. Globalization challenges, for example, the balance between the regional, national and supra-national. Which structural components and which functions can be expected to be served at which spatial level?[11] Is it reasonable to anticipate clusters to emerge within, for example, biomedicine at a multitude of locations around Europe? And what does this imply for national policy actors striving to support such clusters? Another governance issue related to globalization is what types of actors take what roles in pushing the regional, national or supra-national agenda.

All in all, governance issues are central when it comes to innovation systems. However, even given the possibilities of the functional approach, critiques assert that, while the institutional set-up is used to delimit the systems considered, institutional changes and influences, as well as governance aspects, have more seldom been fully captured by IS analyses. In fact, current theoretical development has not coherently shown how to handle these issues (Jordan, 2008). In particular, there is a lack of integration with established governance approaches (Treib *et al.*, 2007; Newell *et al.*, 2008) and the IS framework. A recent attempt along this line was presented by Hillman *et al.* (2009), merging IS theories with the multi level perspective (Geels, 2002; 2004) and governance understanding to form a comprehensive framework.

The governance issue can also be read in the more policy-focused IS documents that, in particular, have been published by the OECD (OECD, 1999; 2002; 2005a; 2005b; 2005c). Already in the early OECD work on innovation systems, it was felt among policy-makers that the institutional set-up for policy was not adequate for the transformations in innovation systems that were taking place, in particular the growing importance of innovations and the decline of the linear model (OECD, 1999; 2002). Later OECD reports have extended that analysis and, based on scanning state-of-the-art practices in several – primarily small – OECD countries, identified the tensions in the innovation systems and the potentials for policy formulation and integration (OECD, 2005a; 2005b; 2005c). Explicitly, the growing importance of sustainability and information society policy is identified as new areas to consider in the re-orientation of IS governance.

Concluding remarks

The aim of this chapter was to enhance our understanding of the theoretical basis on which much of today's Swedish innovation policy rests. While we did not aspire to synthesize or develop new theory, or to move beyond the borders of Sweden, our discussion merits some general reflections.

Modern innovation theories rest on an evolutionary view where learning processes and path dependencies are key, and where variety creation, selection and retention mechanisms are central. The innovation system concepts in its various variants, and the triple helix concept alike, have spun out not only as theoretical constructs, but also as interpreted policy tools. For Sweden, these frameworks are ways to address issues of industrial dynamics. Indeed, while several countries and supra-national units as OECD or the EU have utilized the innovation system approaches to varying degrees during the last decades, Sweden is perhaps at the extreme end of this spectrum, in that the IS concept has been allowed to set the agenda for national and regional policies and program details, and even to name a governmental organization.

The system approaches have, as governance tools – perhaps better than earlier equilibrium-based theories – captured not only a static innovation input or output, but come closer to its dynamic nature. In particular, recent focus on actions (functions) rather than only actors has strengthened the analytical frameworks. In practice, it seems that one important factor in making the frameworks useful for governance of transformation and innovation processes has been the continuous interaction between policy, practice and theory. Nevertheless, there are certainly several flaws in how the frameworks manage to depict a complex reality: openness and intentions in systems as well as delineation of systems borders remains problematic tasks for the analyst. Our review has aspired to highlight such difficulties, thereby giving the governing actor a better ground on which to stand. It is true that governance of IS has not been sufficiently described or penetrated in the research or policy literature.

Notes

1. This section is short as the TH concept is in focus, also in Chapter 3 following this one.
2. Neither is there in the Schumpeterian world a need for the concept "innovator" – distinguished from the "entrepreneur" – as used in modern Swedish language.
3. The IS approach may perhaps not be seen as a formal "theory." Instead, as Edquist (1997) expresses, the IS concept should primarily be understood as a "conceptual framework."
4. This is, in itself, a concept originally formulated by Wittgenstein (1953/1989) in his *Philosophical Investigations.*
5. This notion, as defined by Carlsson and Stankiewicz (1991), should not be confused with Hughes' (1992) term "technological systems." Some literature now refers to the type of innovation systems called "technological systems" as "technological innovation systems" to avoid such confusion (see e.g. Hillman *et al.*, 2009).
6. The diamond-based analysis and the cluster concept are perhaps the most well known, being cornerstones in Porter's analysis of the "competitive advantage of nations" (and regions).
7. In contrast, organizations are "formal structures with an explicit purpose" (Edquist and Johnson, 1995: 11). This means that we do not use the term institutions when we mean the organization as such. For example, a university may be referred to both as an institution influencing societal interactions, and as a specific organization.
8. Note that the biological analogues should not be extended in all directions.
9. And this is particularly problematic to combine with an evolutionary perspective, a problem with which both natural scientists (biologists) and social scientists have struggled. The usefulness of a functions concept differs between physical systems and biological ones, as well as between natural science-related systems, on the one hand, and social systems, on the other (cf. Nagel, 1956).
10. The Almond (1960) input functions are (1) political adaption and recruitment; (2) articulation of interest; (3) aggregation of interest; and (4) political communication. The output functions are (a) rule making; (b) rule application; and (c) rule adjudication.
11. See, for example, the discussions in the chapters by Gustavsson *et al.*, and by Laage-Hellman *et al.* in this volume.

References

Alexander, C. (1964) *Notes on the Synthesis of Form*, Harvard, MA: Harvard University Press.

Almond, G. (1960) "A functional approach to comparative politics," in G. Almond and J. Coleman (eds) *The Politics of Developing Areas*, Princeton, NJ: Princeton University Press.

Amin, A. and Cohendet, P. (2004) *Architectures of Knowledge*, Oxford: Oxford University Press.

Amin, A. and Thrift, N. (eds) (1994) *Globalization, Institutions, and Regional Development in Europe*, Oxford: Oxford University Press.

Asheim, B.-T. and Coenen, L. (2006) "Contextualising regional innovation systems in a globalising learning economy," *Journal of Technology Transfer*, 31 (1).

Asheim, B.-T. and Gertler, M. (2005) "The geography of innovation: Regional innovation systems," in J. Fagerberg, D. Mowery and R. Nelson (eds) *The Oxford Handbook of Innovation*, Oxford: Oxford University Press.

Asheim, B.-T. and Herstad, S. (2003) "Regional innovation systems, varieties of capitalism and non-local relations: Challenges from the globalising economy," in B.-T. Asheim and Å. Mariussen (eds) *Innovation, Regions and Projects*, Stockholm: Nordregio, R 2003:3: 241–274.

Asheim, B.-T. *et al.* (2007) "Constructing knowledge-based regional advantage: implications for regional innovation policy," *International Journal of Entrepreneurship and Innovation Management.*

Benner, M. (ed.) (2005) *Innovationer*, Lund: Studentlitteratur.

Benner, M. and Sandström, U. (2000) "Institutionalizing the triple helix: Research funding and norms in the academic system," *Research Policy*, 29 (2): 291–301.

Bergek, A. (2002) *Shaping and Exploiting Technological Opportunities: The Case of Renewable Energy Technology in Sweden*, Ph.D. thesis. Department of Industrial Dynamics, Chalmers University of Technology, Gothenburg.

Bergek, A. and Jacobsson, S. (2003) "The emergence of a growth industry: A comparative analysis of the German, Dutch and Swedish wind turbine industries," in S. Metcalfe and U. Cantner (eds) *Change, Transformation and Development*, Heidelberg: Physica-Verlag, 197–227.

Bergek, A., Jacobsson, S., Carlsson, B., Lindmark, S. and Rickne, A. (2008) Analyzing the functional dynamics of technological innovation systems: A scheme of analysis, *Research Policy*, 37 (3): 407–429.

Bijker, W.E. (1995) *Of Bicycles, Bakelites and Bulbs: Toward a Theory of Sociotechnical Change*, Cambridge, MA: MIT Press.

Boschma, R. (2005) "Proximity and innovation: A critical assessment," *Regional Studies*, 39 (1): 61–74.

Breschi, S. and Malerba, F. (1997) "Sectoral innovation systems: Technological regimes, Schumpeterian dynamics, and spatial boundaries," in C. Edquist (ed.) *Systems of Innovation: Technologies, Institutions and Organizations*, London: Pinter.

Brulin, G. (1998) *Den tredje uppgiften: Högskola och omgivning i samverkan*, Stockholm: SNS Förlag and Arbetslivsinstitutet.

Brusco, S. (1982) "The Emilian model: Productive decentralization and social integration," *Cambridge Journal of Economics*, 6: 167–184.

Bush, V. (1945) "Science the endless frontier," A Report to the President July 1945, Washington, DC: Government Printing Office.

Carlsson, B. (2000) "Teknologiska system som analytisk ram: en introduction," in B. Carlsson *et al.* (eds) *Innovationssystem, kluster och kompetensblock: fyra essäer om innovationer, tillväxt och sysselsättning*, Stockholm: Rådet för Arbetslivsforskning.

Carlsson, B. (2006) "Internationalization of innovation systems: A survey of the literature," *Research Policy*, 35: 56–67.

Carlsson, B. and Jacobsson, S. (2004) *Dynamics of Innovation Systems Policy-Making in a Complex and Non-deterministic World.* Paper presented at the International Workshop on Functions of Innovation Systems at the University of Utrecht, 23–24 June.

Carlsson, B. and Stankiewicz, R. (1991) "On the nature, function and composition of technological systems," *Journal of Evolutionary Economics*, 1 (2): 93–118.

Carlsson, B. and Stankiewicz, R. (1995) "On the nature, function and composition of technological systems," in B. Carlsson (ed.) *Technological Systems and Economic Performance: The Case of Factory Automation*, Dordrecht: Kluwer.

Carlsson, B., Jacobsson, S., Holmén, M. and Rickne, A. (2002) "Innovation systems: Analytical and methodological issues," *Research Policy*, 31 (2): 233–245.

Cerin, P. (2004) *Turning Corporate Behaviour Into Sustainability*, Ph.D. thesis, Stockholm: Kungliga Tekniska Högskolan.

Coenen, L., Moodysson, J. and Asheim, B.-T. (2004) "Nodes, networks and proximities: On the knowledge dynamics of the Medicon Valley Biotechnology Cluster," *European Planning Studies*, 12 (7): 1003–1018.

Cooke, P. (2005) "Regionally asymmetric knowledge capabilities and open innovation: Exploring "Globalisation 2" – a new model of industry organisation," *Research Policy*, 34: 1128–1149.

Cooke, P., Boekholt, P. and Tödtling, F. (2000) *The Governance of Innovation in Europe: Regional Perspectives on Global Competitiveness*, London: Pinter.

Dahmén, E. (1950) *Svensk industriell företagarverksamhet*, band 1, (IUI), Uppsala: Almqvist and Wiksell.

Dertouzos, M., Lester, R. and Solow, R. (1988) *Made in America: Regaining the Productive Edge*, Cambridge, MA and London: The MIT Press.

Dosi, G. *et al.* (eds) (1988) *Technical Change and Economic Theory*, London and New York: Pinter.

Edquist, C. (ed.) (1997) *Systems of Innovation: Technologies, Institutions and Organizations*, London: Pinter.

Edquist, C. and Johnson, B. (1995) *Institutions and Innovations: A Conceptual Discussion*, Presented at Söderköping Conference within the System of Innovation Research Network 1994–95.

Emery, F. (ed.) (1969) *Systems Thinking*, Harmondsworth: Penguin.

Etzkowitz, H. (2005) *Triple Helix: den nya innovationsmodellen*, Stockholm: SNS Förlag.

Etzkowitz, H. and Leydesdorff, L. (eds) (1997) *Universities and the Global Knowledge Economy: A Triple Helix of University–Industry–Government Relations*, London: Cassell.

Etzkowitz, H. and Leydesdorff, L. (2000) "The dynamics of innovation: From national systems and "Mode 2" to a triple helix of university–industry–government relations," *Research Policy,* 29 (2): 109–123.

Etzkowitz, H., Webster, A. and Healey, P. (eds) (1998) *Capitalizing Knowledge: New Intersections of Industry and Academia*, Albany, NY: State University of New York Press.

Fagerberg, J., Mowery, D. and Nelson, R. (eds) (2005) *The Oxford Handbook of Innovation*, Oxford: Oxford University Press.

Faulkner, W. (1994) "Conceptualising knowledge used in innovation: A second look at the science-technology distinction and industrial innovation," *Science, Technology, and Human Values*, 19 (4): 425–458.

Freeman, C. (1979) "The determinants of innovation: Market demand, technology and the response to social problems," *Futures* (June): 206–215.

Freeman, C. (1983/2004) "Technological infrastructure and international competitiveness," *Industrial and Corporate Change*, 13 (3): 541–569 (reprint of old OECD report).

Freeman, C. (1987) *Technology Policy and Economic Performance: Lessons from Japan*, London and New York: Pinter.

Freeman, C. (2002) "Continental, national and sub-national innovation systems: Complementarity and economic growth," *Research Policy*, 31: 191–211.

Freeman, C. and Lundvall, B.-Å. (eds) (1988) *Small Countries Facing the Technological Revolution*, London and New York: Pinter.

Friedman, T. (2005) *The World is Flat*. New York: Farar, Straus and Giroux.

Galli, R. and Teubal, M. (1997) "Paradigmatic shifts in national innovation systems," in C. Edquist (ed.) *Systems of Innovation: Technologies, Institutions and Organizations*, London: Pinter, 342–370.

Geels, F.W. (2002) "Technological transitions as evolutionary reconfiguration processes: A multi-level perspective and a case-study," *Research Policy*, 31 (8–9): 1257–1274.

Geels, F.W. (2004) "From sectoral systems of innovation to sociotechnical systems: Insights about dynamics and change from sociology and institutional theory," *Research Policy*, 33: 897–920.

Gibbons, M. *et al.* (eds) (1994) *The New Production of Knowledge*. London: Sage.

Granovetter, M. (1985) "Economic action and social structure: The problem of embeddedness," *American Journal of Sociology*, 91 (3): 481–510.

Gustavsson, L. (2009) *Creating Advantage: On the Complexity of Industrial Knowledge Formation in the Knowledge-Based Economy*, Ph.D. thesis, Kungliga Tekniska Högskolan Stockholm.

Hallin, G. (2005) *Sektor möter tillväxt: men hur och var?* Stockholm: Institutet för tillväxtpolitiska studier.

Hekkert, M.P. *et al.* (2007) "Functions of innovation systems: A new approach for analyzing technological change," *Technological Forecasting and Social Change*, 74: 413–432.

Hillman, K., Nilsson, M., Rickne, A. and Magnusson, T. (2009) "Fostering sustainable technologies: A framework for analysing the governance of innovation systems." *Science and Public Policy* 38 (5): 403–415.

Hodgson, G. (1998) "The approach of institutional economics," *Journal of Economic Literature*, 36 (1): 166–192.

Howells, J. (1999) "Regional systems of innovation?" in D. Archibugi *et al.* (eds) *Innovation Policy in a Global Economy*, Cambridge: Cambridge University Press.

Hughes, T. (1992) "The dynamics of technological change: Salients, reverse salients and critical problems," in G. Dosi, R. Gianetti and P.A. Toninelli (eds) *Technology and Enterprise in Historical Perspective*, Oxford: Clarendon Press.

Ingelstam, L. (2002) *System: att tänka över samhälle och teknik*, Eskilstuna: Energimyndigheten.

Isaksen, A. (2005) "Regional clusters building on local and non-local relations: A European comparison," in A. Lagendijk and P. Oinas (eds) *Proximity, Distance and Diversity: Issues on Economic Interaction and Local Development*, Aldershot: Ashgate, 129–152.

Johnson, A. (1998) "Functions in innovation system approaches," Unpublished working paper. Department of Industrial Dynamics, Chalmers University of Technology, Gothenburg.

Johnson, A. (2001) "Functions in innovation system approaches," Electronic Paper at the Proceedings of the Nelson and Winter Conference, Aalborg.

Johnson, A. and Jacobsson, S. (2001) "Inducement and blocking mechanisms in the development of a new industry: The case of renewable energy technology in Sweden," in R. Coombs, K. Green, V. Walsh and A. Richards (eds) *Technology and the Market: Demand, Users and Innovation*, Cheltenham, UK: Edward Elgar.

Jordan, A. (2008) "The governance of sustainable development: Taking stock and looking forward," *Environment and Planning C: Government and Policy*, 26: 17–33.

Kline, J. and Rosenberg, N. (1986) "An overview of innovation," in R. Landau and N. Rosenberg (eds) *The Positive Sum Strategy*, Washington, DC: National Academy Press, pp. 275–305.

Laestadius, S. (1980) *Produktion utan gränser*, Stockholm: Sekretariatet för framtidsstudier.

Laestadius, S. (1992) *Arbetsdelningens dynamik*, Lund, Sweden: Arkiv förlag.

Laestadius, S. (1998) "Technology level, knowledge formation, and industrial competence in paper manufacturing," in G. Eliasson, C. Green and C. McCann (eds) *Microfoundations of Economic Growth*, Ann Arbour, MI: Michigan University Press, 212–226.

Laestadius, S. (1999) *Marshallian Dynamics and New Growth Theory*, Working Paper, Stockholm: Kungliga Tekniska Högskolan, TRITA-IEO R 1999:1.

Laestadius, S. (2000) "Biotechnology and the potential for a radical shift of technology in forest industry," *Technology Analysis and Strategic Management*, 12 (2): 193–212.

Laestadius, S. (2005) "Det dahménska utvecklingblocket: några reflektioner," in M. Benner (ed.) *Innovationer*, Lund, Sweden: Studentlitteratur.

Laestadius, S., Gustavsson, L. and Long, V. (2008) "High-tech innovations in catching-up countries: Conditions and perspectives," in H. Hisch-Kreinsen and D. Jacobson (eds) *Innovation in Low-Tech Firms and Industries*, Cheltenham, UK: Edward Elgar.

Leydesdorff, L. and Etzkowitz, H. (1998) "The triple helix as a model for innovation studies," *Science and Public Policy*, 25 (3): 195–203.

Liu, X. and White, S. (2001) "Comparing innovation systems: A framework and application to China's transitional context," *Research Policy*, 30: 1091–1114.

Lundvall, B.-Å. (ed.) (1992) *National Systems of Innovation: Towards a Theory of Innovation and Interactive Learning*, London: Pinter Publishers.

Lundvall, B.-Å. (1994) "The learning economy," *Journal of Industry Studies*, 1 (2): 23–42.

Lundvall, B.-Å. and Edquist, C. (1992) "Comparing the Danish and Swedish systems of innovation," in B.-Å. Lundvall (ed.) *National Systems of Innovation: Towards a Theory of Innovation and Interactive Learning*, London: Pinter.

Lundvall, B.-Å. and Maskell, P. (2000) "Nation states and economic development," in G.L. Clark *et al.* (eds) *The Oxford Handbook of Economic Geography*, Oxford: Oxford University Press, 353–372.

Malerba, F. (ed.) (2004) *Sectoral Systems of Innovation*, Cambridge: Cambridge University Press.

Malmberg, A. and Power, D. (2005) "(How) do (firms in) clusters create knowledge?" *Industry and Innovation*, 12 (4): 409–431.

Markusen, A. (1996) "Sticky places in a slippery space: A typology of industrial districts," *Economic Geography*, 72: 293–313.

Marshall, A. (1890/1990) *Principles of Economics: An Introductory Volume*, 8th edition, London: Macmillan.

Maskell, P. *et al.* (eds) (1998) *Competitiveness, Localised Learning and Regional Development*, London: Routledge.

Merton, R.K. (1949/1968) *Social Theory and Social Structure: Toward the Codification of Theory and Research*, New York: The Free Press.

Metcalfe, S. (1995) "The economic foundations of technology policy: Equilibrium and evolutionary perspectives," in P. Stoneman (ed.) *Handbook of the Economics of Innovation and Technological Change*, Oxford and Cambridge, MA: Blackwell, 409–512.

Mowery, D.C. and Rosenberg, N. (1993) "The US national innovation system," in R.R. Nelson (ed.) *National Innovation Systems: A Comparative Analysis*, New York: Oxford University Press.

Nagel, E. (1956) "A formalization of functionalism," in F.E. Emery (1969) *Systems Thinking*, Harmondsworth: Penguin.

Nelson, R. (ed.) (1993) *National Innovation Systems: A Comparative Analysis*, New York: Oxford University Press.

Nelson, R. (1994) "The co-evolution of technology, industrial structure, and supporting institutions," *Industrial and Corporate Change*, 3: 47–63.

Nelson R. and Winter, S. (1982) *An Evolutionary Theory of Economic Change*, Cambridge, MA: Belknap Press of Harvard University Press.

Newell, P., Jenner, N. and Baker, L. (2008) *Governing Clean Development: A Framework for Analysis*, Norwich, University of East Anglia.

Nilsson, J.-E. and Uhlin, Å. (2002) *Regionala Innovationssystem: en fördjupad kunskapsöversikt*, VINNOVA Rapport, VR 2002:03.

North, D. (1990) *Institutions, Institutional Change and Economic Performance*, Cambridge: Cambridge University Press.

OECD (1997) *National Innovation Systems*, Paris: OECD.

OECD (1999) *Managing Innovation Systems*, Paris: OECD.

OECD (2002) *Dynamising National Innovation Systems*, Paris: OECD.

OECD (2005a) *Governance of Innovation Systems: Vol 1: Synthesis Report*, Paris: OECD.

OECD (2005b) *Governance of Innovation Systems: Vol 2: Case Studies in Innovation Policy*, Paris: OECD.

OECD (2005c) *Governance of Innovation Systems: Vol 3: Case Studies in Cross-sectoral Policy*, Paris: OECD.

Parsons, T. and Smelser, N. (1956) *Economy and Society*, Glencoe, IL: The Free Press.

Pavitt, K. and Patel, P. (1991) "Large firms in the production of worlds technology: An important case of non globalization," *Journal of International Business Studies*, 22: 1–21.

Perroux, F. (1950) "Economic space: Theory and applications," *Quarterly Journal of Economics*, 64: 89–104.

Polanyi, M. (1967) *The Tacit Dimension*, New York: Double Day Anchor.

Porter, M. (1990) *The Competitive Advantage of Nations*. London: Macmillan.

Porter, M. and van der Linde, C. (1995) "Green and competitive," *Harvard Business Review*, (September–October): 120–134.

Prop. (1996/1997:5) *Forskning och samhälle* (regeringens forskningspolitiska proposition, gvt. proposal of R&D policy), Stockholm: Riksdagen.

Putnam, R.D. (1996) *Den fungerande demokratin: medborgarandans rötter i Italien*, Stockholm: SNS Förlag.

Radcliffe-Brown, A.R. (1952) *Structure and Function in Primitive Society*, London: Cohen and West.

Rickne, A. (2000) *New Technology-Based Firms and Industrial Dynamics: Evidence from the Technological System of Biomaterials in Sweden, Ohio and Massachusetts*, Doctoral thesis, Department of Industrial Dynamics, Chalmers University of Technology, Gothenburg.

Rickne, A. *et al.* (2009) *Conceptualizing and Measuring Operative Interaction and Collective Action in a Region*, The Dahmén Institute, Report to VINNOVA.

Saxenian, A. (1994) *Regional Advantage: Culture and Competition in Silicon Valley and Route 128*, Cambridge, MA: Harvard University Press.

Schumpeter, J. (1934/1962) *The Theory of Economic Development*, Cambridge, MA: Harvard University Press.

Schumpeter, J. (1942/2000) *Capitalism, Socialism and Democracy*, London and New York: Routledge.

Sellenthin, M. (2006) *Beyond the Ivory Tower: A Comparison of Patent Rights Regimes in Sweden and Germany*, Linköping, Sweden: Linköping University.

Sharp, M. (1983) *Europe and the New Technologies*, London: Frances Pinter.

Shinn, T. (2002) "The triple helix and new production of knowledge: Prepackaged thinking on science and technology," *Social Studies of Science*, 32 (4): 599–614.

Simon, H. (1969/1996) *The Sciences of the Artificial*, 3rd edition, Cambridge, MA: MIT Press.

Teece, D. (1986) "Profiting from technological innovation: Implications for integration, collaboration, licensing and public policy," *Research Policy*, 15: 285–305.

Torre, A. and Gilly, J.P. (2000) "On the analytical dimension of proximity dynamics," *Regional Studies*, 34 (2): 169–180.

Treib, O., Bähr, H. and Falkner, G. (2007) "Modes of governance: Towards a conceptual clarification," *Journal of European Public Policy*, 14 (1): 1–20.

Utterback, J. *et al.* (eds) (2006) *Design Inspired Innovation*, Singapore: World Scientific.

von Bertalanffy, L. (1950) "An outline of general system theory," *British Journal for the Philosophy of Science*, 1: 139–164.

von Krogh, G., Kazuo, I. and Nonaka, I. (2000) *Enabling Knowledge Creation*, New York: Oxford University Press.

Wittgenstein, L. (1953/1989) *Philosophical Investigations*, Oxford: Basil Blackwell.

3 "Spaces"

A triple helix governance strategy for regional innovation

Henry Etzkowitz and Marina Ranga

Introduction: From "spheres" to "spaces"

Capitalising on a region's knowledge resources has become an increasing priority for those who are interested in regional economic development, whether at the national or local level. This growing focus on the knowledge production, diffusion and capitalisation capacity of a region has led to a rethinking of regional policy and practice. In this regard, the triple helix model of university–industry–government relations has been found to be of increasing salience to regional innovation policy and practice (Etzkowitz, 2008). The university–industry–government institutional spheres may be variously aligned, with one serving as a gravitational centre around which the others rotate, with different actors taking the lead. For instance, in a *statist regime (triple helix I)*, government plays the lead role, driving academia and industry. In a *laissez-faire regime (triple helix II),* industry is the driving force, with the other two spheres as ancillary support structures. In a knowledge-based society, university and other knowledge-producing institutions play an increasing role, acting in partnership with industry and government, and even taking the leadership in joint initiatives, in a balanced model *(triple helix III)*.

A triple helix approach to regional development originated in Boston during the Great Depression of the 1930s (Etzkowitz, 2002), and has since spread across the US and further afield to Europe, Asia, Africa and Latin America. A seemingly unique historical experience was subsequently found to characterise innovation and regional development dynamics more broadly around the world. The boundary-spanning interaction among the institutional "spheres" of university–industry–government generates a series of "spaces," a physical or virtual bounded context constituted on various criteria that is essential for the process of knowledge-based regional economic and social development. The concept of "triple helix spaces" derives from the experience of New England Council from the 1920s to 1950s as a summit organisation representing regional leadership that included academia, next to industry and government, because this region had an early and unique concentration of academic resources including MIT, Harvard and a wide range of other academic institutions – "the Knowledge Space." The remit of the Council, created by the Governors of the six New England states, was to develop a strategy for the renewal of a region that had been in economic decline from the early twentieth century due to departure of industries and firms to regions with sources of raw materials and possessing cheap labour. After initial attempts to attract branch plants and renew SMEs in dying industries, the Council turned to the region's unique resource and comparative advantage, its high concentration of academic resources. They focused upon the start-up phenomenon of firms emanating from MIT and Harvard in scientific instruments from the turn of the century and in the newly emerging radio industry in the 1920s, and invented the venture capital firm to expand and intensify this process – "the Innovation Space."

This paper discusses the creation and development of the triple helix spaces of knowledge, innovation and consensus, showing how they contribute to the governance of regional development. We draw upon the first author's interviews with members and observers of various projects in Silicon Valley, Rio de Janeiro and Amsterdam, participation in meetings at the New York Academy of Sciences during the mid-1990s, expert interviews in Niteroi, Brazil in 2002 and archival research in San Diego and Boston to better understand the generalisation of the New England experience into a triple helix spaces model with policy implications. The following three sections discuss the dynamics of the knowledge, innovation and consensus spaces. After that we examine recent events in Sweden as an instantiation of triple helix spaces. Before concluding we delineate the non-linear dynamics of the spaces and possible transitions among them. Key factors, including a time dimension, are introduced to better understand the governance process of regional innovation.

The knowledge space

The knowledge space is primarily the realm of R&D in science and technology, located in universities, firms and government labs, as well as its functional equivalent in the arts, which generates artistic and cultural activities created through an organised collective process, similar to scientific R&D, but with their own distinct discovery, validation and dissemination procedures. However, non-R&D activities, such as technology adoption, combining existing knowledge in various ways, organisational change, etc., which have been increasingly recognised in recent years to play an important role in improving productivity and competitiveness and fostering innovation, can also be included in the knowledge space.

The creation of a knowledge space may occur through a variety of paths involving different local and regional resources, such as relocation or aggregation of existing research organisations and formation of new ones, networking, etc., as exemplified below:

- *Relocation of some national public research organisations from the capital to less research-intensive regions.* This was the case of government research laboratories moved from Mexico City to other regions of Mexico after the mid-1980s earthquake, and inserted into new surroundings where they took on a new potential (Casas *et al.*, 2000). The researchers started to use their skills and the lab resources to address problems in their new locality. This policy was eventually broadened from a diffusion of research resources to an explicit regional development policy, and more research institutes were transferred from the capital to other regions to give them a research capacity that had heretofore been lacking. Another example of moving existing research resources is North Carolina's Research Triangle – a classic case of using the state's political clout to gain second level government labs as an attractor of second-level corporate labs as the initial framework for hi-tech development (Hamilton, 1966).

- *Foundation of a science-based university*, as in San Diego, where a new branch of the University of California was gestated in the 1950s and eventually became the basis for a leading hi-tech complex. The attraction of leading researchers in fields with commercial potential, such as molecular biology, was early recognised as an economic development strategy by the coalition of academic, business and political leaders that called for the founding of this campus. The transformation of San Diego from a naval base and military retirement community to a knowledge-based conurbation followed. The University of California, Merced has recently been established as an "entrepreneurial university" to promote high-tech development in an agricultural region, with the intention of replicating the San Diego experience of several decades earlier. The strategy, following the classic endogenous model, albeit with significant exogenous support, was to create and then leverage location-specific knowledge assets to induce new investment and create new value.
- *Congregation of geographically dispersed groups from university and industry around common themes, with government support,* such as the Canadian Networks of Centres of Excellence (NCE)[1]. Highly qualified personnel work co-operatively through NCE-sponsored initiatives of industrial relevance. Their work involves everything from leading-edge investigations into the causes of and potential cures for cancer to the use of robots and intelligent systems to perform tasks in difficult environments. Large government sponsorship originally motivated widely dispersed academic and firm research units to work together on a proposal, typically dividing up the funds to extend existing local projects, with a suitable overlay of collaborative rhetoric. Through the interaction and discussion necessary to get a proposal, new research ideas were produced that led to the transmutation of the sponsor's rhetoric and conditions of funding into a new reality in which projects created for the sake of capturing funds and splitting the resources became genuine intellectual collaborations.
- *A hybrid format that networks existing knowledge-based organisations and creates new ones through collaboration among existing players.* Several examples of this phenomenon may be found in Sweden, including

the founding of the Stockholm School of Entrepreneurship as a joint initiative of Stockholm University, Royal Institute of Technology (KTH), and more recently also including the Royal Art College. The Oresund project linking southern Sweden (Skåne) and Copenhagen included the creation of "Oresund University," an organisation that encourages collaboration and joint projects between universities on both sides of the strait that previously divided this cross-border region. Karolinska Institute initiated a university-building strategy of incorporating a series of small schools in the biological sciences, nursing and other loosely related fields scattered across Sweden, and even across the Norwegian border, in order to create a greater "critical mass" of research, training and "third mission" activities. The aggregation of local resources into a critical mass is an important driver of collaboration in small regions and countries that, sooner or later, realise its necessity in order to be internationally competitive.

- *Re-organisation of research funding from a linear research council model to an interactive model.* Sweden has undergone a revolution in its research funding system that, to a greater or lesser degree, can be seen in many other countries. There has been a shift in emphasis from funding research as an end in itself, or for military or other specific purposes, to encouraging institutional spheres to work more closely together to promote innovation. A group of foundations were founded in the early 1990s to focus on filling gaps in the country's innovation system. One of their main objectives is to encourage internal academic reform, as well as assist universities to move out of the governmental sphere, engage more closely with industry and become more involved in their regions. Heretofore, most research councils were oriented to the older universities and traditional academic disciplines. An industrially oriented research council was focused on the country's traditional mid-tech industries and a limited range of higher-tech fields. The foundations opened up a rigid innovation system both by making available alternative sources of funds and by their willingness to seek out new research providers, beyond creating a diversity of funding sources in a research funding system that had become out of phase with the country's innovation needs. The foundations are a "wild card" in the Swedish innovation system, with the resources to pursue new directions. The logic of politics has provided Sweden with a "war chest" to assist its transition to a new innovation regime. One question is what direction to take: whether to emphasise forming local interdisciplinary "critical masses" of researchers, whether academic, governmental or industrial, as virtual institutes, versus geographically decentralised networks of sectoral R&D activity. Of course, various combinations of strategies resulting in a four-fold matrix are also possible.

Research, or more specifically, R&D, is thus a significant factor in the knowledge space, but it can often be under-recognised in the development of other industries, such as the creative industries, simply because it does not take place in labs with white-coated personnel.[2] For example, the arts have a functional equivalent of R&D that performs similar roles, such as *The Kitchen*[3] in New York City's Soho District that invent new forms of conceptual art, new artistic formats and modes of performance that serve as models for other artists and are disseminated through international performance tours. Although *The Kitchen* members and followers did not explicitly view themselves from an innovation perspective, they instigated an innovation process in their domain.

Another example is the fashion industry, which is typically viewed as a cluster of firms, with educational institutions playing a peripheral training role. However, educational institutions such as the Fashion Institute of Technology in New York and the Costume Institute at the Metropolitan Museum of Art also constitute "R&D Centres" in a non-traditional sense. The Institute is the cultural memory of the industry that is regularly utilised as a source of ideas in the form of historical styles that may be reinterpreted in new ways with new materials or hybridised into new formats. The Fashion Institute's student training is a forcing ground of new ideas that help new entrants to the field make their mark. Thus, the design class in the Fashion Institute serves a similar purpose to the teaching laboratory, the classic engine of chemical research, in utilising students as an R&D resource.

After-hours clubs have also been recognised as providing venues for artists, fashion designers and other creative individuals to develop new projects across the disciplines, warranting permission to keep late hours in recognition of their economic development mission (Currid, 2007). Movement of artists into abandoned industrial buildings and preserving them as low-cost space for qualified persons through creative use of municipal zoning authority regulated the transition of a declining manufacturing district into Soho, the arts equivalent of a science city project based on advanced academic research (Etzkowitz and Raiken, 1980).

Public arts funds and more general public employment schemes supported the work of many of these artistic and cultural organisations that provided the infrastructure that led to the transformation of abandoned industrial sites into chic neighbourhoods. Ironically, they often became too expensive for their pioneers, driving them to new frontiers of declining neighbourhoods elsewhere in the city and region. The phenomenon has been recognised to the extent that municipalities seeking renewal often invite arts groups to relocate to subsidised space in order to jump-start a gentrification process that has become an overlay on arts-based economic renewal. An interesting example in this sense is Barcelona's *@22* urban science park project to recycle an old industrial district into a platform for knowledge-based enterprises.[4] Here, in a first stage, artists who had taken over abandoned factories and utilised them for ateliers, galleries and living spaces were evicted by the police, but several years later, *@22's* leadership invited artists to return in order to revivify an increasingly sterile urban environment, where many of the early twentieth-century factories, documented in an architectural survey volume, had been torn down and replaced by anonymous modern structures. *@22* is a success in terms of attracting national and multinational firms to locate in Barcelona. However, its top-down design failed to take account of and incorporate spontaneous bottom-up developments, such as the artists' influx, that could have made it an even greater success as a hybrid technology/arts district, with a greater potential to spawn creative industries at the interface.

A similar promising opportunity was lost in Brazil, when the Porto Digital software technology park project to recycle an under-utilised old commercial district allowed the construction of a shopping mall at the edge of the district that sucked the life out of a street of cafes, galleries and boutiques that had taken root as a nascent cultural district before the founding of the urban science park. A decade ago, it appeared that the members of the software start-ups would take advantage of these cafes to socialise, much as Silicon Valley bars provided a space for interaction among firms during the early days of Silicon Valley (Saxenian, 1994) and thereby expand the firm-formation activities of the Porto Digital Project by informal means.

Regional strategies to develop the knowledge space could focus on: mapping regional actors such as public and private research labs, firms, universities, arts and cultural organizations, etc. and analyzing their evolution and future trends, understanding their priority-setting and the design of their agendas, scope of operations (regional, national, international) and regional impact. Policy initiatives may also be directed at developing human resources for R&D in the sciences and arts at national/regional level, including labour market aspects for researchers, employment, education and training, immigration to attract world-class researchers, making research careers more available to various categories of the local population, especially women and minorities, reducing brain drain and improving brain gain at various stages of education and research career. When R&D and educational resources are weak, the strategy focus should be on their development; when they are strong, the focus will be on how best to utilise these capacities to disseminate the knowledge and enhance the innovation-based economic growth.

An essential condition for the creation and consolidation of the knowledge space is the existence of a '*critical mass*' of academic research and education resources on a particular topic in a local area, as well as other R&D and non-traditional R&D actors. The university's integrating and convening ability in providing a neutral site for other knowledge actors to collaborate increasingly makes it the key actor in the knowledge space (Etzkowitz, 2002). This comes from the unique characteristic of the university, in contrast to R&D units of firms and government-sponsored research institute, to have a high flow-through of human capital that occurs through its original mission of teaching. By admission and graduation of students, this flow-through continually brings new people and new ideas into the academic research group, improving its chances of creativity. As students are newly trained in research, they are typically less bound by received ideas and thus more likely to try new approaches. When they graduate and gain employment, they perforce act as knowledge transfer carriers and by their departure open up room for new entrants into the group on a continuous basis. This enhances the chances of organisational renewal, in comparison to research institutes and R&D units of firms, which are more static organisational models in comparison to academic research groups. Only by continuously reviewing and rethinking its specialties, both at the level of individual positions, departments and even faculties, can universities reinvent themselves to be at the knowledge frontier.

The innovation space

The innovation space is the venue for recombining elements of existing organisational models together with new concepts for organisational functioning in order to invent better ways to encourage and promote innovation. Activities in the innovation space include the aggregation of resources to create new organisational formats, induction of people into newly conceptualised roles and the creation of legitimating themes to justify the enterprise by linking it to both old and new societal goals. At the local level, this space involves the adaptation or creation of new organisations to fill a gap in the regional innovation environment, as well as in other social systems in need of renewal. Such new organisations typically have hybrid formats, synthesising elements from different institutional spheres, e.g. the origin of the venture capital firm from a university–industry–government collaboration in New England during the 1930s and 1940s to fill a seed capital and business support gap in that region (Etzkowitz, 2002).

The appropriate organisational format is a function of the strengths and weaknesses of the region. For example, a region with informal networks of angel investors and serial entrepreneurs would be less likely to need formal incubator facilities than one lacking these capabilities. On the other hand, a region lacking a university may want to create one, either as a means of raising the technological level of existing clusters or as a source of new ones. Other typical developments involve linking one mechanism to another in a sequence of steps, e.g. incubation may lead to the expansion of entrepreneurial training to increase the number of candidate firms. However, the order of steps is not predetermined, and depends on local circumstances. Thus, the need for an analysis of the strengths and weaknesses of a region as an initial step to formulating a strategy rather than simple adoption of a current example of best practice drawn from a successful region.

The common characteristic of the innovation space organisational formats is the *knowledge-based entrepreneurship* that is enacted as an institutional phenomenon encompassing all the triple helix actors. They can undertake collaborative projects to enhance the regional innovation environment and improve its abilities to serve as the take-off platform for business ventures. Lacking such an environment, a firm will be deprived of "oxygen" and be forced to move to a more supportive local context. For example, a Northeast UK inventor with a technology similar to that of SUN Microsystems found that he was unable to move his technology forward locally, because the local shortage of competitors and collaborators made it difficult for the firm to flourish in its original environment, and sold his firm to SUN and moved to Silicon Valley for a period of time to transfer his technology.

The concentration of entrepreneurial talent, intellectual capital and tacit knowledge in relatively few world-class regions gives them a clear competitive advantage in drawing talent and innovative firms into their orbit from emerging regions. For instance, Silicon Valley is noted for its relative strength in semiconductors and software, while Boston is widely recognised formerly for minicomputers and presently for biotechnology. This clustering reflects the economies of scale generated by locating in the same area as other firms working on similar technologies or products (Feldman and Francis, 2004). Location advantages of successful hi-tech regions are a result of a long-term development process, based on local entrepreneurial capacities and institutional support for new firm formation and innovation (Mason and Harrison, 1992). Other regions with agglomerations of research resources, such as New York and Chicago, largely failed to develop commensurate hi-tech clusters, to date. Nevertheless, in the wake of the downturn of New York's financial industry, steps are being taken to build alternative economic drivers on the region's extensive research base.

Entrepreneurial innovation is a change-inducing force and a driver of regional development that can be built from various cultural foundations,[5] and relies on values, attitudes and behaviours that can be combined with various policy measures to foster a strong entrepreneurial orientation. It is also influenced by the motivation, aptitude and level of economic and technological performance of the region (Thwaites and Wynarczyk, 1996; Lee and Peterson, 2000), which can be strengthened by effective entrepreneurial training programmes and business plan competitions. Indeed, such support measures are increasingly found worldwide, in a variety of localities, regions and nations (Morris, 1998). Even as knowledge-based regional development is taken up in a wide variety of contexts, a relatively common series of stages and phases may be identified.

The diachronic co-evolution of the triple helix institutional spheres in the innovation space is exemplified by the constant expansion and adjustment taking place in this space. When a university takes an initial step in establishing a liaison or technology transfer office, it soon realises that a much broader range of services and support structures are required in order to market intellectual property and create spin-off firms. Sometimes, this involves inserting the new innovation mechanism into a broader institutional environment such as a national or regional-wide network of transfer offices in order to identify market opportunities and partners. Other times, an alliance with local city and regional governments may pave the way to funding an incubator facility to assist in spinning off firms. Innovation is a never-ending process and the innovation space is insatiable in its demands. This explains why the state of California recently passed a $3 billion stem cell research initiative to strengthen the biomedical research capacities of its universities, combined with programmes directed at enhancing the availability of venture funds and assistance to start-up firms. Debt-funding mechanisms, previously utilised to develop moderate-income housing, were adapted to generate research funds and public venture capital (Etzkowitz and Rickne, 2009).

An emerging model of knowledge-based regional economic development is not one tied to any specific mechanism, whether science park, incubator or a venture capital firm. It typically takes shape within a geographical region, with an idea of bringing the governmental, academic and industrial resources into a common organisational framework where an analysis is made of regional strengths and weaknesses. Then some thought is given to how these gaps can be filled. In some instances, they can be filled by extending the traditional role of the university or by new government programmes. Sometimes, industry fills the organising role. The framework for looking at these activities is not to see them as either rooted in extending academia into new functions or industry extending itself by moving into increasing the knowledge intensive activities, or government, by itself, instigating programmes but looking at these three actors and their relations with each other as being the source of knowledge-based regional economic development.

The consensus space

The consensus space is, at one and the same time, the physical space, convening platform and virtual framework that brings together the triple helix actors to brainstorm, discuss and evaluate proposals for advancement towards a knowledge-based regime. Through cross-fertilising diverse perspectives, arising from different standpoints, ideas may be generated and results may be achieved that actors could not likely have accomplished individually. In the consensus space, triple helix actors come together to formulate strategies to initiate projects for realising and enhancing the local growth potential. This is essentially a governance space, in the sense of governance defined by Kuhlmann as:

> a process through which a socio-political community achieves binding decisions in the face of conflicting interests. The processes of consensus-building, decision-making or even implementation of decisions are not merely determined by state actors or formal governments. Rather, due to growing complexity and segmentation of modern societies and issue areas, it is the interaction of societal and state actors that defines problems, builds up the necessary degree of consensus on problems and solutions, consolidates conflicting interests, and (pre-) determines political decisions.
>
> (Kuhlmann, 2001: 957)

The consensus space is characterised by the following:

- interdependence between organizations – rather than seeing themselves as isolated entities, firms, universities and local government actors begin to see themselves as part of a larger whole: in some cases, a newly invented identity such as Oresund; at other times, a reviving traditional locality such as Norrköping, Sweden;
- broader coverage of the governance concept, including government, as well as non-government actors;
- changing boundaries of the state, with shifting and more transparent boundaries between public, private and voluntary sectors;
- continuing interactions between network members, caused by the need to exchange resources and negotiate shared purposes;
- game-like interactions, rooted in trust and regulated by the rules of the game negotiated and agreed by network participants;
- networks are not accountable to the state – they are self-organizing; and
- although the state does not occupy a privileged, sovereign position, it can indirectly and imperfectly steer networks.

The diachronic co-evolution of the consensus space is exemplified by the dynamics arising from the application of "multi-level governance" (Hooghe, 1996; Marks *et al.*, 1996; Cooke *et al.*, 2000), at the supra-national (e.g. EU, NAFTA), national, regional or urban levels. For example, in Europe, we assist at an increasing process of devolution of central powers to regional and local administrations, in parallel with an increasing role of EU policies at national and regional level, which influences the co-evolution of "political systems" and "innovation systems" in Europe (Kuhlmann, 2001).

The consensus space is typically a mix of top-down and bottom-up processes to create leadership through collaboration rather than diktat, a neutral ground where the different actors in a region, from different organisational backgrounds and perspectives, can come together to generate and gain support for new ideas promoting economic and social development. A process of "cross-institutional entrepreneurship" spanning the triple helix spheres is thus initiated and aims at improving the conditions for knowledge-based regional development.

The process of getting relevant actors to work together often includes the creation or modification of an organisation to provide a home for brainstorming, analysis of problems and formulation of plans. Examples include the Pittsburgh Hi-tech Council or the Petropolis Technopole in Rio de Janeiro State (Mello and Rocha, 2004), or the Board of the Recife Brazil Science Park, explicitly representing key actors in regional innovation and playing a "quasi-political" role for enhancing innovation capacity in the region. In addition, the very process of including actors from different backgrounds in the strategy review and formulation process may provide access to the resources required to implement a project. The New England Council representing university, industry and government leadership worked out a political strategy to make the venture capital firm a viable entity by lobbying to change laws that prevented large financial institutions from investing in risky ventures. Another example is an informal group of academics and entrepreneurs in Rio de Janeiro that came up with the idea of starting a PhD-oriented new campus of the State University of Rio de Janeiro in Friburgo to provide a neighbouring declining industrial region with high-level knowledge inputs, rather than simply training support personnel for existing firms as might have happened in an undergraduate campus. Similarly, the Knowledge Circle of Amsterdam meets regularly to brainstorm ideas for enhancing knowledge-based development.

Government and its agencies may play an important role in shaping the consensus space, as, for example, the Governors of the New England States' founding of the New England Council in the 1920s. They encouraged a range of public, semi-public, civic and private actors to bring their resources into the system of governance and shape the consensus space. However, when government structures are less active, universities, industry associations and NGOs may take the leadership role in creating a venue that brings relevant actors together to address innovation issues and gaps. Thus, Birmingham University provided an ideal neutral site for the deliberations of a consortium of triple helix actors in projecting the post-Rover, post-automotive future of the West Midlands, UK as a future technology corridor, including a biomedical complex based on area research resources. A boundary-spanning leadership, capable of transcending entrenched local interests, was found to be crucial to this endeavour (Gibney *et al.*, 2009).

The consensus space, like the other spaces, is closely related to the concept of *leadership,* but in this case, the concept gets a broader dimension. Who shall assume a leadership role at the regional level is a frequently asked question, even in countries with strong regional governments. In countries with no or only weak regional governments, there may not be a governmental actor available to take the lead at the regional level, but other actors, such as universities and firms, may come forward to take the innovation organiser role to set forth a future achievable objective. The innovation organiser typically occupies a key institutional position, enunciates a vision for knowledge-based development and has sufficient respect to exercise convening power to bring the leadership of the institutional spheres together. Once the vision is accepted, a broader group representing the spheres may take steps to aggregate and commit resources to implement a project. In the 1930s New England case, MIT's President Compton was the innovation organiser; in the 1990s Silicon Valley downturn, a respected CEO took the lead. In New York in the mid-1990s, the head of the New York Federal Reserve Bank initially took the lead, but did not persist and passed the mantle to the head of the New York Academy of Sciences. In this instance, failure to set forth a compelling vision, with consistent leadership, resulted in a gradual withering away of the consensus space of university, business and government leaders that had originally been convened. Nevertheless, a positive collateral effect was realised through the organisation of university technology transfer offices into a regional group to discuss common issues and problems.

Energising a nascent triple helix

The development of the Swedish innovation system provides an opportunity to study the moving strands of the triple helix. In the *Research 2000* debate, some Swedish academics argued that the two worlds of industry and academia should be kept apart. As one put it, "The academy should be the academy, carry out basic research, open up understanding between worlds rather than changing it. The alternative perspective is that academic research should change dramatically, open up to business." Academic differences are often "academic" until an economic crisis occurs and a decision is made to upgrade from a mid-tech economy to one more closely based on knowledge and academic research. This transition has led to a re-evaluation of the role of the university in society.

Sweden suffered severe decline in the early 1990s as corporations outsourced production and moved headquarters to other countries, threatening the national consensus. In response, universities that were creatures of the government, limited to human capital formation and research, fundamental investigation and serving national firms' interests, were expected by some policy-makers to reproduce US academic entrepreneurship. Following the model of separation, the university's "third mission" is either interpreted as saying the university should be involved in spreading knowledge of its activities or they might say it is a waste of time to do that. They might say the university should have an office to translate its research or be opposed to it. In either event, it would be seen as an activity that would be carried out discretely by different agents of the university.

An alternative view, taken by a few elements in the academic system, such as the regional colleges or the Karolinska Institute, have arrived at a more integrated perspective: to pursue these various activities in tandem as a coherent strategy for conjoint academic and high-tech development, building a critical mass in a particular sector. They see the teaching, research and economic development activities as becoming part of an integrated whole. Certainly, in some of the regional colleges, their natural path to development is to argue that resources should be placed with them so that they may pursue these activities in tandem. At the Karolinska Institute, the network model of university of incorporating various pieces of academia from parts of the country to create a critical mass in order to pursue these various activities represents another possible model for academic hi-tech development. It transcends the focus on a local region, but tries to build critical mass in a particular sector across a broader geographical span by creating a virtual knowledge region.

Some elements in the Swedish academic system see the teaching, research and economic development activities of the university as the basis for a new role as regional leaders. Certainly, at some of the regional colleges, their natural path to institutional development is to argue that resources should be placed there so that they can pursue this new mission. What is especially interesting is that older universities such as Uppsala in collaboration with non-elite agricultural universities are also exploring a new role in regional innovation. Networks are generated from collaborations between large firms and academic researchers, e.g. Pharmacia and Uppsala University, that left in place a substrate of ties that became the basis for new firm formation in biotechnology. It also appears informally among firms in a common area of activity that then may be formalized into a "valley" through the organisation of an association, e.g. Radio Valley in Gothenburg, Sweden.

Sweden has undergone a revolution in its research funding system that, to a greater or lesser degree, can be seen in many other countries. There has been a shift in emphasis from funding research as an end in itself, or for military or other specific purposes, to encouraging institutional spheres to work more closely together to promote innovation. A group of foundations were founded in the early 1990s to focus on filling gaps in the country's innovation system. One of their main objectives is to encourage internal academic reform, as well as assist universities to move out of the governmental sphere, engage more closely with industry and become more involved in their regions. As each institutional sphere "takes the role of the other," the stage is set for a new series of closer interactions and collaborations. The foundations thus play a key role in facilitating and incentivising university–industry collaborations, accelerating a process underway for other reasons.

The premise of governmental activism is that the conditions for hi-tech economic growth are not spontaneous creations; rather, they can be identified and put in place by explicit measures. As regions formulate knowledge-based innovation strategies, the constellations of actors, and their relative importance in the local political economy is transformed. With knowledge assuming increased significance as a factor of production, in both high-technology and older manufacturing industries, the traditional elements of land, labour and capital reduce in importance with various political consequences, including the inclusion in regional growth coalitions of knowledge-producing institutions such as universities.

The foundations revitalised the traditional research funding system. Heretofore, most research councils were oriented to the older universities and traditional academic disciplines. An industrially oriented research council was focused on the country's traditional mid-tech industries and a limited range of higher-tech fields. The foundations opened up a rigid innovation system both by making available alternative sources of funds and by their willingness to seek out new research providers. Beyond creating a diversity of funding sources in a research funding system that had become out of phase with the country's innovation needs, the foundations typically have taken a more proactive role in R&D management as "public venture capitalists."

Top-down incentivisation of bottom-up and lateral interactions characterise Sweden. Counties were required to develop growth strategies in order to access cluster development funds. VINNOVA, the national innovation agency, encouraged regions to support knowledge-based development in collaboration with universities. Through grants competitions such as the VINNVÄXT programme, requiring evidence of triple helix collaborations in developing the project, consensus spaces were induced across Sweden. Beyond simple financial incentive, VINNOVA also introduced training programmes in triple helix interactions to facilitate collaboration among regional industrial academic and governmental actors. Some universities were recruited into these regional efforts, while others played a leading role in organising them.

Swedish universities are changing from an ivory tower stance and support structure to the existing industrial infrastructure to play a more prominent role in the creation of a new industrial base. Entrepreneurship is expected to be an integral part of academia and government, as well as industry. The Linköping case involves both the creation of the necessary conditions for a multi-linear model, the establishment of a broad-based research university but also the sufficient conditions, the specific mechanisms to translate knowledge into useful economic activity. Only some of these projects are internal to the university; most involve a collaboration process with industry and government actors. At Stanford, the university was built from an industrial fortune; at Linköping, government-industry collaboration initiated the university development project. Both instances involved projection of a new economic base for the region based on firm-formation from academic knowledge.

Conclusion: The new dynamics of regional innovation

How can aspiring knowledge regions best learn from world leaders? A synthesis of contemporary best practice is often relied upon to create technopoles, innovative milieus, learning regions, science cities, etc. However, contemporary practice may not always be the most productive starting point for an aspiring region, as it sets the bar very high and often ignores the early developmental phases that may be more relevant to an emerging region. Transferring a particular mechanism, such as venture capital or incubation, from a success case may not "take," unless an appropriate local context has also been created to support its insertion. Just as an organ transplant may take or be rejected, depending upon surrounding conditions in the organism, so the successful transplantation of social mechanisms is similarly dependent upon local conditions. Thus, the same mechanism may "take" in one set of circumstances and be rejected in another. Identifying and/or creating conditions appropriate for transfer is thus an important part of the regional policy-making remit. Emerging regions may benefit more by adopting a model abstracted from the formative stages of success cases, rather than imitating features of late development stages of successful knowledge regions. The fallacy of taking the end for the start, ignoring the long-term nature of endogenous development, may short-circuit development.

The dynamics of triple helix spaces is important in the transition among stages of regional development. Elsewhere (Etzkowitz and Klofsten, 2005) we outlined a four-stage model of regional growth and renewal, as follows:

1. *Genesis*: creating the idea for a new regional development model.
2. *Implementation*: starting new activities and developing infrastructure.
3. *Consolidation and adjustment*: integration of activities to improve the efficiency of the infrastructure.
4. Self-sustaining growth and renewal of the system by identifying new areas of growth.

At the genesis stage, a key issue is the state of the knowledge space, which can already exist or must be created or enhanced. A comparison among the New England, Linköping and Freiburg cases is instructive. In the first, the knowledge space was significant; in the second, moderate; and in the third, it was virtually lacking. In these cases, different development decisions were taken: the New England Council moved directly to the innovation space to promote commercialisation. In Linköping, a move to strengthen the knowledge space was made: in order to extend mechanical expertise into emerging areas, an interdisciplinary university was started to promote interrelations and interconnections, as well as development of new fields such as computer science. A similar approach was also adopted in Freiburg, where a tightly focused advanced academic programme was created in order to infuse new knowledge into traditional industries.

The knowledge space is also especially salient at the stage of self-sustaining growth and renewal phase. This was especially noteworthy in Silicon Valley in the mid1990s, where many successful firms had outgrown their university links, or were spin-offs of an early generation of firms and had never developed extensive academic links. Indeed, by this time, many of the Valley's high-tech firms tended to view themselves as a self-generated phenomenon, a cluster of interrelated firms, rather than as part of a broader university–industry–government complex. However, in the mid-1990s downturn, such firms felt the need to connect or reconnect to academic institutions and local government in order to move the region forward. A new organisation, Joint Venture Silicon Valley, was established for this purpose and a public brainstorming process was initiated in the form of a series of open meetings focused on generating ideas for the future technological candidates. A venture capital approach was taken, with a few promising ideas, such as computer networking, selected from a larger collection (Miller, 1997).

The innovation process folds back in on itself when one space becomes the basis for the development of another. For example, science parks created at a university as a mechanism to retain firms originated from that university, as well as provide an income stream to the region, could subsequently be replicated by other universities to assist the firm-formation process and enhance links between local R&D units and university researchers. Stand-alone science parks were also established, primarily as a site for large corporate R&D units and branch R&D units of multinational corporations. In recent years, science parks, such as Stockholm's Kista, which have lost many of their original clients, have taken steps to renew themselves by expanding their knowledge space, founding entrepreneurial universities as a source of future start-ups.

Innovation can no longer be assumed to take a conventional linear path, whether from research through development or from identification of market opportunities to product introduction. In some countries, there is a movement away from an assumption that there is single starting point of research and an end point of the economy: an autonomous linear model based on laissez-faire assumptions in which innovation takes its own course. Innovation was expected to largely take place within industry with other institutional spheres playing only a limiting contributing role, government, for example, acting only when clear market failures could be identified. In countries that, to one degree or another, relied on central planning, it has become accepted that government programmes have an important role to play, not only from the national level – top down – but also from the local level – bottom-up, often in collaboration with other organisations in civil society.

Innovation policy can no longer be conceived only as a "top-down" initiative of national government but should also be seen as the cumulative result of the creation of regional governance structures through interaction among governments at various levels, businesspersons, academics and NGOs comprising membership from all of these spheres. A triple helix regional innovation system, with self-renewing capabilities, is created through interaction among spheres and spaces. When a knowledge space exists without a consensus space, full advantage is unlikely to be taken of its potential due to the lack of a convening and organising process to create the intermediary and transfer organisations and networks – the innovation space – that are the breeding ground of new knowledge-based clusters. This multi-dimensional model of innovation may be used as a benchmark to evaluate existing projects, as well as a theoretical framework to guide new policy initiatives. Triple helix spheres and spaces are the "Lego building blocks" of the knowledge society.

Notes

1. See www.nce-rce.gc.ca/index_eng.asp for further details on the Canadian Networks of Centres of Excellence.
2. Notable exceptions, albeit without the "uniform," include MIT's Media Lab and its worldwide progeny at IIT Kanpur and Tokyo University and Newcastle University's Culture Lab.
3. *The Kitchen* is a non-profit, interdisciplinary organisation that provides innovative artists working in the media, literary, and performing arts with exhibition and performance opportunities to create and present new work. Using its own extensive

history as a resource, the organisation identifies, supports, and presents emerging and under-recognised artists who are making significant contributions to their respective fields, as well as serves as a safe space for more established artists to take unusual creative risks (www.thekitchen.org).
4. See www.22barcelona.com for details.
5. For example, Sombart (2001), in his analysis of the Catholic merchants of Bruges, found similar entrepreneurial characteristics to those that Max Weber (2003) identified in his classic work *The Protestant Ethic and the Spirit of Capitalism.*

References

Casas, R., R. de Gortari and M.J. Santos. (2000). "The building of knowledge spaces in Mexico: A regional approach to networking." *Research Policy* 29 (2): 229–241.

Cooke, P., P. Boekholdt and F. Todtling. (2000). *The Governance of Innovation in Europe*. London: Pinter.

Currid, E. (2007). *The Warhol Economy: How Fashion, Art, and Music Drive New York City*, Princeton: Princeton University Press.

Etzkowitz, H. (2002). *MIT and the Rise of Entrepreneurial Science*. London: Routledge.

Etzkowitz, H. (2008). *The Triple Helix: University–Industry–Government Innovation In Action*. London: Routledge.

Etzkowitz, H. and M. Klofsten. (2005). "The innovating region: Towards a theory of knowledge based regional development." *Research Management* 35 (3): 243–255.

Etzkowitz, H. and L. Reiken (1980). *Artists Social Movements of the 1960's and 70's: From Protest to Institution Formation*, Conference Paper, 23 March 1980, Eastern Sociologial Society, Boston, MA.

Etzkowitz, H. and A. Rickne. (2009). "Science policy and direct democracy: Proposition 71, California's experiment in stem cell innovation." Paper presented at the Triple Helix 7 Conference, Glasgow, 17–19 June 2009.

Feldman, P.M. and J.L. Francis. (2004). "Home-grown solutions: Fostering cluster formation." *Economic Development Quarterly* 18 (2): 127–137.

Gibney, J., S. Copeland and A. Murie. (2009). "Toward a 'new' strategic leadership of place for the knowledge-based economy." *Leadership* 5 (1): 5–23.

Hamilton, W.B. (1966). "The research triangle of North Carolina: A study in leadership for the common weal." *South Atlantic Quarterly.*

Hooghe, L. (ed.) (1996). *Cohesion Policy and European Integration*. Oxford: Clarendon Press.

Kuhlmann, S. (2001). "Future governance of innovation policy in Europe: Three scenarios." *Research Policy* 30: 953–976.

Lee, M.S. and S.J. Peterson. (2000). "Culture, entrepreneurial orientation and global competitiveness." *Journal of World Business* 35 (4): 401–416.

Marks, G., F.W. Scharpf, P.C. Schmitter and W. Streeck. (1996). *Governance in the European Union.* London: Sage.

Mason, C. and R. Harrison. (1992). "Strategy for closing the small firms finance gap." In: K. Caley, D. North, R. Leigh and D. Smallbone (eds) *Small Enterprise Development*. London: Paul Chapman.

Mello, J.M.C and F.C.A. Rocha. (2004). "Networking for regional innovation and economic growth: The Brazilian Petrópolis technopole." *International Journal of Technology Management* 27 (5): 488–497.

Miller, W. (1997). "Interview with Henry Etzkowitz." Stanford University Business School and Joint Venture Silicon Valley.

Morris, M.H. (1998). *Entrepreneurial Intensity: Sustainable Advantages for Individuals, Organisations, and Societies*. Westport, CT: Quorum Books.
Saxenian, A.L. (1994). *Regional Advantage*. Cambridge, MA: Harvard University Press.
Sombart, W. (2001). *Economic Life in the Modern Age*. Piscataway, NJ: Transaction Publishers.
Thwaites, A. and P. Wynarczyk. (1996). "The economic performance of innovative small firms in the South East region and elsewhere in the UK." *Regional Studies* 30 (2): 135–149.
Weber, M. (1904/2001). *The Protestant Ethic and the Spirit of Capitalism*. London: Routledge Classics.

4 Regional dynamics in non-metropolitan hi-tech clusters

A longitudinal study of two Nordic regions

Jukka Teräs and Håkan Ylinenpää

Introduction

This chapter is an empirical journey into the development of knowledge-based clusters in two Nordic non-metropolitan regions that, despite a similar history and seemingly similar development conditions, reveal different development trajectories: the Oulu region in Finland and the Luleå region in Sweden.[1] The purpose is to investigate important factors related to the dynamic development of knowledge-based clusters utilizing longitudinal data in order to develop a more in-depth understanding of the often non-trivial and multi-faceted factors and interdependencies in play. Building on historical secondary data and data from studies in 1998 and 2008, we compare and analyze the dynamic development of knowledge-based (hi-tech) clusters for these two contrasting cases. Which factors and conditions are decisive for understanding why one region will experience a dynamic development of its knowledge-based industries while a twin region located to a non-metropolitan area with seemingly similar development conditions will experience a far less favorable development path? Are the answers to this quiz stable and given when recognizing that the "rules of the game" for developing regional competitiveness may be affected by a new competitive logic formed by, for example, a more globalized economy?

In order to address these questions, we will first outline the theoretical underpinnings that we will build on in this chapter. In a second section, we will then describe the methodological approach we have utilized in our empirical and longitudinal study of our two case regions, followed by a third section where we present some basic data on the selected non-metropolitan regions under study. In our fourth section, we report on the results from our empirical study of these two regions and especially compare findings from a previous, comparative study from 1998 with a more recent 2008 study on the same two regions. In a concluding section, we summarize our findings as a base for suggested "lessons to be learned" for academic research and practice.

Theoretical points of departure

The fact that different regions have diverging trajectories continues to spur interest in the issue of what constitutes factors and drivers for a dynamic regional development. With the introduction of a more knowledge-based economy, the importance of understanding high-technology industries has been especially underlined, paving the way for concepts and theories employed in innovation system research. Examples of new or reused concepts often utilized in research depicting regional dynamics are, for example, different kinds of innovation systems, clusters and critical mass, or theories and concepts related to the dynamic evolution of regions.

The blessings and dynamics emanating from the knowledge-based economy are unevenly distributed in space, since knowledge organizations as well as business innovation activities tend to concentrate to a few specific regions (Cooke *et al.*, 2007). Regional decision-makers and regional development experts in, especially, non-metropolitan areas therefore face the challenge of how to stimulate and maintain competitiveness and innovation in regional industries, and how to achieve an alert and dynamic responsiveness to a rapidly changing environment (Kautonen, 2006). Some guidance related to this challenge may be found in previous research in the field, e.g. in research related to the concept of regional innovation systems (RIS) with its roots back to Marshall's early studies of Italian industrial districts (Marshall, 1916). Similar to research on national innovation systems, this stream of research regards the territorial aspect of innovation systems as crucial. Fertilized by Saxenian's (1994) studies of Silicon Valley and Route 128 and Porter's (1998a, 1998b) studies of clusters as a territorial phenomenon, research in this stream has highlighted the importance of localized learning, and how social interaction based on cultural and spatial closeness facilitate trust, exchange and innovative activities in local or regional systems of interacting actors (cf. Maskell *et al.*, 1998; Cooke *et al.*, 2000; Asheim and Coenen, 2006).

The territorial aspect is, however, not the only rationale for defining and investigating innovation systems. As pointed out by Carlsson *et al.* (2002), innovation systems may be understood and analyzed by utilizing a number of different analytical and methodological approaches. Laestadius (2007) identified – besides regional innovation systems – three other (occasionally competing) perspectives on how innovation systems are structured and constituted: sectoral innovation systems, technological innovation systems, and development blocks. Sectoral innovation systems depart from the understanding that innovations develop in or with an application for a specific line of industry or industry sector. As examples, new process technology may be related primarily to process industry, new ICT innovations to ICT industry, and new biotech innovations to biotech industry. These industries are normally operating under global competition and are often, to a significant degree, represented by global multinational corporations. While the concept of national innovation systems hence has a main focus on the nation as the primary unit of analysis (and an interest in analyzing the competitiveness of this national unit as compared to other national units), advocates for the sectoral innovation system approach devote their main interest into understanding how, for example, information and knowledge are transferred, adopted and used for innovative work in different lines of industries regardless of national borders (Malerba, 2005). The concept of a technological innovation system (Carlsson and Stankiewicz, 1991) share this basic understanding that technology innovation diffusion and transformation into innovations are not delimited by spatial dimensions such as national borders. Instead they are understood to emerge and develop on the basis of a specific technology, but not necessarily delimited to a specific line of industry. Laestadius (2007) and Johansson *et al.* (2007) provide illustrative examples of how information technology is not only delimited to ICT industries, but serves as the base for innovations in several other lines of industries. Similarly, Coenen *et al.* (2007) describe how knowledge developed on the basis of biotech research is capitalized on in traditional industry sectors such as the food industry. Development blocks, finally, an important Swedish contribution to the theory on innovation systems (Dahmén, 1950; 1984), may be understood as a concept that includes both sectoral and technological aspects of an innovation system, but in a more explicit way links development of an innovation system to Schumpeterian dynamics by recognizing the structural tensions that create new opportunities and thus enable or serve as drivers for innovation.

In the wake of a more knowledge-based economy and a more globally intertwined economy, the importance of university-based research and global networking is often highlighted in recent research (cf. Laestadius *et al.*, 2007). Universities have, therefore, during recent decades been assigned a more important role as actors in different kinds of innovation systems. That modern innovations are not undertaken in isolation has moreover laid the basis for more extended policy-related research, notably in the innovation systems literature, but also paved the way for theoretical concepts such as open innovation (Chesbrough, 2007). Recognizing, however, that proximity still matters (cf. Maskell *et al.*, 1998), clusters have in recent years dominated the theoretical discussion and the practical operations of decision-makers in the field of regional development. Adopting Porter's (1998a) classic definition, a cluster is a geographical concentration of interconnected companies, specialized suppliers, service providers, firms in related industries, associated organizations (e.g. universities, different kinds of agencies, and trade associations) in particular fields that co-operate but also compete. Porter states that clusters affect competition by increasing the productivity of companies located to a specific area, by driving the direction and pace of innovation, and by stimulating the formation of new businesses (Porter, 1998a). The cluster phenomenon has attracted remarkably increased attention during the 1990s and 2000s, resulting in an avalanche of published academic papers (Maskell and Kebir, 2006). Porter's "diamond model" from 1980 has often been used to address the development of industrial dynamics – first on the national level (Porter, 1990) but later also on a regional level (Ylinenpää and Lundgren, 1998).

The relationship between clusters and different innovation systems concepts has been subject to a number of studies. Most of the literature on clusters regards the phenomenon as nested within, and impacted by, other spatial scales of analysis, including innovation systems (Wolfe and Gertler, 2004). There are, however, differences in the way we understand and define the regional cluster concept in the literature, ranging from a narrow definition (firms only) to broader definitions (firms, research organizations, governmental organizations, research organizations, financial institutions, and institutions for collaboration; see, for example, Sölvell *et al.*, 2003).

Compared to the literature on the cluster structure, the research literature on the dynamic evolution of regional clusters is not too abundant. Porter (1998a) states that self-enforcing cycles promote the growth of a regional cluster, especially when local institutions are supportive and local competition is vigorous, and where the success stories of a growing cluster attract the best talent. Porter, however, also states that clusters can, and over time do, lose their competitive edge due to both external and internal forces (cf. Sternberg, 1995 – a study of Japanese clusters from a life cycle perspective). Adopting a life-cycle perspective, it has been stated that the development of a dynamic cluster could be seen as a process containing four stages, where embryonic clusters represent those at the early stages of growth, established clusters those perceived as having room for further growth, mature clusters those that are stable or will find further growth difficult, and declining clusters as those that have reached their peak and are failing or declining (DTI, 2004). Clusters at this stage are, however, sometimes able to reinvent themselves and enter the development cycle again. Westerberg *et al.* (2007), in an attempt to conceptualize the content and drivers of such "positive spirals," suggested the degree of actor commitment and leadership as an important mechanism driving the development of clustered innovation systems. In such systems, locomotive companies located at the market-end of a regional or local value-processing chain have often been found to play a significant role (cf. Saxenian, 1994 – classic studies of Silicon Valley and Route 128). These often larger corporations serve as an interface to the outer world, introducing new market requirements and new technology to their (smaller) local partners. They also serve as a role model for the growth and future development of their smaller partnering companies. Especially in hi-tech regions, locomotive companies such as Hewlett Packard, Digital Equipment and Nokia have had an important driving function for the development of the regions in which they operate.

Large cities and metropolitan regions hold key positions in the emerging knowledge economy (Cooke *et al.*, 2007). There are, however, numerous examples of smaller, non-metropolitan regions that have been capable of building up and sustaining hi-tech concentrations. Non-metropolitan regions, however, face different challenges compared to metropolitan regions. Tödtling and Trippl (2005) hence state that metropolitan regions, regarded as centers of innovation, often face the problem of "fragmentation": a lack of networks and interactive learning representing an innovation barrier, and resulting in the development of new technologies and the formation of new firms below the factual potential of the region. More peripheral regions may instead often face and suffer from "organizational thinness," where important RIS prerequisites such as specialized expertise or knowledge centers are weakly developed, where there is a lack of dynamic clusters, and where the critical mass for a dynamic cluster is not reached. The concept of critical mass is especially relevant for the discussion of clusters and regional innovation systems in metropolitan versus non-metropolitan regions (Doloreux *et al.*, 2007).

Purpose, research questions and methodological approach

The brief odyssey into relevant research literature above constitutes an important base for formulating the purpose of this chapter: to investigate important factors related to the dynamic development of knowledge-based clusters utilizing longitudinal data from two contrasting non-metropolitan cases; the regional industrial and scientific concentrations in Oulu, Finland, and Luleå, Sweden. In order to avoid the in-built analytical problems related to choosing metropolitan (capital) regions to study (where causes and effects normally are difficult to identify), we have chosen to focus on non-metropolitan regions with a similar geographical location and history (cf. Doloreux *et al.*, 2007). More specifically, we are interested to learn more about the importance of regional cross-sector collaboration, the importance of locomotive company functions and cluster formations, the impact of knowledge-hubs such as universities and science parks, and if and how internationalization and globalization affects regional dynamics and regional development in our two contrasting cases. This, then, is our contribution to this book: to learn from confronting longitudinal data from two contrasting cases (the two twin cities/regions of Oulu and Luleå) manifesting different development trajectories during the recent decades. Oulu is commonly regarded as a hot spot for hi-tech and knowledge-based industries and an undisputed regional metropolis of Northern Finland, while Luleå, with its dependency of traditional industries, is only one of several cities and regional centers in the northern part of Sweden. For our chosen purpose, statistics on demography and job development serve as an important input and a point of departure. They do not, however, tell us very much about why development occurs and why development paths differ between regions. To develop such an understanding we need to also address reasonable explanations behind different regional development trajectories in terms of strategies, activities and actors that may be related to these differences. This chapter focuses on addressing two research questions:

1. What kind of dynamic evolution did the Oulu and Luleå regions experience in the period 1998–2008, especially referring to their hi-tech sectors?

2. Which factors are important for understanding the development of hi-tech concentrations in the studied non-metropolitan areas, as well as the differences occurring between these regions?

The methodological approach in this chapter is grounded in the following data sources:

1. Statistical data and other secondary data from national statistical authorities, regional governments, local communities and other sources.
2. Previous studies of the two regions, especially national studies and comparisons (e.g. Männistö, 2002; Hyry, 2005) and studies comparing regions in the Nordic countries (Dahl *et al.*, 2006).
3. A multiple case study comparing Luleå and Oulu regions based on both quantitative, secondary data and qualitative data retrieved from in-depth interviews with 14 key actors representing business, society and academia in the two studied regions (Ylinenpää and Lundgren, 1998). Interviews were conducted utilizing both unstructured and semi-structured interview guides.
4. A qualitative multiple case study (Teräs, 2008) on the structural characteristics and dynamic evolution of regional science-based clusters, including 21 in-depth interviews with key cluster actors in Oulu and Luleå in 2007–8. A snowball sampling method (Scott, 1991) was selected to identify the relevant people in the case study regions Oulu and Luleå. All the major cluster actor groups defined by Sölvell *et al.* (2003) – companies, research community, financial institutions, government, and collaboration organizations – were represented in the interviews. Moreover, several interviewees represented and/or were connected to more than one category of cluster actors.

It is relevant to address the issues of validity and reliability of the study. As a multiple case study of two contrasting, non-metropolitan regions with hi-tech clusters, there are limitations referring to the degree to which our results may be (statistically) generalized to a broader population. This is neither the purpose with this specific study. Instead, we aim for a better understanding of the forces in play in knowledge-based HT concentrations. Studying such knowledge-based concentrations in highly urbanized metropolitan areas involves a greater risk for erroneous conclusions due to the number of different variables in play in such regions. Through in-depth studies of hi-tech clusters in non-metropolitan regions where interdependences between different variables are more evident, we should therefore have greater possibilities to build a better understanding of the phenomenon (cf. Doloreux *et al.*, 2007). Several case studies suggest that clusters require a decade or more to develop depth and real competitive advantage (Porter, 1998a). The bulk of studies in this field are based on analyzing success stories offering "snapshots" of regional success rather than the capability of specific regions to sustain a dynamic development over time (cf. Chapman *et al.*, 2004). In order to reduce this risk, we use a longitudinal approach with data from two regions/hi-tech concentrations covering several decades. In order to secure the study's reliability and validity, we utilize a multitude of data sources as a base for our analyses, such as documents, archival records, interviews, and direct and participant-observations. The study involves a number of different key actors and decision-makers, researchers, and other experts with whom tentative research findings, rival explanations, and interpretations have been discussed and documented. The selection of respondents has, in both the study of Ylinenpää and Lundgren (1998) and Teräs (2008), utilized a snowball approach in order to ensure identification of informants with different perspectives and relevant insights into the cases we here analyze (cf. Miles and Huberman, 1994).

The next section, addressing this chapter's first research question, will give some brief data on the development of hi-tech concentrations in Luleå and Oulu.

The evolution of the hi-tech concentrations in Luleå and Oulu

The Luleå region in Northern Sweden and the Oulu region in Northern Finland have significant similarities regarding history and location. Both cities were founded in the seventeenth century by the Swedish king as centers in the far north for trade and exploitation of natural resources. The anticipated growth of these regions did not, however, start until the second half of the eighteenth century, when a dynamic regional economy emerged based on exploitation of wood products and international trade. Oulu developed as an administrative center for Northern Finland, manifested also by the establishment of the second government-funded university in Finland in 1958. In Luleå, a university of technology was established in 1971, but the role as an administrative center for the north part of the country was – and still is – challenged by several other cities located to the north of Sweden. During the 1990s, it was obvious that the twin-cities/regions of Oulu and Luleå, despite a similar history and location (city centers are located at a distance of 150 kilometers on the Gulf of Bothnia), experienced very different development paths. The Oulu region was regarded as a modern and booming center for economic development and as "the Silicon Valley of the North" (Hyry, 2005), while Luleå and the Norrbotten county in which Luleå is located followed a different and less favorable development trajectory (Ylinenpää and Lundgren, 1998).

A basic and common "bottom-line indicator" depicting the dynamic status and development path of a region is the number of inhabitants in a specific region. Table 4.1 depicts basic demographic indicators of the Oulu and Luleå regions between 1970 and 2006, including population numbers in each region's regional center, where Oulu is the leading city of the Northern Ostrobothnia region and Luleå is the leading city of the Norrbotten region.

Table 4.1 reveals that the positive population growth has continued in both cities in the period 1998–2006, but that the growth rate is markedly higher in Oulu than in Luleå during this period. When comparing regions, the Northern Ostrobothnia region reveals a continuous and steady population growth, while Norrbotten – despite a booming economy based on demand for the region's natural resources during recent years – has experienced a loss of inhabitants.

Ylinenpää and Lundgren (1998) identified the evolution of hi-tech sectors in the two regions as one key factor behind these demographic differences, and concluded that, while hi-tech firms in the Oulu region almost doubled their employment from 4,590 to 8,141 between 1987 and 1996, an already marginal hi-tech employment in the Luleå region decreased to 700 employees in 1996. During 1997–2005, however, both Oulu and Luleå experienced a significant growth in HT (hi-tech) sectors, especially in ICT-related industries. Table 4.2 depicts the development of hi-tech jobs in the two regions between 1997 and 2005.

In Oulu, the rapid positive development of the HT sector in 1958–2001 ("the Oulu phenomenon") was followed by a stabilization phase in 2002–6. According to Teräs (2008), there has not been any significant slowdown period in Oulu HT cluster in 1958–2006. In Luleå, the HT sector developed at a much slower pace compared to Oulu. The HT sector experienced a significant

Table 4.1 Demographic data on Oulu and Luleå regions, 1970–2006

	Population 1970–2006 (1,000 inhab.)			*Population change (%)*	
	1970	*1997*	*2006*	*1970–2006*	*1997–2006*
Oulu city	71	114	130	+ 83	+ 14
Northern Ostrobothnia	300	360	380	+ 27	+ 6
Luleå	59	71	73	+ 24	+ 3
Norrbotten	255	262	252	– 1	– 4

Source: Ylinenpää and Lundgren (1998), Norrbotten County Council (2007), National Institute of Statistics (2011).

Table 4.2 Hi-tech jobs in the Oulu and Luleå regions, 1997–2005

	No. of hi-tech jobs	
	1997	*2005*
Oulu region (city of Oulu and 9 neighbouring municipalities)	11,294	18,426
Luleå region/Norrbotten	700	2,800

Source: Ylinenpää and Lundgren (1998), Oulu Growth Agreement Programme (2006), Regionfakta (2008).

growth period from 1997 until the year 2001, peaking at 3,800 jobs in 2001. This upturn was followed by a slowdown caused by turbulence in the ICT sector. This downturn, however, affected the Luleå region much harder. According to Nordin (2008), 40 percent of Luleå-based ICT companies went bankrupt around the millennium shift. In the years 2004–6, the Luleå HT sector then recovered from this ICT sector downturn. The amount of people working in the ICT sector in Norrbotten was, in 2006 2,855 persons, out of which 1,818 were located to the city of Luleå.

The development of the Oulu and Luleå regions and their hi-tech concentrations has been analyzed previously by several researchers. Männistö (2002) lists the following characters of major importance in the development of Oulu region: the foundation of the University of Oulu; the birth of the cable industry; the redirection of the electro-technical education and research of the University of Oulu to electronics; the diversification of the traditional industry in the region to electronics industry; the foundation of the VTT electronics research center in Oulu; the foundation of the Technopolis technology park; the Technology City Oulu project; the substantial investments of Nokia into the region; the expansion of the markets for local telecommunication companies; and the Oulu Centre of Expertise program developed by local government, industry and academia. According to Hyry (2005), the most important elements in the Oulu phenomenon process are as follows: the active and strategic role of national government forcing cities to put strategic plans into effect in conjunction with their partners; the contributions of University of Oulu, VTT and Tekes to technological development; a joint development strategy; the role of Nokia, encouraging technology development and generating employment; and the roles of specific key individuals.

Dahl *et al.* (2006) compare the role of universities in Oulu, Luleå and Aalborg, noting differences referring to spillover effects from universities to existing and new industries. In Oulu (and Aalborg) knowledge spillover effects instrumentally contributed to "the creation of entire new industries in wireless communications, while the Luleå case is more dominated by spillover and renewal of the existing industry structure" (ibid.: 141).

From previous studies, we may hence learn that sectoral, structural and institutional factors such as policy, the role of leading (locomotive) companies or the presence of spillover effects seem important when trying to understand the different development trajectories in our two contrasting cases, but also that specific actors per se play a critical role. Thus, moving from sectors into the domain of actors, one may question whether different development trajectories in fact may emanate from differences in the configuration, roles and functions of different actors operating in each regional innovation system. The key actors operating in the HT clusters in Oulu and Luleå regions identified by Teräs (2008) are listed in Table 4.3.

Table 4.3 depicts actors considered to be significant players in each region's innovation system. The most important *company* in the Oulu HT cluster is Nokia, with two major units in Oulu: Nokia Mobile Phones and NSN Nokia

Table 4.3 Major actors of Oulu and Luleå HT clusters

Actor group	*Oulu*	*Luleå*
Companies	Information technology cluster (Nokia employing 4,500 people in Oulu region), wellness technology cluster, content and media cluster, environment cluster, biotechnology cluster	ICT cluster as the dominant hi-tech cluster (international ICT companies such as TeliaSonera, TietoEnator, CapGemini and Google; government-owned ICT companies; hi-tech SMEs)
Government	City of Oulu, Council of Oulu region, TEKES regional office, Northern Ostrobothnia TE Center	City of Luleå, Norrbotten County Council
Financial institutions	Banks, regional venture capital companies	Banks, regional venture capital companies and special funds
Research community	University of Oulu, VTT Technical Research Centre of Finland, Oulu University of Applied Sciences	Luleå University of Technology
Institutions for collaboration	Oulu Innovation Ltd, Technopolis Technology Park	Aurorum Science Park, Internet Bay, LTU Innovation, LNAB, IUC

Source: adopted from Teräs, 2008

Siemens Networks. In 2007, Nokia employed 4,500 people in the Oulu region (Kaleva, 2007). The Oulu HT cluster includes several companies listed on OMX Helsinki Stock Exchange, e.g. EB Group, formerly Elektrobit, which specializes in demanding embedded software and hardware solutions for automotive and wireless industries. The leading wellness technology company is Polar, manufacturing, for example, heart rate monitors. In Norrbotten, ICT consultancy and services represented 41 percent, telecommunication 32 percent, electronics industry 15 percent, and ICT trade and renting 12 percent of the ICT jobs in the region (Regionfakta, 2008). The Luleå HT cluster does not have any major locomotive companies such as Nokia in Oulu. The companies can instead be divided into four major categories: units of internationally operating ICT companies (e.g. TeliaSonera, Tietoenator, CapGemini, Google), government-owned companies (e.g. Tulldata, Metria), niche companies having already experienced rapid development (e.g. Avantra, Marratech, NordNav, Upzide), and new spin-off companies (e.g. new firms hosted in Aurorum Business Incubator).

In Oulu, the major regional *governmental authority* is the Council of Oulu region. TEKES (the Finnish Funding Agency for Technology and Innovation) has a regional office in Oulu. The TE Employment and Economic Development Centre office in Oulu provides regional services for the Ministry of

Employment and Economy, for example. The City of Oulu provides services to businesses directly and via special units such as the Oulu Region Business Agency. In Luleå, the major regional authority in Norrbotten is traditionally Norrbotten County Administrative Board, but to an increasing degree also the County Council. The City of Luleå promotes business development directly and via the business development company LNAB.

In Oulu and in Luleå, *financial institutions* include banks and venture capital companies (public and private). Oulu hosts, for example, the regional office of Finnvera Plc., a specialized financing company owned by the state of Finland. In Luleå, there are several financial institutions in Luleå active in the field of hi-tech including, for example, ALMI, Norrlandsfonden, Emano, Lunova and Längmanska Företagarfonden.

The major actors of the *research community* in Oulu include the University of Oulu, with more than 16,000 students and 2,800 staff members; the state-owned independent research organization VTT (Technical Research Centre of Finland); and the University of Applied Sciences of Oulu, formerly known as Oulu Polytechnic, with approximately 9,000 students and a staff of 800. Luleå University of Technology (LTU) has 16,000 students and a staff of 1,500. There are several research and knowledge centers connected to LTU, such as CDT (Centre for Distance-Spanning Technology).

In Oulu, the *organizations for collaboration* or *IFCs* include Oulu Innovation Ltd, a development company that promotes hi-tech-related businesses; and Technopolis Ventures, the business incubator and business service provider and a subsidiary of the technology park company Technopolis Plc. In Luleå, Internet Bay is an industrial network established in Luleå in 1999 to promote ICT-based business and co-operation activities in Northern Sweden. ProcessIT is an initiative launched by Internet Bay in 2003 to develop new products and services based on the needs of the process industry in Northern Sweden for new innovative ICT solutions, with the focus on measurement technology, mobile workplace, optimization of automated processes, and maintenance systems. Aurorum Science Park is located near the Luleå university campus area. Furthermore, IFCs in Luleå include, for example, the Chamber of Commerce, IUC, Export Development Norrbotten, Innovationsbron, and the Swedish Federation of Business Owners.

Both the Oulu and Luleå regions hence expose a multi-faceted list of actors operating in each regional innovation hi-tech system or cluster. There are, however, differences referring to both size and actor configurations. If and how such differences may be related to the differences referring to development paths that we here are interested in is an important theme in our next section. To conclude, secondary data such as statistics, web-based information and previous studies hence provide some information, enabling a better understanding of how these two regions and their hi-tech sectors have developed over the years. In order to find answers to how and why these development trajectories have occurred, we must however rely on other data sources. This issue, addressed by this study's second research question, is therefore the theme in the following section.

Factors affecting the hi-tech development

Critical development factors

Ylinenpää and Lundgren (1998) systematically compared the development in Oulu and Luleå regions and searched for answers to the question why the development of these regions differed. The authors suggested a number of different reasons explaining why the Oulu region had experienced a more favorable development path when compared to the Luleå region:

- the role of the locomotive company Nokia in Oulu, serving as a demanding customer and role model for other companies in the region;
- the presence of dynamic clusters in emerging and knowledge-based sectors of the economy in Oulu, while Luleå and its region, to a high degree, was characterized by company constellations in mature and often declining business sectors;
- a dynamic mode of co-operation between industry, the university/research sector and local and regional government;
- a significant role in the national hierarchy of cities for Oulu as a natural center for the northern part of the country;
- differences related to the universities (Oulu with a complete university structure while Luleå was focusing mainly on technology) and the science parks (where Technopolis in Oulu, at this time, was regarded as one of the most successful parks in the world);
- the Finnish leadership and management style (which could be advocated to work better than a "consensus-seeking" Swedish ditto during turbulent and dynamic development periods); and
- a more marked international orientation among the Finnish SMEs.

A follow-up study by Teräs (2008) included interviews with a total of 21 key cluster actors in Oulu and Luleå in 2007–8, following the Sölvell *et al.* (2003) categorization of major cluster actors. Teräs (2008) concludes that the regional HT cluster in Oulu experienced a long period of exceptional growth until the beginning of the 2000s, when the growth stagnated. The Oulu HT cluster has been propelled by the Nokia-driven ICT sector and "the Oulu Spirit" – a willingness to collaborate across sectors and cluster actors and involving strong leadership. The previous cornerstones of the Oulu HT cluster – the research community and the Nokia-driven ICT industry – are still in place but the insufficient amount of hi-tech entrepreneurship and a low degree of internationalization of the SMEs have been raised as crucial challenges for the future. The Oulu HT cluster currently attempts to diversify its activities into several fields outside the dominant ICT sector although a critical mass of actors in these other sectors has not been reached yet. The mindset of the key actors of Oulu HT cluster and decision-makers in the Oulu region has, despite the fact that the Oulu HT cluster today is stronger than ever in many aspects given e.g. its accumulated know-how in global hi-tech business, changed from optimism into uncertainty concerning the future of hi-tech development. In Luleå, hi-tech companies based on ICT-driven research and expertise with an origin in Luleå University of Technology experienced a significant growth period in the late 1990s, but ran into severe difficulties when the dot-com crash hit the economy. Today, this sector has recovered and entered into a new growth period. The Luleå University of Technology remains a cornerstone of the Luleå HT environment. Bottlenecks hampering further expansion include a relatively low level of internationalization of its SMEs and a lack of internationalization expertise, and a lack of critical mass of actors and competences offering a sufficient infrastructure and support services to companies. Also, the ability of the Luleå University of Technology to attract and keep its best talent and to build up credible research units, and the weak motivation of financial institutions to set up offices in Luleå, are regarded as challenges for the future.

The results from the interviews by Teräs (2008) regarding the key actors (companies, government, financial institutions, research institutions, and institutions for collaboration) are summarized below.

Regarding *company actors* in Oulu, several interviewees expressed concerns about the fact that decision-making in companies such as Nokia now takes place outside Oulu. Some interviewees also claimed that the success of Nokia has not necessarily been favorable to the formation of hi-tech start-up companies in the Oulu region: it has been too easy for many companies and entrepreneurs to remain as subcontracting companies with Nokia as the dominating, if not the only, customer. In Luleå, respondents frequently commented on the lack of large locomotive companies and even medium-sized companies. Some interviewees stated, however, that there are new, promising companies rapidly developing in the field of ICT, some of which have already been bought up by international companies. Worries were expressed, however, regarding some of the most promising ICT companies moving out to, for example, the Stockholm region at a relatively early stage.

In Oulu, some interviewees, regarding *governmental actors*, stated that the national decision-makers are located in the Helsinki region, too far away. The regional funding organizations were criticized for not being able to prioritize development projects, for example. The city of Oulu ought to play a more significant role in the Oulu HT development, according to many interviewees. In Luleå, the regional government was considered as more important for the Luleå HT cluster than the national government by most interviewees. Moreover, some interviewees also stated that, although the regional government rhetorically emphasizes hi-tech branches as important in their strategies, a great deal of funds is allocated to traditional industries. The role of the City of Luleå was largely seen as important, although many interviewees pinpointed the limited funding resources available through the city for hi-tech development in the region.

In Oulu, the majority of the interviewees regarded the role of *financial institutions* as limited. The amount of active business angels, re-investing money gained from previous hi-tech ventures, was also considered to be limited, although growing. The amount of potential hi-tech growth companies is simply not sufficient to motivate the long-term presence of venture capital actors in the Oulu region. In Luleå, the majority of the interviewees stated that there is lack of venture capital in the Luleå region. Many interviewees mentioned the lack of business angels in Luleå. Some interviewees claimed, however, that the availability of venture capital is not any major bottleneck in the Luleå HT development.

In Oulu, the *research sector* was considered a critical element in the Oulu HT cluster. According to some respondents, the Nokia activities in the Oulu region would never have taken place without a strong technical university and research community in the region. The role of the University of Oulu in building up the Oulu HT cluster was widely recognized, as well as the roles of Technical Research Centre of Finland (VTT) in identifying and promoting applications of the latest technologies and Oulu University of Applied Sciences in delivering a well-educated labor force. Also in Luleå, the strength and competitiveness of Luleå University of Technology was considered to be one of the critical success factors behind the Luleå HT cluster. There were, however, worries among the interviewees regarding the ability of LTU to continue to produce top-class experts in the future. The occurring trend in 2007–8, when the university was not able to fill all its vacant student places, was seen as a warning signal of students not being attracted enough to studying hi-tech branches in Luleå. Also, insufficient university–industry contacts were clearly expressed by the informants.

In Oulu, the informants claimed that nobody seems to have taken over the clearly dominant role as an *organization for collaboration (institution for collaboration, IFC)* and as a support organization for the hi-tech cluster after the technology park Technopolis broadened its activities into other cities in Finland and turned into a publicly listed stock company in 1999. In Luleå, almost all the interviewees stated that there are too many IFCs in the Luleå region, which makes the IFCs' activities fragmented. Some interviewees argued that the importance of the IFCs to the Luleå cluster development, therefore, is only marginal. The co-operation between the IFCs was seen as insufficient, and the lack of trust between the IFCs was mentioned as a problem by several interviewees.

Changes in the factors

The following analysis "revisits" some of the hypotheses and findings proposed by Ylinenpää and Lundgren (1998) on the basis of the new information provided by the study of Teräs (2008). Moreover, the analysis employs the opportunity to put the findings into a broader context. The primary focus of the analysis is not limited to finding similarities and differences between Oulu and Luleå hi-tech clusters. The aim is to provide new, value-added information on cluster dynamics in non-metropolitan regions, where the findings of Ylinenpää and Lundgren (1998) are compared with and complemented by more recent data (Teräs, 2008). Addressing our second research question – which factors are important for understanding the development of hi-tech concentrations in the studied non-metropolitan areas, as well as the differences occurring between these regions? – we then summarize our findings. The factors are presented in terms of their importance to the regional development in the following main categories: "important," "of some importance," and "less important."

Table 4.4 summarizes the findings by Ylinenpää and Lundgren (1998), suggested as *important differences* or *differences of some importance* when comparing Oulu and Luleå HT clusters, and contrasts these findings to the new findings by Teräs (2008) "ten years after." It also depicts findings from Teräs (2008) that were not anticipated in the 1998 study (impact from globalization and critical mass of actors/competences).

The role of the regional center in the national hierarchy of cities was one factor that Ylinenpää and Lundgren found important for explaining the different development paths of Oulu and Luleå regions. Moving ten years ahead from 1998, we find that the population of Oulu has grown markedly faster than the population of Luleå in the period 1998–2007. The Finnish Ministry of the Interior (2006) categorizes Oulu as a "versatile university region" with a ranking number four in Finland, overtaken only by the Helsinki, Tampere, and Turku regions. The national importance of Luleå is absolutely and relatively speaking lower than the Finnish case, and less interesting for companies, organizations and people looking for an attractive location. The role of the regional center is, therefore, also considered an *important* factor and difference in 2008 when comparing Oulu and Luleå.

The degree of co-operation between key actors was, in 1998, regarded as another important factor explaining the different development trajectories. The interviews in 2007–8 revealed that key decision-makers still trust each other and that cross-sector collaboration still is viable in Oulu. However, the "Oulu Spirit" has partly diluted. The role of locomotive companies such as Nokia has changed with decision-making power and the relative importance of local managers transferred to company headquarters and/or overseas locations. Also, the role of local subcontracting companies has diminished. Even if Teräs (2008) concludes that co-operation between local actors in Luleå (with the exception that intermediaries or IFCs are not co-operating in any optimal way) is well-functioning, and that both Oulu and Luleå today seems to suffer from a lack of leading persons in their hi-tech clusters, co-operation between key actors is still considered as a stronghold in the Finnish region and a factor of *some importance* comparing Oulu and Luleå regions in 2008.

A third important explanatory factor in Ylinenpää and Lundgren (1998) was related to the role of sophisticated local customers. In 1998, Nokia had a significant role in shaping the electronics and software industry in the Oulu region as a buyer, technology transfer catalyst, and a stimulator and role model for smaller firms. According to Teräs (2008), the overall importance of Nokia

Table 4.4 The 1998 study re-visited: Important findings, findings of some importance, and findings not anticipated in 1998

Ylinenpää and Lundgren (1998)	*Teräs (2008)*
Oulu has a significantly more important role than Luleå in its national context (**important**)	Oulu more important nationally, exposing a population growth faster than in Luleå. Both regions suffer from remote location **(important)**
Co-operation between key actors is more developed in Oulu than in Luleå (**important**)	"Oulu Spirit" partly diluted but more developed than in Luleå. The relative importance of local co-operation however diminished **(of some importance)**
Nokia is more important as a demanding local customer to SMEs in Finland than, e.g. SSAB or LKAB are in Sweden (**important**)	Nokia important for Oulu. Co-operation Nokia-local SMEs diminishing. ProcessIT increases co-operation in Luleå **(of some importance)**
Oulu hosts clusters of firms in expanding or emerging industries, while Luleå has mature branches of industry. The dominating clusters in Oulu are to a higher degree characterized by early home demand and early saturation (**important**)	Oulu still promotes cluster activities outside ICT branch despite the disappointments with regional growth programme in the early 2000s. No major coordinated clustering initiatives outside ICT and ProcessIT in Luleå **(of some importance)**
The role of university is more significant in Oulu than in Luleå (**of some importance**)	Oulu has additional research community strongholds. University-related special units important in both regions **(of some importance)**
The Finnish management style and practices are more favourable during times of rapid change and turbulence (**of some importance**)	International management style, rather than Finnish or Swedish management style, gaining ground **(less important than in 1998)**
Firms in Oulu have a more international orientation than firms in Northern Sweden have (**of some importance**)	Rapid ICT growth in Oulu resulted in the increase of internationalized SMEs. Less internationalized SMEs in Luleå but the situation has improved. **(of some importance)**
The role of the science park is more significant in Oulu than in Luleå (**of some importance**)	The relative importance of the science park diminished in Oulu region **(less important than in 1998)**
Not specifically anticipated in 1998	Globalisation affects both Oulu and Luleå HT clusters. Oulu has however more global business experience **(of some importance)**
Not specifically anticipated in 1998	Critical mass of (especially ICT) competences and other resources more developed in the Oulu HT cluster **(important)**

is still important for the Oulu region but the diminishing relative importance of local Nokia managers is slowing down the pace and intensity of the co-operation between Nokia and local SMEs. In Luleå, large companies such as SSAB (steel) and LKAB (mining) have developed their co-operation with their local university and its research centers. The ProcessIT Innovation initiative is a good example of this increased co-operation between large companies, research and education, and SMEs in the Luleå region. There is, however, currently a lack of a locomotive companies in the Luleå hi-tech cluster, and companies such as TeliaSonera and TietoEnator have even cut down the number of jobs in Luleå between 1998 and 2008. To conclude, the more significant role of sophisticated local customers in the Oulu region compared to the Luleå region is still considered a factor of *some importance* in 2008.

The role of emerging and expanding industries was, in 1998, considered to be a fourth relevant and highly important factor for explaining the different development paths in Luleå and Oulu. In Oulu, the success period in the late 1990s resulted in attempts to promote and support several new hi-tech clusters in addition to the dominant ICT cluster – a strategy that, at least up to date, has not paid off as expected. In Luleå, the rapid development of especially Internet applications in 1998–2002 took an abrupt end caused by the global "dot-com" downturn. Also, the possibility to cultivate other hi-tech sectors has only to a limited degree been exploited. The role of emerging and expanding industries is therefore still considered a factor of *some importance* for the dynamic development of these non-metropolitan regions and for the different development trajectories they reveal.

Ylinenpää and Lundgren (1998) perceived the role of universities as a factor of some importance. Together with the University of Oulu of Applied Sciences and the VTT Technical Research Centre, the university, the City of Oulu and key companies in the region today form the Oulu Triple Helix Alliance (City of Oulu, 2007) – an alliance that according to Teräs (2008) represents a strong tie between research, hi-tech companies and local government. In Luleå, the interviews by Teräs (2008) emphasize the great potential of the university know-how for the development of technology-based companies, but also advertise for a more close collaboration between business and research. The role of universities – and the entire research and educational community – is therefore still, in 2008, considered a factor of *some importance.*

The 1998 study concluded that the Finnish *management style* is more oriented towards alert decision-making, while the Swedish management style is characterized by team-building and establishing decisions firmly in the own organization. The increasing globalization has, however, according to Teräs (2008), also had an effect on national management styles and cultures, and accordingly also affected prevailing management styles and cultures in regions and clusters. The preference for Finnish or Swedish management styles has hence withered away in favor of an international or global management style, with guidelines and role models often coming from a foreign owner. Difference in management styles/cultures is therefore considered a factor of *minor importance* comparing Oulu and Luleå regions in 2008.

The degree of international (export) orientation of firms was, in the 1998 study, yet another factor considered having some relevance for explaining the dynamics and the differences between Oulu and Luleå. The 1998 study concluded that the degree of internationalization of the SME sector (to a significant degree driven by corporations such as Nokia) was higher in the Oulu region than in Luleå. In Luleå of today, the number of internationally oriented persons in hi-tech SMEs is markedly higher than it was 10 years ago, but the number of internationally oriented professionals in the Luleå HT cluster is still low. The international orientation of firms hence still, in 2008, is considered a factor of *some importance* for understanding the hi-tech dynamics in the Oulu and Luleå regions.

The role of the science park (in the study from 1998 identified as a factor of some importance) has, in Oulu, changed remarkably during 1998–2007. Technopolis Technology Park first broadened its geographic scope of operations into other cities in Finland and was listed at the Helsinki Stock Exchange in the year 1999. Today, Technopolis is also operating outside Finland. The role of Technopolis thereby has changed from being a local technology park with the City of Oulu as its largest shareholder into a publicly listed company. This change has also had an effect on regional development in the Oulu region, when Technopolis, previously a creative and important locomotive for regional development, has become more of a private real estate company with added-value consultancy services. In Luleå, Aurorum Science Park as an IFC has so far not played any critical role in the Luleå HT development, although, for example, a functioning incubator and spin-out resources such as LTU Innovation have been established in recent years. The current role of the science parks is hence considered a factor of *less importance* for the understanding of current hi-tech concentrations and for explaining the differences between Oulu and Luleå.

Ylinenpää and Lundgren (1998) did not anticipate globalization per se – beyond international/export orientation – as a factor of importance in the comparison between the Oulu and Luleå hi-tech clusters. The ongoing globalization has, however, in a marked way affected both the Oulu and Luleå HT clusters during the period 1998–2007. Many respondents in the study of Teräs (2008) stated that a global marketplace and global trends today largely drive hi-tech development. The importance of attracting foreign experts to Oulu region was emphasized. In Luleå, the interviewees argued that the global trends drive especially the ICT sector. One of the implications of the globalization concerns the development path of new hi-technology start-ups: instead of following the traditional stages from operating on a local to national and international marketplaces, many start-ups enter as "born globals" and sometimes also leave Luleå at an early stage. Globalization generally strongly affects both the Oulu and the Luleå HT clusters, Oulu currently being better prepared and competitive mainly due to their broader experience from globally oriented activities by the ICT cluster. Globalization is thus, by Teräs (2008), considered a factor of crucial importance for the current and future development of both Oulu and Luleå HT clusters. Oulu HT cluster's stronger position may, however, be a factor of *some importance* in explaining the current dynamics and the differences between the Oulu and Luleå regions in 2008.

The importance of critical mass was not predicted by Ylinenpää and Lundgren (1998). Critical mass representing actors, competences and other kinds of resources, however, play an important role in the development of knowledge-based HT clusters in non-metropolitan regions (Teräs, 2008). Lack of such resources implicates a low degree of absorptive capacity (Cohen and Levinthal, 1990) and increases the risks of missing attractive development opportunities. The demographic statistics for 1998–2008 indicate that Oulu still is a growing capital of northern Finland. Furthermore, the Nokia-driven rapid ICT sector development in the Oulu region in the late 1990s and early 2000s increased the amount of ICT experts in the Oulu region by thousands of professionals, developing a critical mass and serving as a buffer against exogenous shocks such as the dot.com downturn in the 2000s. At the same time, the Luleå ICT sector suffered severely from the dot-com crisis and lost hundreds of professionals, many of whom moved to the Stockholm region. The critical mass of the HT cluster and the region is therefore considered an *important* factor when comparing the dynamics and the recent development of Oulu and Luleå regions.

Conclusions and implications

This chapter has focused on addressing two research questions: the dynamic evolution of the Oulu and Luleå hi-tech regions in 1998–2008 and the most important factors underlying the development of the Oulu and Luleå hi-tech regions. The dynamic evolution of the Oulu and Luleå hi-tech regions has been addressed in the third section of this chapter. The answer to the second research question has been addressed in the fourth section. Based on the "re-visit" of the Oulu and Luleå regions in 2007–8, 10 years after the first visit, the following major conclusions may be drawn.

When comparing the development of hi-tech clusters in the Oulu and Luleå regions over time, co-operation between regional cluster actors, the cluster leadership itself and the role of locomotive or flagship companies have partially lost their explanatory power as key factors for understanding different development trajectories. Although several policy initiatives have been launched in both our regions during the studied period of time, the impact from such top-down initiatives and the adoption of trendy, best-practice examples and concepts such as attracting "the creative class" (Florida, 2003) or building "triple helix collaboration" (Etzkowitz and Leydesdorff, 2000) through local or regional cross-sector growth agreements may be questioned. As illustrated by our empirical cases, there is a risk when developing elegant top-down innovation systems and policies that the entrepreneurial elements and commercial driving forces vital for the implementation of such concepts will be left on the sidelines. Top-down initiatives may, at best, support a healthy, market-driven development but can hardly by themselves generate any dynamic cluster development (cf. Laestadius *et al.*, 2007). Both Oulu and Luleå currently suffer from a lack of leading persons from different sectors in their hi-tech clusters although the existing co-operation between key actors is still considered a stronghold. Especially in Oulu, the role of the locomotive company Nokia has undergone a remarkable change due to the continuous globalization of the company, resulting in an erosion of its previous role as a regional locomotive. Together with the new role for the science park Technopolis, this has had obvious implications for the functionality of the leadership and collaboration between the region's HT cluster actors. As pointed out already by Dahmén (1950, 1984), however, restructuring of economies and the tensions thereby created also represents the soil for Schumpeterian dynamics to emerge and develop. Such entrepreneurial drivers (regardless if this talent is developed locally or attracted to the region) are vital for a functional and sustainable regional development. From this perspective, the Swedish example of ProcessIT Innovations represents an interesting initiative to build a new entrepreneurial arena based on a historic tension between mature basic industries and new hi-tech industries developing in the region.

The adaptation to the challenges emanating from a more globalized economy for regions such as Oulu and Luleå has only begun. The international hi-tech business environment has changed remarkably in the last 10 years. As highlighted by this study, operating on a local or regional arena only is not sufficient to achieve competitiveness in a modern economy. In order to achieve international or global competitiveness, a more open rather than regional and localized mindset is required when selecting target markets, collaboration partners, and future employees. Functional regional development strategies may, in other words, be understood as modes of behavior that today combines "local buzz and global pipelines" (Bathelt *et al.*, 2004), which in our empirical study is especially evident when comparing the situation in 2008 with the situation 10 years earlier. For non-metropolitan areas, the challenge hence is not only to develop a functional regional expertise, collaboration and leadership by attracting and retaining competencies and resources required, but also to develop "global pipelines" that interact with this "local buzz" of regional/cluster actors.

From a more practical and applied point of view, our study highlights the importance for regional decision-makers to counteract regional lock-in effects by actively and proactively adjusting development strategies to an increasingly global reality. The competitive environment in which regional and company-based world-class competitiveness is embedded is constantly changing. This requires both persistent development investments and sensitive adjustments to emerging "new rules of the game." In this endeavor, prescriptions for success are to a significant degree time – and situation – dependent, and former success formulas (or copying other regions' historical success formulas) are – as illustrated by our contrasting cases – often a bumpy and risky road to heaven. Our study also pinpoints the important role a brave regional (and often collective) leadership has for cluster development anchored in strengths and specific resources/competences in the region. Finally, our study should have contributed with underlining the importance of developing a critical and optimal base of resources and competences as a fundament for regional competitiveness instead of a too diversified, low-risk strategy. Such development efforts should be anchored in both a local/regional critical mass of actors and a more global mindset and global partnerships.

Although based on two specific Nordic non-metropolitan regions, this empirical study should also be of interest and relevance to a broader context. As noted by, for example, Laestadius *et al.* (2007), regional development, in general, and development of knowledge-based clusters, in specific, is an endeavor that, to an increasing degree, is affected by a global economy and a global knowledge community. The development of functional knowledge-based clusters hence, to an increasing degree, requires not only development of local and regional recourses and capabilities, but also a combination with resources and capabilities that are located outside the region itself.

Note

1. We gratefully acknowledge the comments and suggestions that we received from the editors of this book.

References

Asheim, B.T. and Coenen, L. (2006) "Contextualising regional innovation systems in a globalising learning economy: On knowledge bases and industrial frameworks," *Journal of Technology Transfer*, 31: 163–73.

Bathelt, H., Malmberg, A. and Maskell, P. (2004) "Clusters and knowledge: Local buzz, global pipelines and the process of knowledge creation," *Progress in Human Geography*, 28: 31–56.

Carlsson, B. and Stankiewicz, R. (1991) "On the nature, function and composition of technological systems," *Journal of Evolutionary Economics,* 1: 93–118.

Carlsson, B., Jacobsson, S., Holmén, M. and Rickne, A. (2002) "Innovation systems: Analytical and methodological issues," *Research Policy*, 31: 233–45.

Chapman, K., MacKinnon, D. and Cumbers, A. (2004) "Adjustment or renewal in regional clusters? A study of diversification amongst SMEs in the Aberdeen oil complex," Institute of British Geographers.

Chesbrough, H. (2007) *Open Business Models*, Boston, MA: HBS Press.

City of Oulu (2007) "Oulu triple helix: Oulun korkeakoulutoiminnan kehittäminen osana innovaatioympäristöä," City of Oulu.

Coenen, L., Asheim, B., Ljung, B. and Moodysson, J. (2007) "VINNVÄXT i Skåne," in S. Laestadius, C. Nuur and H. Ylinenpää (eds) *Regional växtkraft i en global ekonomi*, Stockholm: Santérus Academic Press Sweden, 219–36.

Cohen, W.M. and Levinthal, D.A. (1990) "Absorptive capacity: A new perspective on learning and innovation," *Administrative Science Quarterly*, 35: 128–52.

Cooke, P., Boekholt, P. and Mayer, F. (2000) *The Governance of Innovation in Europe: Regional Perspectives on Global Competitiveness*, London: Pinter.

Cooke, P., De Laurentis, C., Tödtling, F. and Trippl, M. (2007) *Regional Knowledge Economies,* Cheltenham, UK: Edward Elgar.

Dahl, M.S., Dahlum, B. and Ylinenpää, H. (2006) "Regions with comprehensive universities: The cases of Alborg, Luleå and Oulu," in J.-E. Nilsson (ed.) *The Role of Universities in Regional Innovation Systems: A Nordic Perspective*, Copenhagen: Copenhagen Business School Press.

Dahmén, E. (1950) *Svensk industriell företagsverksamhet*, Uppsala, Sweden: Almqvist & Wiksell.

Dahmén, E. (1984) "Schumpeterian dynamics: Some methodological notes," *Journal of Economic Behavior and Organization*, 5 (1): 25–34.

Doloreux, D., Dionne, S. and Lapointe, D. (2007) "Institutional structure and modes of governance in non-metropolitan innovation systems," *International Journal of Entrepreneurship and Innovation Management*, 7: 405–23.

DTI (2004) "A practical guide to cluster development," A report to Department of Trade and Industry and the English RDAs' DTI/Ecotec Research & Consulting.

Etzkowitz, H. and Leydesdorff, L. (2000) "The dynamics of innovation: From national systems and 'mode 2' to a triple helix of university–industry–government relations," *Research Policy,* 29: 109–23.

Florida, R. (2003) "Cities and the creative class," *City and Community*, 2 (1) (March): 3–19.

Hyry, M. (2005) "Industrial growth and development in Northern Finland: The case of Oulu 1970–2002," A thesis for the Degree of Doctor of philosophy, Coventry Business School, Coventry University.

Johansson, J., Wincent, J. and Ylinenpää, H. (2007) "Triple-Helix-konceptet som funktionell modell eller retorisk återvändsgränd?" in S. Laestadius, C. Nuur and H. Ylinenpää (eds) *Regional växtkraft i en global ekonomi*, Stockholm: Santerus Academic Press, 77–98.

Kaleva (2007) "Nokia työllistää Oulun seudulla noin 4 500 ihmistä," Newspaper article on February 15, 2007.

Kautonen, M. (2006) "The regional innovation system bottom-up: A Finnish perspective," Academic dissertation, University of Tampere.

Laestadius, S. (2007) "VINNVÄXT programmets teoretiska fundament," in S. Laestadius, C. Nuur, and H. Ylinenpää (eds) *Regional växtkraft i en global ekonomi*, Stockholm: Santerus Academic Press, 27–56.

Laestadius, S., Nuur, C. and Ylinenpää, H. (eds) (2007) *Regional växtkraft i en global ekonomi*, Stockholm: Santerus Academic Press.

Malerba, F. (2005) "Sectoral systems: How and why innovation differs across sectors," in J. Fagerberg, D.C. Mowery and R.R. Nelson (eds) *The Oxford Handbook of Innovation*, Oxford: Oxford University Press, 380–406.

Männistö, J. (2002) "Voluntaristinen alueellinen innovaatiojärjestelmä," University of Lapland, Acta Universitatis Lapponiensis 46.

Marshall, A. (1916) *Principles of Economics*, London: Macmillan.

Maskell, P. and Kebir, L. (2006) "What qualifies as a cluster theory?" in B.T. Asheim *et al.* (eds) *Clusters and Regional Development*, London: Routledge.

Maskell, P., Eskelinen, H., Hannibalsson, I., Malmberg, A. and Vatne, E. (1998) *Competitiveness, Localised Learning and Regional Development*, London: Routledge.
Miles, M.B. and Huberman, A.M. (1994) *Qualitative Data Analysis: An Expanded Sourcebook*, Thousands Oaks, CA: Sage.
Ministry of the Interior (2006) "Large city-regions," available at: www.intermin.fi/intermin/biblio.nsf/7397BE88A5D8FDF1C225721A004686D3/$file/452006.pdf (accessed on May 20, 2010).
National Institute of Statistics (2011) "Population statistics," available at: www.stat.fi/tup/seutunet/ouluseutu_vaesto.html (accessed on August 5, 2011).
Nordin, H. (2008) "Bubblan som sprack: IT-kraschens inverkan på konkurser bland IT-företag i Luleå," Bachelor thesis, Luleå University of Technology.
Norrbotten County Administrative Board (2007) *Fakta om Norrbotten*, Publication Series nr 3/2007, Luleå.
Oulu Growth Agreement Programme (2006) "Annual report," City of Oulu.
Porter, M. (1990) *The Competitive Advantage of Nations*, London: Macmillan.
Porter, M. (1998a) "Clusters and the new economics of competition," *Harvard Business Review*, November–December: 77–90.
Porter, M. (1998b) *On Competition*, Harvard Business Review Book, Boston, MA: Harvard Business School Publishing.
Regionfakta (2008) "Statistics on Luleå and Norrbotten," available at: www.regionfakta.com (accessed on February 22, 2008).
Saxenian, A. (1994) *Regional Advantage: Culture and Competition in Silicon Valley and Route 128*, Cambridge, MA: Harvard University Press.
Scott, J. (1991) *Social Network Analysis*, London: Sage.
Sternberg, R. (1995) "Supporting peripheral economies or industrial policy in favour of national growth? An empirically based analysis of goal achievement of the Japanese Technopolis Program," *Environment and Planning*, 12: 425–39.
Sölvell, Ö., Lindqvist, G. and Ketells, C. (2003) *The Cluster Initiative Greenbook*, Stockholm.
Teräs, J. (2008) "Regional science-based clusters: A case study of three European concentrations," Thesis No. C 302 for the degree of Doctor of Technology, ACTA Publication series, University of Oulu.
Tödtling, F. and Trippl, M. (2005) "One size fits all? Towards a differentiated regional innovation policy approach," *Research Policy*, 34: 1203–19.
Westerberg, M., Wincent, J. and Ylinenpää, H. (2007) "Triple Helix och utvecklingen av innovationssystem," in S. Laestadius, C. Nuur and H. Ylinenpää (eds) *Regional växtkraft i en global ekonomi*, Stockholm: Santerus Academic Press.
Wolfe, D. and Gertler, M. (2004) "Clusters from the inside and out: Local dynamics and global linkages," *Urban Studies*, Special Issue.
Ylinenpää, H. and Lundgren, N.-G. (1998) "Regional dynamics: A comparison of two Nordic regions," Paper presented at the conference "SMEs and Districts," LIUC, Castellanza, Italy, November.

5 Regional strength in global competition

Collaborative patterns for life science firms in Western Sweden

Jens Laage-Hellman, Annika Rickne and Emelie Stenborg

Introduction

All over the globe, policy initiatives aim to create regional clusters within life science.[1] Indeed, in addition to the existing hot spots such as those in California and Massachusetts, several regions in the US, Asia and Europe compete on becoming a world-leading hub in a certain niche. There are several underlying reasons for this aim to create strong regional agglomerations. First, following the acclaimed success of regions such as Silicon Valley and Route 128 it is a common view that success lies in forming strong regions or clusters. Also, for the case of life science, the importance of strong clusters has been thoroughly documented (Cooke, 2001). However, in what ways such regions could or should be formed or strengthened is still up for debate.

Second, regional initiatives to develop industry, stimulate economic growth and create new jobs often build on the assumption that collaboration among *regional* actors (firms and others) is of crucial importance (see e.g. Asheim and Isaksen, 2002; Gertler, 2004). In particular, this is claimed to be true for science-based or knowledge-intensive industries, such as life science and ICT (see e.g. Powell, 1998; McKelvey *et al.*, 2004; Tidd *et al.*, 2005). In such industries, where development and growth to a large extent is driven by the creation of new knowledge and technological innovation, especially research and development (R&D) collaboration between firms, as well as between industry and university is regarded as important. For example, there is widespread belief among policy makers in many countries that local universities can play a role as an engine of growth, for example, by generating new knowledge that can be transferred to and commercialized by companies.

One issue of particular difficulty is the balance between the regional, national and international levels. What resources and competencies could or should be formed within the region? This paper takes the proposition of the importance of regional knowledge interaction to court, and does so for a highly research-intensive sector, namely that of life science. We pose the question whether it is true that biomedical firms are dependent on regional knowledge networks, and if so, what the collaborative patterns look like. We perceive that any such regional cluster formation should take as its starting point the needs of the actors within that region. Our analysis then concerns to what extent the life science firms, in one such region aiming to be a European hub, in practice are interested and engage in regional interaction, and to what extent they turn to the national or international arena for their knowledge and resource needs. Our research question then relates to what types of regional interaction matter for life science firms. The answer to this question tells us something about the role of regional knowledge nodes and how policy actors could approach regional cluster formation.

To discuss these issues, we draw on the example of Western Sweden, where more than 200 firms deal with life science applications related to health. The analysis is based primarily on a survey to 222 firms, with complementary interview data from selected companies.

The chapter is structured in the following way. First we discuss the importance of an open mode of innovation, and presents some underlying arguments in the literature for regional versus national or international interaction. After that, the research design is motivated and the details of the sample and the manner of analysis are presented. The following two sections contain the gist of the analysis and present findings on patterns of regional interaction. In the final and sixth section, we focus on conclusions for open innovation processes within the region.

Open innovation and the region

Open innovation

Innovation is an interactive and highly collective process. It is facilitated by positive externalities creating benefits for a wide set of actors, and it requires active learning processes, where such necessity for interaction and learning between organizations long has been stressed by writings on innovation systems, clusters, and industrial districts. Our focus in this paper is the intentional processes to create externalities and learning between organizations (or individuals): here called an 'open innovation' mode, an idea gaining ground in the research literature as well as in policy and strategy debates in the recent decade (Chesbrough, 2003). While it is not a new concept to firms that, in order to compete, knowledge and the ability to use and transform such knowledge are key assets, firms today seem more conscious to address the implementation of an open innovation mode. Companies strive to use all available sources and paths to gain access to and explore knowledge pools in terms of science, technology, markets, design, logistics, etc. For example, through mobility of skilled people – moving in and out of small and large firms, industry and academia – knowledge transfer is made possible, as well as by being a sharing organization the firm may pool its resources with others, thereby creating a larger whole.

Not least, the life science area has been subject to many empirical studies covering network aspects related to innovation (see e.g. Pisano, 1991; Powell, 1998; Haanaes, 1999; Orsenigo *et al.*, 2001; Oliver, 2004; Waluszewski, 2004; Gilsing and Nooteboom, 2006; Roijakkers and Hagedoorn, 2006). Due to its science-based character, relationships between firms and research-performing organizations are of particular importance and collaboration between large firms and small, research-based firms plays a key role for the commercialization of science and the development of new products (drugs, medical devices, etc.).

Regional knowledge flow

Despite increasing globalization of the economy, it is in some literature argued that regions are the locus of innovation (Lundvall and Borrás, 1997; Asheim and Isaksen, 2002; Gertler, 2004). One reason for regional or local collaboration being of particular importance is the tacit nature of much of the knowledge involved in innovation. Such tacitness makes it difficult to transfer this knowledge in a codified form, and therefore personal interaction is often required – both for the transmission of information and for the generation of new knowledge. With unprompted meetings, recurrent exchanges in person, the possibility to observe the actions of competitors, and the diffusion of what has been termed the ‘local buzz’, knowledge is diffused at a lower cost (Scott, 1998; Bathelt *et al.*, 2004; Storper and Venables, 2004). As a consequence, geographical proximity is supposed to facilitate knowledge diffusion and learning. On a local/regional arena with a critical mass of actors there may also be access to inexpensive R&D skills and other complementary resources from other actors.

From the individual firm’s point of view, there are many advantages associated with locating in strong clusters, or differently put, well-developed sectoral regional innovation systems (RIS). Then, in addition to the possibility of finding local partners, the region can offer mobilization of external resources, formation of a local market for the firm’s products, and various spillover effects, such as access to a pooled labor market and availability of local suppliers (Bergek *et al.*, 2008).

This implies that localized learning processes can be seen as a means to develop ‘effective’ or ‘functional’ RIS, where firms build competitive advantages through interaction with localized capabilities (Maskell and Malmberg, 1999). These include knowledge and skills available in the region, as well as the region’s natural resources, infrastructure and institutional endowments. While many industrial clusters historically have been based on the local availability of traditional inputs and resources, more recent clustering of firms tend to be based more often on localized knowledge bases. For example, life science firms tend to be situated in regions characterized by agglomeration of technology-intensive firms, universities and research institutes. Thus, much of the knowledge development and commercialization takes place in those biotech clusters that have emerged around the world (see e.g. Cooke, 2001; 2002).

However, while the region matters for innovation, other geographical units may matter equally much, or more. In fact, the very researchers previously arguing for the region as the main focus now incorporate arguments of globalization of knowledge flows. For example, Maskell and Malmberg (1999: 173) point to the fact that the resources available in a certain cluster comprise both the region's own resources and the ones available through import from other parts of the world. Thus, a region – and its actors – needs to build relationships with actors located outside its own borders. Also, the region, as such, may need to build relationships with other regions.

Today, governments and policy-makers in various ways want to induce both 'regional interaction' – i.e. interaction among regional actors – and regional actors' collaborative patterns with actors outside the region.[2] Nevertheless, in several regional initiatives much focus still lies on promoting regional interaction: not least has this been true in Sweden. Our study scrutinizes, from the firm perspective, this balance between the regional, national and international.

Research design

Characterizing the RIS for biomedicine in Western Sweden

The region of Western Sweden – and, in particular, the city of Gothenburg – has a long and proud tradition in the health-related life science sector, even though it has only recently been recognized as a national 'cluster' (such as Uppsala-Stockholm or Medicon Valley in Southern Sweden/Denmark).

There are in the region more than 200 companies that can be said to belong to a biomedical cluster. This includes firms that develop, manufacture and/or market pharmaceuticals, diagnostics, medical devices, biotechnology tools for research and production, and contract or clinical research. From an economic or employment point of view, a limited number of large firms dominate the cluster: AstraZeneca, Astra Tech, Getinge, Nobel Biocare, Mölnlycke Health Care and SCA Incontinence Care. Together, they employ some 5,000 people. While AstraZeneca's large R&D facility (with some 2,500 employees) works in the pharmaceutical field, the other five firms are all medical technology (medtech) companies. This observation exemplifies and illustrates that, in the pharmaceutical sub-sector, the region's biomedical industry is completely dominated by one firm. There are very few small drug discovery firms – one probable reason being AstraZeneca's long-standing and close collaboration with the university's medical research, which over the years has resulted in successful commercialization of several innovative drugs. By contrast, in the medtech sub-sector, there exist a significant number of small and medium-sized companies. However, the industry is quite fragmented in terms of technologies, products and applications. It means that, from an operational point of view, the potential for collaboration among the region's companies is often limited. One important exception is the area of biomaterials, including, for example, dental implants. In fact, in the biomaterials field, there has for a long time been an internationally strong 'sub-cluster', which comprises large and small firms, as well as leading academic research units. Most of these companies are, in one way or another, based on research carried out at regional universities. Some companies are direct university spin-offs, while others are corporate spin-offs or have entered the field from other origins.

In the third major sub-sector, biotech tools, there are no large companies but rather a growing number of small R&D-based firms. A couple of them work with stem cells. For example, the company Cellartis – a spin-off from the University of Gothenburg – has a unique resource in the form of embryonic stem cell lines making it an internationally attractive research partner both for universities and companies. Existing activities in biomaterials and stem cells are now converging making the whole area of biomaterials and cell therapy a particularly strong profile area for the region.

The University of Gothenburg and Chalmers University of Technology perform prominent research in life science and have made important contributions to the development of a now thriving biomedical industry.[3] The health science faculty at the University of Gothenburg (Sahlgrenska Academy) with 1,500 employees (850 of whom are researchers or teachers) is organized into six departments, conducting basic pre-clinical research, experimental studies of disease mechanisms and research on public health care and epidemiology. The pharmaceutical research is clinical in nature and undertaken in close collaboration with Sahlgrenska University Hospital. Examples of strong research fields are obesity, diabetes, cardiovascular diseases, biomaterials, pharmacology, neuroscience, pediatrics, epidemiology, rheumatology, microbiology and odontology (e.g. dental implants). At the Faculty of Sciences relevant research involves e.g. the Department of Cell and Molecular Biology. At Chalmers, technically oriented research of relevance to biomedicine and health care is performed at different departments, altogether employing more than 100 life science researchers. This includes groups working on, for example, biomedical engineering, systems biology, bio-nanotechnology, physical chemistry and polymer materials for medical applications. Chalmers has, for many years, supported entrepreneurial activities and has spun off some life science-related companies.

The region is also home to three publicly owned research organizations with R&D activities in the field of, or of relevance to, biomedicine: SP, SIK and IMEGO. Furthermore, the region hosts Sahlgrenska University Hospital – one of Northern Europe's largest hospitals – and a few other relatively large hospitals. Also, there is a unit providing health care services, Region Västra Götaland, responsible for financing and managing public health care in the region.

Finally, the region is home to a range of support, bridging, or policy-making organizations that are involved in biomedicine, such as Sahlgrenska Science

Park, Chalmers Innovation, and Business Region Göteborg. Among important regional financiers of biomedical R&D or other innovation activities, one finds the state-owned Innovationsbron Väst, Chalmers Invest, and a couple of private venture capital firms.

The study

The life science industry is, largely, science-based and in many countries a sector prioritized by policy-makers at the national and regional levels as it is perceived to offer attractive growth potential. As highlighted above, Western Sweden hosts a fairly large number of life science firms. Also, the sector is now one of those prioritized by the regional and municipal authorities. In fact, there is a regional development initiative – GöteborgBIO – with the idea of stimulating growth by developing the RIS for life science. One of the central aims is to strengthen the linkages among regional actors, thereby contributing to build a stronger and more dynamic innovation system.

We chose to conduct a survey designed to capture the companies' acquisition of knowledge and their collaborative patterns. The target population consisted of all life science firms active in the region: in essence, 222 firms including those involved in development, manufacturing or marketing of pharmaceuticals, diagnostics, medical technology, biotech supply and contract/clinical research services. As much as 55 percent of the population (123 firms) carry out R&D activities in the region, and we will denote these as 'research intensive firms' (see Table 5.1).[4] Many of the others are service-oriented firms or sales companies that market life science products in the Swedish or regional market, and may play a role in the transfer or use of new R&D-based knowledge.

Using a web-based survey tool, a comprehensive questionnaire was distributed in 2006 and sent to the local branches of global firms.[5] In total, 78 firms completed it (35 percent).[6] The response rate is higher for the R&D-performing firms (46 percent) than for those that are not doing any R&D (21 percent). The response rates reflect the profile of the region's life science industry in general, with a predominance of medtech firms in comparison to pharmaceutical and biotech supply firms. Based on the firms' self-descriptions, less than one fifth (17 percent) of the responding companies are involved in

Table 5.1 Response rate

	Number of firms	*Response rate*
Total sample	222	35% (78 firms)
Firms with R&D, 'research intensive firms'	123	46% (57 firms)
Firms without R&D	99	21% (21 firms)

pharmaceuticals, while the same number of firms (13 firms) have diagnostics marked as their main product area. Medical technology, including biomaterials, is the dominant product area (59 percent of the firms), while biotech supply accounts for only one tenth of the responding firms. In a non-response analysis, three variables were investigated: number of employees, age and main activity (i.e. whether the firms work with pharmaceuticals, medical technology or other products). As regards 'research intensive firms', the results obtained were not proved to be fully representative for the whole population of such firms.[7] Testing for 'all responding firms' showed that there was less difference between respondents and non-respondents, but still some differences with regards to activity, and this should generate some caution as to generalizing the results to the whole population.[8] To verify and complement the results in the survey, personal interviews were carried out with 13 firms representing different sub-sectors (Andersson *et al.*, 2007).

Evidence of open innovation

Almost all the firms here investigated consider themselves to be innovative, and 6 out of 10 firms report that they had introduced 2–10 new products in the past 5 years. Such innovation processes (from idea to finished product on the market) are of course by no means linear, i.e. they are not foreseeable procedures where input (e.g. in terms of R&D) always leads to the desired output. Instead, it is complex routes involving a set of different inputs and competencies. Refuting a linear relationship between R&D investments and innovative output, it is nevertheless the firms that invest heavily in R&D that consider themselves most innovative.

Of the 'research intensive firms', 60 percent carry out scientific work or develop basic technologies, in addition to product and service development. These firms tend to spend a comparatively large amount of their resources on R&D, when measured as a percentage of total costs. Indeed, in 2005, about 60 percent of the firms spent more than 10 percent of their total costs, and more than one third spent over 20 percent. However, the R&D expenditures in terms of absolute numbers are, in many cases, less impressive, where three quarters of the firms spent less than 10 man-years on R&D in 2005, and only four firms invested more than 20 man-years. This is, above all, a reflection of the fact that most life science firms in the region are relatively small.

The heavy focus on R&D is also mirrored in that the R&D-performing firms consider technology and intellectual property as the most vital assets for their competitive advantage. Interesting, however, is that they acknowledge market knowledge, the firms' customer relationships, and their human resources as nearly as important. The firms thus speak about their need for a set of different types of knowledge: technology and market-related ones, as well as application-specific knowledge, scientific knowledge, production knowledge, and also, to some extent, distribution/logistics and design. There is thus a multitude of knowledge types required for innovation, and we will below discuss if and how these are developed in an open innovation mode.

Our first question is to what extent the Swedish life science firms engage in their innovation process in an 'open mode', i.e. where external knowledge sources are consulted and used and where knowledge flows in and out of the company in order to speed up the development. An important finding is that many firms in our study view their innovation activity to be mainly an internal process. In fact, two thirds of the 'research intensive firms' say their R&D activities are carried out completely in-house, without the involvement of external partners. This may be seen as positive in the sense that it illustrates that companies have the internal strength and capability to run such complex projects as life science development requires. Conversely, given the crucial importance of external interaction for successful innovation, do the firms rely too much on internal knowledge, leading to the potential risk that they be less open to knowledge impulses and learning from 'the outside'? We will dwell on this issue in some depth below.

Turning the figures around, one third of the R&D activities in 'research intensive firms' entail external partners. This corresponds well to, for example, the findings of Enkel and Gassmann (2008). The extent of open innovation is thus on an 'average level' in this sample of firms.

In essence, the respondents acknowledge the need for external knowledge sourcing. An important such 'channel' to identify or acquire new knowledge alludes to the firms' participation in various situations where knowledge is diffused, such as conferences, trade associations or EU projects. Out of the many sources of codified knowledge – be they patents, licenses or scientific publications – the firms denote the acquisition of licenses to be of only moderate interest, while patent analysis and the analysis of scientific reports and journals are seen as more momentous.

The respondents also point to the impact of university relations on knowledge identification and acquisition. For example, the respondents claimed that recruitment from universities was a vital channel for knowledge acquisition. These findings confirm the scientific character and the important role played by academic research for biomedical firms. Furthermore, the firms considered *direct* interaction with universities the most effective way for both identifying and acquiring knowledge, as compared to more indirect means of technology transfer.

Clearly, then, innovation processes in these firms are truly open, where the firms identify knowledge needs and source knowledge trough a range of 'channels', be they based on codified knowledge such as patent analysis or conferences, or be they in the form of interaction with, for example, universities or firms. It is evident that firms have a wide range of partners: customers, suppliers, producers of complementary goods and services, competitors, health care providers, universities, research institutes and public organizations. Naturally, the companies have a number of motives for collaborating with others: seeking corporate partnerships for co-operation in the marketing of products and services, co-manufacturing products or outsourcing manufacturing, and co-developing or acquiring new technologies, among others. Our interviews indicate that the differences in interaction patterns, in terms of partner type and frequency, and in terms of purpose of co-operation, are very much related to the stage of development the company has reached.[9] As one would expect, early-phase companies tend to be more focused on interaction with universities, university hospitals and other research organizations. When a company starts growing, it becomes more important to establish collaborative relationships with other types of partners, including customers/users, suppliers, contract manufacturers, authorities, distributors and financiers. These relationships are needed in order to access the complementary resources and competencies the company needs for its further development and growth. Some of the resources the company needs are of a rather general character, such as skilled labor, competent managers and money. Other resources are of a more technological or product-specific nature. It can be specialized equipment for research or production, suppliers of dedicated goods or services, or potential customers with specific applications. Companies may also need to broaden their scientific bases by establishing collaborative relationships with research groups in other parts of the world to access complementary technologies or to promote their own solutions and products, for example.

In our study, the firms see their customers, significantly, as their most important and frequent type of partner. Most of the firms in our sample are well established on the market. A quarter of the firms state that they currently collaborate with one to four customers, 45 percent have 4–20 such partnerships, while almost a quarter have more than 20 interactions ongoing.[10] It is a well known fact that, within the life science industry – as in many other sectors – customer input is key to accomplishing timely and successful innovations. Interestingly, while much of this interaction naturally has a market focus, 73 percent of the firms that collaborate with customers sometimes, or often, engage them in scientific projects and publications. Even in patenting activities, one out of five firms sometimes involve customers. Plainly, customers are crucial to firm innovation.

Also, suppliers take on a prominent role as partners, while providers of complementary products and services are seen as somewhat less important, and competitors – as might be expected – rarely function as key partners. This indicates that, in the life science industry, customers and suppliers are not only actors in the transaction chain, but they may also be seen as potential collaborative partners who can make valuable contributions to the development and commercialization of new products. This rhymes well with the literature, where interaction with customers and suppliers is seen as a central feature of innovation in many industries. This includes, for example, broad economics-based studies of innovation such as Freeman (1982), Rosenberg (1982) and Dosi (1988), as well as the more focused literature on industrial networks and marketing, exemplified by von Hippel (1988), Håkansson (1989), Biemans (1992), and Håkansson and Waluszewski (2004).

To verify the survey findings, 13 selected companies from the sample were interviewed (Andersson *et al.*, 2007). These companies displayed varying patterns of interaction, with regard both to the number and type of partners. On one extreme, we found a biotech supply firm having some 150 collaborative relationships. On the other extreme, one of the medtech companies said that it has no R&D interaction at all.[11]

Open innovation within the region

As argued in the introduction, open innovation is a phenomenon quite without boundaries, implying that knowledge flows between various types of partners and across geographical borders. However, sometimes regional collaboration is preferred by firms – or perhaps preferable from an innovation and growth perspective – and therefore sometimes a national or foreign partner is chosen. The intriguing question is: in what situations and for what reasons? Therefore, let us now take a look at open innovation from a spatial perspective.

The first overall finding is that, to the responding life science firms, partners are considered of equal general *value* to the firms' innovation process, independent on being located in, respectively, the region, elsewhere in Sweden, Europe or North America. Fewer firms, however, saw interactions with partners in Asia as influential, possibly indicating a lack of networks in this part of the world.[12] To give an example from the interview study, the managing director of one pharmaceutical firm, which is a university spin-off, says that, for their firm, proximity is not a major criterion in its selection of R&D partners. Thanks to modern communication technology, it has become less significant where the partners are located. Instead, partners are chosen on the basis of competence and price. Another university spin-off reports that all its R&D partners but one are situated abroad. What is surprising with these findings, therefore, is not that the firms have a wide geographical distribution of partners, but that, in this globalized world, regional interactions are still equally dear. In what follows, we will delve into the details of why regional partnering retains its value.

Industrial and clinical interaction

Our question for this section is thus which of the interactions presented in the previous section that are with regional, national or foreign partners, and why. We start with analyzing the (non-academic) partners, i.e. health care providers, clinical research organizations (CROs) and firms.

It becomes readily apparent that it is the health care providers, CROs and pharmaceutical firms that are of foremost frequency at the regional level (see Figure 5.1).[13] In fact, for pharma and health care firms, partnering in the region is about equally common as in the nation as a whole. This is not surprising given that the region holds the dominant pharma firm, AstraZeneca, and that Sahlgrenska University Hospital is well renowned and eager to enter into firm partnerships.

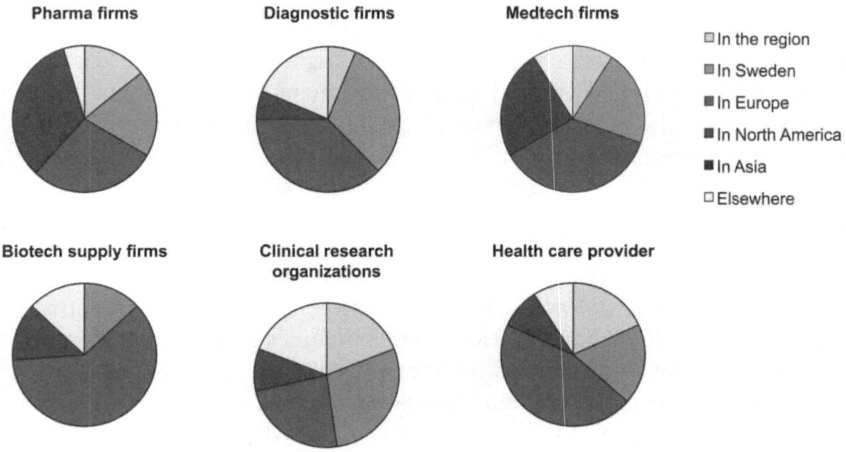

Figure 5.1 Overview of the location of the firms' non-academic partners (percentage of non-academic partners in each location).

Somewhat surprisingly, there is little intra-regional co-operation with medtech or diagnostics firms, and none whatsoever with the biotech supply sector, even though these three groups together make up the bulk of the regional life science industry. One of the problems – contributing to explain this observation – is the heterogeneous and fragmented nature of the industry. Not least in the medtech and biotech supply sub-sectors, there exists a broad range of technologies, products and applications, and the firms tend to be highly specialized. This means that, from the perspective of an individual firm, it is often difficult to find common points of interest with other regional life science firms. One exception in our focal region is the field of biomaterials, where there exists a group of interrelated firms with various types of technological or market-related connections. Here, we can in fact observe how intra-regional interaction has contributed to strengthen certain firms. However, the number of interactions among these firms is still rather limited.

Even though regional partnerships are present as seen above, it is more frequent to have a national firm or health care partner outside of the region than within it.[14] About a third of the respondents collaborate with Swedish (non-regional) diagnostic firms and CROs, and about a fifth with such medtech or pharmaceutical firms or with health care providers. In general, Swedish partnerships are thus relatively frequent, but they are primarily established with non-regional actors. In fact, while pharma and health care partners may very well be regional, it is not primarily within the region that one can find other firms or health care organizations to partner with – for that, one needs rather to turn to other regions within Sweden.

Overall, the dominant type of partnership is the European (non-Swedish) one, especially with biotech supply firms, health care providers, diagnostic and medtech firms, pharmaceutical companies and CROs (see Figure 5.1). Thus, even though many firms in the sample are small, they are international in their innovation processes, seeking collaboration in various countries. It seems that, for the younger or smaller firms, European partners are relatively more important. Generally speaking, it is probably easier for these firms to start developing their businesses in Europe, and then grow later on by entering the larger, more distant and challenging North American market.[15]

Non-European co-operation is predominantly with pharmaceutical companies and medtech companies, but also with other types of partners. Once we cross the boundaries of Europe, partners from North America are the most pronounced. This is quite logical since the US is the dominant market worldwide for life science products. The interview study showed that many of the somewhat larger medtech and biotech supply firms regard North America as their key market. This also means that it is crucial for them to collaborate with American partners and customers/users.

Regional open innovation with universities and institutes

As for the location of partnering with *universities and research institutes*, the pattern here is different in that these are more often local relationships. Thus, the region plays a more essential role for these types of collaboration, especially when compared to partnerships with firms, CROs and health care organizations, as discussed above (see Figure 5.2).

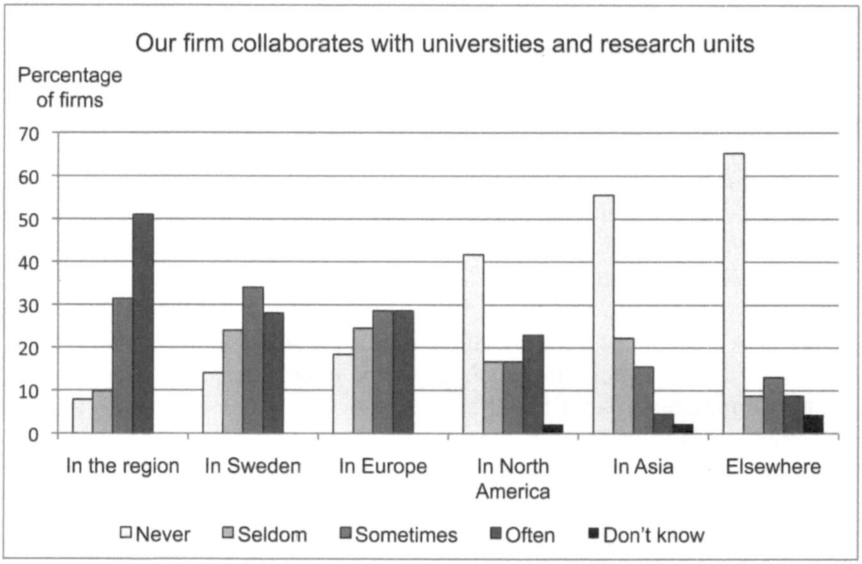

Figure 5.2 Location of partnering universities and research units.

Note: N = 123, $n_{mean} = 44$

Interestingly, then, the *regional* university and institute partners are relatively more frequent than what was the case on the business side. From the interviews, we see several reasons to these patterns.

First, research – in contrast to later stage product development – usually involves a higher amount of tacit knowledge, which presumably contributes to the explanation as to why geographical proximity and face-to-face interaction are relatively more essential when it comes to academic collaboration (as compared to customer and supplier collaboration).

Second, many of the analyzed biomedical firms in Western Sweden are spin-offs from the two regional universities. The typical pattern in Sweden is that the academic founders of start-up companies stay in academia while being involved in parallel in the company (e.g. as part-time employees, advisors or board members). It is natural for the company to maintain a close relationship with the founding research group at the university. It is not uncommon that the original invention manifested in a product concept needs to be substantially refined in order to meet the requirements from the market or production. The role of the academic researchers may be to transfer technology to the company and assist in the development of products and applications. The continued research at the parent university may also bring about new results that are valuable to the company.

Third, the Swedish system with intellectual property ownership resting with the individual researcher, rather than with the university as such,[16] paves way for an informal manner of initial collaboration. In fact, the tradition of collaboration is strong between firms, Sahlgrenska University Hospital, the research groups at the universities, and the institutes in the region. While there certainly are collaborations starting with a formal contact at the central university level, it is not seldom that an interaction springs out of a personal relation leading to initial, rather informal discussions, which in time may lead to more formalized projects. Such informal networks are therefore both deeply rooted in the regional soil and hard to copy to other spatial settings.

Fourth, one aspect may be that former undergraduate and PhD students keep contact with their old universities. That informal relationship should not be underestimated but is also probably more of an externality than related to intentional open innovation strategies.

Fifth, in addition to the possibility of acquiring knowledge, the local universities constitute particularly vital resource providers for the smaller firms. Some interviewees representing such firms emphasized the great value of being able to use unique research facilities owned by the universities. This mainly concerned the use of advanced scientific instruments that the companies cannot afford to buy – at least not in the short term. To support the development of small firms, regardless of whether they are university spin-offs or not, it is thus imperative that the universities pursue a policy that means they are open to this kind of interaction. For these firms, it is an obvious advantage if these resources are available in their local environment. To go outside of the region to gain access to them may prove to be too costly or complicated.

One may argue that university and institute collaborations – and regional collaborations in particular – have their limits. For example, some firms primarily connect to universities in early phases of their development. Other firms, however, continue to collaborate with universities throughout their life. Due to the science-based nature of the biomedical industry, university- and institute-based research may be very important for the firm's technological development. Traditionally, the principal role of these research organizations is as co-developers of science and basic technology. It can be, for instance, inventions in the form of new concepts for diagnosis or therapy or novel techniques used in the R&D process. Conventionally, then, business partners such as customers and suppliers contribute to the more applied phases of product, service and process development. However, while this is still true in the general case, such demarcations between what different partners may bring seem gradually to disappear. For example, university researchers who work as medical doctors may take a key role in defining the market as they daily see the patients' needs. In fact, clinical researchers at university hospitals can both be customers and involved in clinical trials and application development.

New organizational forms for regional firm-university/institute collaboration within joint centers may be one way to organize such ongoing exchange. The two regional centers of excellence BIOMATCELL and SuMo are good examples of such a coordination platform. Both centers relate to life science, and have set up collaborative schemes between industry, academia, institutes, health care providers, and regional and national policy actors. The firms – small as well as large ones – stress the benefits of university interaction as, for example, identification of new knowledge areas, access to a network of researchers and experts, technology and method development, and use of high quality equipment (Brink and Johansson, 2009). In particular, as the firms cannot afford to keep in-house the wide range of knowledge areas needed, the university connections both ensure high-quality development and speed up knowledge acquisition.

Discussion

We started this chapter by asking what role regional knowledge nodes may have for life science firms, and how policy actors could approach regional cluster formation. In this last section, the discourse focuses around what conclusions we may draw as regards open innovation processes, particularly within the region, and what this may imply for policy.

The analysis shows clearly that interaction with other actors – such as firms (including customers) in the health care sector, service providers, clinics and research organizations – is a key prerequisite for gaining access to critical resources needed in the innovation process. These firms truly engage in an open innovation mode.

The empirical data illustrates that, while such interaction sometimes takes place regionally, working with national or international partners predominates, especially for the large and well-established companies. This is what one would expect to find in a globalized business such as life science. The established firms, even those that are not so large, operate in international markets and tend to see themselves as players on a global arena. To the extent that they need or wish to establish R&D interaction with, for example, customers, suppliers or other life science firms, it is relatively easy for them to find partners outside the region, where most of the business counterparts are situated. In other words, the need for regional interaction is relatively limited for these firms.

Despite this conclusion, it cannot be denied that knowledge interaction is often easier and less expensive when the partners are located nearby, especially for smaller firms. The empirical evidence shows that regional interaction is of high consequence to life science companies.

The firm-to-firm and health care interaction

As regards collaboration with industrial firms, service firms or health care, the survey shows that, while regional partnerships are indeed influential today, they are not the primary type of interaction for most firms. Above all, the regional partners that come into play are health care providers, CROs and pharmaceutical firms, and also universities and research organizations (as we will see below). We conclude that business relationships with customers and suppliers are used not only for purely commercial purposes, but also, in many cases, as a resource for technological development. This type of partner is usually located outside of Western Sweden, which is quite understandable given the relatively small size of the region. The life science firms tend to be focusing on technologically advanced niche products, implying that, in order to make profits on their R&D investments and achieve sustainable growth, the larger international market needs to be penetrated. In fact, it is not unusual that life science firms can be characterized as 'borne globals' – i.e. contrary to the traditional internationalization model, they focus on the export market already from the beginning.

Even if it is true, at the aggregate level, that customer interaction takes place mainly with non-regional partners, this does not mean that the availability of regionally based customers is trivial. We know from the research on business-to-business markets that the introduction of new products into new geographical markets often requires technological co-operation with local partners/customers.[17] There may, for example, be a need to make adaptations of products or applications to local demands (e.g. due to country-specific regulations, traditions or other local circumstances). Furthermore, in order to be willing to buy, local customers may require results from clinical trials or other types of references from their own country. This means that innovating firms may need to establish strategic partnerships with customers (e.g. 'lead-users') in several different geographical markets.

In life science, many of the product ideas developed by small firms have their origin in technical or pre-clinical research. It is of utmost importance that the product development, preferably at an early stage, is linked to clinical needs and that fruitful feedback is received from pioneering customers. For natural reasons, young life science firms tend to be small and have limited resources. Therefore, it is a great advantage to them if they can establish interaction with nearby potential customers. In our sample, we have many firms that have close interaction with Sahlgrenska University Hospital. Our interviews gave some good examples of this, illustrating how advantageous it is for the industrial development that there is in the region a health care organization that is not only advanced and progressive, but also open to interaction with industry and to testing new ideas and methods.

Suppliers of goods and services are also prominent as partners, as shown by our data. This is in line with general industrial patterns. Involvement of suppliers in product development is not a new phenomenon. Nonetheless, during the last decades, the trend is that companies tend to use suppliers more often as an R&D source.[18] The increased outsourcing of production is one of the explanations for this trend. When firms purchase more and more complex products (such as complete modules or sub-systems), the increasing interdependency between buyers and sellers in many cases creates a need for technological co-operation. Previous research on the medtech industry in Sweden has shown that new, research-based companies often choose to 'outsource' the production – possibly with the exception of certain key components on which the innovation is built or the final assembly and quality control (Laage-Hellman, 1993). For these small firms, with their limited resources, it is often an advantage if they can find regional suppliers – especially for essential components or materials that need to be adapted. The larger and well-established firms appreciate interaction with local suppliers, for the reason of convenience, but for them it is easier to work with more distant partners, and they also do that to a large extent.[19]

Clearly, the firms in this study acknowledge the importance of a competent life science industry within the region. However, as to the region of Western Sweden, there is a perceived lack of a sufficient critical mass of firms. To alleviate this problem is one of the drivers for the firms to engage in the regional initiative GöteborgBIO. Firms think there could be more knowledge spillover and synergies from the co-location with other life science companies. Indeed, in the survey, many respondents pointed to the current lack of a critical mass as an inhibitor to their own development and growth. They believe that the entry of additional life science firms would be very positive for the region and for them as companies. First and foremost, they feel that this would most likely lead to more interaction, and contribute to an increased information flow among firms enhancing knowledge spillover. Second, it is believed that, if additional firms would be attracted, the region could be marketed as a 'bio-region' with a vibrant firm base. This could, for example, make the region attractive for firms that can supply specialized intermediate goods and services.

Third, a larger cohort of life science firms would create a pooled labor market for R&D personnel and managers, and make it easier to attract expertise from other countries. Fourth, the companies think the existence of a critical mass would give the sector increased political power, which would lead to an enhanced public awareness and legitimacy of their technologies and products. Being more of a critical mass of firms would allow them greater ease in influencing legislation and policy actors.[20]

The respondents put forward some specific viewpoints on how to increase the critical mass. One point is that, in order to increase the number of firms and enable their growth, there is a need to improve the availability of venture capital (VC), even though several VC firms are present in the region. For example, attracting internationally leading VC firms that specialize in life sciences to the region would bring many benefits, be they in the form of money or expertise in the life science business. Creating a pooled labor market is another means to attract companies and to facilitate for the existing ones. It is true that, today, the labor market functions acceptably for the firms, and there is a sufficient supply of engineers in particular. However, for a majority of firms, it is not easy to recruit managers from the region. Also, measures to increase the mobility of employees, both within the industry and between the industry and academia, as well as assisting small firms to internationalize, can be expected to pay positive dividends. In fact, knowledge spillover can accrue from recruitment processes, further indicating the importance of having a critical mass of life science firms in the region, or nation, within which personnel can circulate.

The university and institute interaction

A chief finding was the importance of collaboration with the universities in the region, for the purpose of identifying and acquiring new knowledge. As obvious from our analysis, the biomedical firms have strong relations to the regional universities. The first reason lies in that tacit research-related knowledge may be more easily accessible in spatial proximity. In addition, many companies are founded on inventions and research findings that originated within the two main regional universities, with whom they maintain close relationships (e.g. for further developing the technology). Moreover, the specific structure of intellectual property gives rise to a culture of informal collaboration, with time developing into more formalized patterns. Also, the possibility for sharing of equipment, access to expert judgment, etc. is valuable, especially for small firms.

The ability to innovate with the aid of regional universities is, in essence, a positive trait of this set of firms, in that variety may be created and innovation spurred. One may also say that the RIS functions well from this point of view, providing access to prominent universities, and having an organizational set-up that makes firm–university–institute interaction possible. This indicates that this is a feature that could, or should, be strengthened further and exploited even more. For example, the stimulation of academic researchers to increase their contacts with existing firms (or vice versa), and the encouragement for other forms of academic entrepreneurship (e.g. spin-off firms) are paths to explore. Here the universities and institutes, but also regional and national policy organizations, may take a role.

From a commercialization point of view, it must be emphasized, though, that for university spin-off companies it is of crucial importance to complement the key relationship with the founding institution with other collaborative relationships. Above, we discussed the need to link up with users/customers in different countries for application development, promotion of new methods, etc. The company may, however, also need to broaden its interface with the academic world – it may, for example, be in order to access complementary technologies needed to develop a complete product offer (which may include various auxiliary equipment, software or services). Regional university and institute relations are thus, most often, not sufficient.

It can be of interest to relate the results reported in this paper to studies of other biomedical innovation systems or clusters. Nummi (2007), for example, has studied the medtech industry in the Oulu region in Northern Finland. Here, close collaboration between companies, regional organizations for higher education and the university hospital has been decisive for rapid growth of the medtech industry. The case thus illustrates how tight industry–university collaboration within a small region – actively supported by local and national policy-makers and exploiting the advantages of geographical proximity – can lead to the emergence of a new, internationally competitive cluster. It seems that, in many respects, the strategy now pursued by GöteborgBIO is quite similar (including, for example, investments in science parks and stimulation of regional networking). One element of the Oulu approach that is missing in Western Sweden, however, is the formation of company associations on the regional level. These are strategic level institutions that, among other tasks, express the companies' interests to regional and national policy-makers. They facilitate interaction within industry but also make it easier to start collaboration between companies and research institutes.

The role of policy

Obviously, regional knowledge nodes are essential for life science firms. The empirical exposé has highlighted numerous ways in which firms benefit from their regional resource environment. The specific pattern discerned as to which types of regional interactions that are of substance depends, of course, on the precise industrial set-up and is not generic in character. Other sets of firms may, most likely, show their unique pattern of when regional, national or international interaction is preferred. A general conclusion is, then, that only by understanding the very details of the interaction patterns and the 'gaps' in the value chains can policies be designed to better support the innovation and growth processes.

Generally, it must be concluded that the room for policy intervention directly affecting firm interaction patterns is relatively limited. However, policy-makers may take certain actions, mainly indirect ones, which have some effects on the collaborative pattern. All measures aiming at increasing the critical mass, as discussed above, would improve the preconditions for finding local partners. For example, firms with certain types of key products or technologies may be identified as missing in the region, and efforts could be made in order to attract such companies from outside to enter the region. Also, supporting the building of infrastructure platforms for multilateral R&D projects of interest to several firms would facilitate regional interaction. One example from GöteborgBIO is the new Institute for Biomaterials and Cell Therapy, where different constellations of firms can come together in joint projects.

A debatable issue is, of course, if innovation policy should actively encourage and support the formation of *regional* interaction specifically, or center more generally on the encouragement of knowledge flows. One argument may be that such focus on the region may be done at the possible expense of less attention on international interaction. It is therefore crucial to balance the respective focus on interactions within the regional arena with that of a global outlook. Perhaps this has been somewhat of a neglected area in much of regional cluster building around the globe. In particular, this may be one critique to direct towards the Swedish VINNOVA program, scrutinized in several chapters in this book. At the onset, efforts were more directed towards regional activities and inter-linkages, rather than on positioning the cluster and its actors in a global perspective. This potential deficit has been alleviated over time, and several of the regional initiatives throughout Sweden today have an international outlook and vision.

Our final comment on the role of policy relates to the concept of the region. While the study takes its empirical offspring in what is, in practice, considered as a region in the Swedish context, this may be up for debate. In fact, the finding that Swedish partnerships are frequent – but not necessarily local ones – indicates that the 'region' of Western Sweden may be too small a geographical unit for this type of innovation and growth activity. Comparing with dominant life science nodes around the globe, Western Sweden has a rather limited population, market and scope. To policy, this may signify that to divide Sweden (with a total population of 9 million) into regions may not be a useful strategy. Instead, Sweden should be seen as one arena, where policy measures should be coherent and where each sub-region should contribute to the larger whole.

Notes

1. Acknowledgements are given to GöteborgBIO for financial support, and to the responding firms for their time and sharing of experiences. We are also grateful to the book's editors and contributors for comments on earlier drafts.

2. As Asheim *et al.* (2003) point out, there are a range of ways in which policy strives to address network-building to enhance innovativeness, including schemes directed at individual actors or larger groups, as well as to provide the needed resource base, and those wanting to accomplish institutional change.
3. Besides these two main universities, there are a few smaller and younger universities located in smaller towns.
4. We classified the firms as performing R&D based on the report by VINNOVA (2005), as well as company information, such as their websites.
5. This may give a bias to the answers, as the role of a local branch of a global firm may be precisely to have close contacts within the Gothenburg region.
6. The questionnaire was sent to managing directors or R&D managers, depending on the size of the firm.
7. For 'research-intensive firms', the analysis shows that, for all three variables (with significance level $p = 0.05$), the firms that have responded differ from those that did not. If we look at *all firms*, we find that, regarding age and number of employees, the firms can be said to come from the same distribution, but that they differ with regard to activity ($p = 0.05$).
8. In addition to the complexity of its structure and processes, the innovation system is constantly evolving over time. It is difficult to catch many of the dynamic mechanisms through which the system evolves with the type of survey we have used. However, it is our intention to carry out a follow-up study after 2–3 years. This will give us an excellent opportunity to address the issue of innovation system dynamics.
9. This also confirms previous studies (Laage-Hellman, 1993).
10. Our data shows that, for most categories, the majority of firms have only 1–3 partners they collaborate with. However, when we look at interaction with customers and suppliers, their partners are, on average, roughly 9 and 7 respectively. There are a significant number of firms collaborating with more than 10 customers or suppliers. Thus, there is a great deal of variation among the firms.
11. It is noteworthy that, so far, this company has not been successful in launching new products onto the market. There may be different reasons for this failure, of course, but one could speculate that this could have something to do with this lack of knowledge flows.
12. Given the rapidly increasing investments being made in biotechnology by several Asian countries – such as Japan, South Korea, Singapore, India and China – there are strong reasons for Swedish firms to become more active in that part of the world.
13. The question posed was where each type of their partner was primarily located, thus they were to give the most important location for each category of collaborative actor. $N = 123$, $n_{mean} = 23$.
14. This may, in fact, indicate that the relevant size of the 'region' in this case is Sweden as a whole.
15. This indicates that helping younger and smaller firms to establish themselves in North America may be a feasible task for regional support organisations.
16. This is called the teacher's exemption.
17. See, for example, Håkansson (1982) and other literature on business-to-business marketing. Laage-Hellman (1996) addresses this issue in a special case, that of the Japanese market.
18. Toyota and other Japanese car manufacturers have served as role models for other firms that have adopted successful Japanese practices. This holds for automakers as well as firms from other engineering industries (Bidault *et al.*, 1998).
19. Several of the large medtech firms in Western Sweden have their manufacturing outside the region. This diminishes the motive for choosing partners from the region, even if they carry out a large part of their R&D activities there.

20. However, a critical mass is not to the same extent believed to create a regional product/service market. Instead, most companies are already well established in several markets, and a regional market is of little significance.

References

Andersson, F., Ang, T., Dertz, M., Hebib, G., Kensfors, H. and Söderström, C., 2007, 'R&D Interaction in the Biomedical Industry', Project work carried out for the Master's programme Management and Economics of Innovation, Chalmers University of Technology (*mimeo*).

Asheim, B.T. and Isaksen, A., 2002, 'Regional Innovation Systems: The Integration of Local "Sticky" and Global "Ubiquitous" Knowledge', *The Journal of Technology Transfer*, 27, 1: 77–86.

Asheim, B. T., Isaksen, A., Nauwelaers, C. and Tödtling, F. (eds), 2003, *Regional Innovation Policy for Small-Medium Enterprises,* Edward Elgar, Cheltenham, UK.

Bathelt, H., Malmberg, A. and Maskell, P., 2004, 'Clusters and Knowledge: Local Buzz, Global Pipelines and the Process of Knowledge Creation', *Progress in Human Geography,* 28, 1: 31–56.

Bergek, A., Jacobsson, S., Carlsson, B., Lindmark, S. and Rickne, A., 2008, 'Analyzing the Functional Dynamics of Technological Innovation Systems: A Scheme of Analysis', *Research Policy*, 37, 3: 407–429.

Bidault, F., Despres, C. and Butler, C., 1998, 'The Drivers of Cooperation Between Buyers and Suppliers for Product Innovation', *Research Policy*, 26: 719–732.

Biemans, W.G., 1992, *Managing Innovation Within Networks*, Routledge, London.

Brink, J. and Johansson, M., 2009, 'Industry and University Collaboration in an Open Innovation Center: Pressure to Change', *Memo*, University of Gothenburg.

Chesbrough, H.W., 2003, *Open Innovation: The New Imperative for Creating and Profiting from Technology*, Harvard Business School Press, Boston.

Cooke, P., 2001, 'Biotechnology Clusters in the UK: Lessons from Localisation in the Commercialisation of Science', *Small Business Economics*, 17: 43–59.

Cooke, P., 2002, 'Regional Innovation Systems: General Findings and Some New Evidence from Biotechnology Clusters', *Journal of Technology Transfer*, 27: 133–145.

Dosi, G., Freeman, C., Nelson, R, Silverberg, G. and Soete, L. (eds), 1988, *Technical Change and Economic Theory*, Macmillan, New York.

Enkel, E. and Gassmann, O., 2008, 'Driving Open Innovation in the Front End: The IBM Case', Working Paper University of St. Gallen and Zeppelin University, St. Gallen and Friedrichshafen.

Freeman, C., 1982, *The Economics of Industrial Innovation*, Frances Pinter, London.

Gertler, M., 2004, *Manufacturing Culture: The Institutional Geography of Industrial Practice*, Oxford University Press, Oxford.

Gilsing, V. and Nooteboom, B., 2006, 'Exploration and Exploitation in Innovation Systems: The Case of Pharmaceutical Biotechnology', *Research Policy*, 35, pp. 1–23.

Haanaes, K., 'Strategic Resources in Action: A Case Study of the Dynal Biotech Joint Venture', *Scandinavian Journal of Management*, 15: 249–267.

Håkansson, H. (ed.), 1982, *International Marketing and Purchasing of Industrial Goods: An Interaction Approach*, John Wiley, Chichester.

Håkansson, H., 1989, *Corporate Technological Behaviour: Co-operation and Networks*, Routledge, London.

Håkansson, H. and Waluszewski, A., 2002, *Managing Technological Development: IKEA, the Environment and Technology*, Routledge, London.

Laage-Hellman, J., 1993, 'Forskningsbaserat medicintekniskt företagande: En studie av kommersialiseringsformer, framgångsfaktorer, utlandsexploatering och europasamarbete', *Nutek B 1993:12*, Stockholm.

Laage-Hellman, J., 1996, 'International Technology Cooperation: The Case of Sweden-Japan', Chalmers University of Technology, Gothenburg (*mimeo*).

Lundvall, B.-Å., and Borrás, S., 1997, *The Globalising Learning Economy: Implications for Innovation Policy*, Office Publications of the European Communities, Luxembourg.

McKelvey, M., Rickne, A, and Laage-Hellman, J., 2004, *The Economic Dynamics of Modern Biotechnology*, Edward Elgar Publishing, Cheltenham, UK.

Maskell, P. and Malmberg, A., 1999, 'Localised Learning and Industrial Competitiveness', *Cambridge Journal of Economics*, 23: 167–185.

Nummi, J., 2007, 'University–Industry Interaction in Medical Devices Development: Case Study of the Oulu Region in Finland', in Lester, R.K. and Sotarauta, M. (eds), 'Innovation, Universities, and the Competiveness of Regions', *Technology Review 214/2007*, Tekes, Helsinki.

Oliver, A.L., 2004, 'Biotechnology Entrepreneurial Scientists and Their Interactions', *Research Policy*, 33: 583–597.

Orsenigo, L., Pammolli, F., and Riccaboni, M., 2001, 'Technological Change and Network Dynamics: Lessons from the Pharmaceutical Industry', *Research Policy*, 30: 485–508.

Pisano, G.P., 1991, 'The Governance of Innovation: Vertical Integration and Collaborative Arrangements in the Biotechnology Industry', *Research Policy*, 20: 237–249.

Powell, W.W., 1998, 'Learning From Interaction: Knowledge and Networks in the Biotechnology and Pharmaceutical Industries', *California Management Review*, 40, 3: 228–240.

Roijakkers, N. and Hagedoorn, J., 2006, 'Inter-firm R&D Partnering in Pharmaceutical Biotechnology since 1975: Trends, Patterns, and Networks', *Research Policy*, 35: 431–446.

Rosenberg, N., 1982, *Inside the Black Box: Technology and Economics*, Cambridge University Press, Cambridge.

Scott, A., 1998, *Regions and the World Economy*, Oxford University Press, Oxford.

Storper, M. and Venables, A.J., 2004, 'Buzz: Face-to-face Contact and the Urban Economy', *Journal of Economic Geography*, 4, 4: 351–370.

Tidd, J., Bessant, J. and Pavitt, K., 2005, *Managing Innovation: Integrating Technological, Market and Organizational Change*, John Wiley, Chichester, UK.

VINNOVA, 2005, 'Regional klusterprofil: företag inom bioteknik, läkemedel och medicinsk teknik: Västra Götaland 2004', Särredovisning och fördjupning av VINNOVA Analys VA 2005:2. Stockholm.

von Hippel, E., 1988, *The Sources of Innovation*, Oxford University Press, Oxford.

Waluszewski, A., 2004, 'A Competing or Co-operating Cluster or Seven Decades of Combinatory Resources? What's Behind a Prospering Biotech Valley?' *Scandinavian Journal of Management*, 20: 125–150.

6 Between the regional and the global

Regional innovation systems policy and industrial knowledge formation

Linda Gustavsson, Cali Nuur and Staffan Laestadius

Introduction

The globalization of the world economy, with for instance the diffusion of the activities of multinational corporations (MNCs) and the increasing "footlooseness" of production witnessed at the beginning of new millennium (see e.g. Castells, 2000; Storper 2000; Archibuigi and Lundvall, 2001; Epstein, 2003; and Freidman, 2005), have led to an increased interest in and need for a deeper understanding of the processes of knowledge formation. On the one hand, globalization appears to reinforce the importance of the territorial dynamics in generating industrial competences and skills development (Porter 1998, 2000; Storper 2000). Several concepts have been developed to capture this spatial dimension of industrial development, such as clusters (Porter, 1990), industrial districts (Marshall, 1890/1920) and regional innovation systems (Asheim and Gertler, 2005; Cooke, 2005). On the other hand, there are indications that knowledge formation and diffusion are not primarily locally coordinated processes (see e.g. Amin and Cohendet, 2005; Boschma, 2005). As the activities of many firms are, to a greater or lesser extent, globally distributed, the argument that the process of knowledge formation is primarily connected to geographical proximity is coming under scrutiny (Malmberg and Power, 2005). The current stage might imply that certain, previously localized resources, will become more generally available, reducing previous competitive advantages.

Still, knowledge formation and geographical proximity have been paired together for a long time, and this coupling has been described in length in theory and applied greatly in policy. There is no doubt that there are distinctive infrastructural, social or institutional features connected to a certain place – leading to concentrations of specific activities in certain locations and varied growth patterns of regions (Malmberg and Maskell, 2005). Examples of such concentrations are software development in Bangalore or telecom firms in Shanghai or manufacturing firms in Suzhou. It has also been argued that globalization and regionalization is a joint development, where increasing specialization leads to regional centers of excellence and a global search for expertise (Sölvell and Birkinshaw, 2000). Consequently, some of the intellectual underpinnings for stating that localized processes of knowledge formation differ greatly from non-local processes are currently being revisited (Lagendijk and Oinas, 2005; Lorenzen 2005), thus also shifting the view on which role the region has for knowledge formation and economic growth. Understanding the spatial as well as the non-spatial components of knowledge formation is therefore essential for understanding how, when and why geography matters for economic activities.

The density of institutional and inter-personal relations assumed to create an industrious climate favoring the location of activities to certain "hot spots" is an important argument for regional development agencies and policy-makers who want to establish – or maintain – the global competitiveness of their regions. However, although clustering to certain locations obviously takes place, the relations between actors within "hot spots" can be far less dense than what may be expected and much of the innovation and knowledge formation processes take place outside these spatial clusters (Laage-Hellman *et al.*, 2007; Malmberg and Power, 2005). Not least do the globally dispersed innovative dynamics of many multinational corporations (MNCs) challenge the notion that geographical proximity is most beneficial for innovations (Bathelt *et al.*, 2004).

This chapter contrasts corporate knowledge formation with a regional innovation policy initiative and explores the interrelated questions of: *How do, on the one hand, a regional innovation systems policy initiative and, on the other hand, a multinational company with strong roots in the same region, organize the knowledge formation processes?* The regional policy initiative is one of several initiatives within the larger policy program, VINNVÄXT (Regional Growth through Dynamic Innovation Systems). Our aim is not to evaluate the role of the initiative in enhancing the competitiveness of individual actors within the region, but to highlight two contrasting approaches to the organization of knowledge formation processes. The context of the chapter is Sweden, a relatively small country with a limited domestic market, which, for a long time, has been dependent on international trade for corporate and national economic growth. For instance, its exports in 2007 amounted to approximately 140,000 MEUR, which equals 52 percent of GDP (Statistics Sweden, 2008). Sweden has an open economy and ranks as number three (behind Belgium and Austria) in the globalization index that analyses the economies of industrialized nations from parameters such as economic integration, trade restrictions and information flows (Dreher, 2007). Another characteristic of the Swedish economy is that large firms have historically dominated the industrial structure. More than 60 percent of the industrial employees work in MNCs, many of which today are dominated by foreign capital (ITPS, 2007).

Swedish regional development policy, which historically has been characterized by regional convergence, has, in the last few years, shifted towards a more endogenous regional growth approach (Nuur *et al.*, 2009; Nuur and Laestadius, 2010). Thus, analyzing corporate knowledge formation processes and contrasting these with a regional innovation systems policy initiative is all the more relevant when put in the context of a country such as Sweden.

Apart from this introduction, this chapter consists of three other sections. First, we discuss the role of regional and extra-regional knowledge links, and their importance for the MNC as well as for regional innovation systems policy. The next section contains the empirical part, which is based on two case studies. The first case study concerns a regional innovation systems initiative, namely "Robot Valley" in the region of Mälardalen in Central Sweden – a policy initiative that aims at making this a leading, and even world-leading, region within the field of industrial automation. The second case study is on one of the large firms within industrial automation in the region. This MNC is also one of the major corporate actors in the Robot Valley initiative. Finally, our fourth section concludes the chapter by suggesting that policy measures on knowledge formation need to include strategies to stimulate extra-regional links.

Knowledge links and policy response

In this section, we discuss the role of regional and non-regional links for knowledge formation in a globalized world. We also revisit the concept of proximity and discuss the need for a more nuanced view on the geography of knowledge formation.

Regional and extra-regional links

The role of location in facilitating social interaction, knowledge formation and innovation has been discussed in the literature for over a century based on concepts such as industrial districts (Marshall, 1890/1920), clusters (Porter, 1990), science parks (Saxenian, 1994) and innovative local environments (Malmberg, 2002). In particular, the concept of *clusters* – denoting the tendency of firms in the same field to locate in the same geographical area – gained significance as an instrument of regional development in the 1990s. Although the process of clustering of economic activities was a key aspect of Marshall's thesis a century ago, it was Porter (1990) who gave it potency in policy circles. The origins of clusters share commonalities with that of agglomerations and include historical circumstances, proximity to sources of endowment, local demand and chance, and the advantages include human resource specialization, innovations, knowledge formation and scale economies. However, although the concepts of agglomeration and clusters are occasionally used synonymously, in many circumstances mere agglomeration might only be based on factor endowment and hence does not necessarily mean the presence of interdependencies (Malmberg, 1998).

Since its introduction in the 1990s, the cluster concept has shaped regional development strategies by giving authorities insights into the factors that drive the economies of particular regions and the challenges that they may face (Waits, 2000). However, the cluster concept has been criticized for being "vague" and contributing to both "conceptual and empirical confusions" (Martin and Sunley, 2003: 10). Despite this, the cluster concept is still potent in the endogenous regional development approach of several industrialized economies (Nuur and Laestadius, 2010).

If the cluster concept was the instrument for regional development in the 1990s, it is the related concept of regional innovation systems (RIS) that has captured the attention of policy-makers in the 2000s. The RIS approach, as a mechanism of regional development, follows scholarly evidence that differences between nations do not solely depend on factor endowment but rather on the extent of knowledge utilization and innovative capabilities, as discussed by the national innovation systems literature (Freeman, 1987; Lundvall, 1992). The literature on the spatial dimension of innovation systems (Asheim and Coenen, 2005; 2006; Asheim and Gertler, 2005) argues that innovations often are the result of social processes in which there is knowledge formation through localized interactions. Defined by Asheim and Gertler (2005: 299) as "the institutional infrastructure supporting innovation within the production structure of a region," the RIS concept differs from the cluster concept in that it includes more actors than firms when analyzing regional dynamics – actors such as research organizations, educational infrastructure, technology transfer agencies, business associations, financial institutions, etc. (Isaksen, 2001).

The increased interest in RIS as a mechanism of development can be derived from several features of the spatial dimension that are often mentioned as beneficial for knowledge formation. For instance, relations with other actors in a region can be facilitated by geographical proximity as this provides actors with shared norms and language; it eases face-to-face interaction, etc. Further, informal exchange of knowledge – and particularly tacit knowledge exchange – is positively correlated with geographical proximity. The cumulativeness and path-dependency of knowledge and innovation can result in an importance of inter-regional interaction (Cantwell and Iammarino, 2003). These features relate to both static and dynamic economies of agglomeration, where the static agglomeration economies involve economies of scale, transaction and transport costs etc., and the dynamic agglomeration economies emphasize the role of knowledge formation and localized learning, as once observed by Marshall (1890/1920).

However, it seems as if the advantages of geographical proximity can be offset by advantages offered by dispersion. In organizational activities, for example, the advantages of decentralization can include the closer connection between production and R&D, knowledge about local markets, customers and suppliers, as well as the access to specialized knowledge in a particular technology cluster (Cantwell and Iammarino, 2003). This issue is discussed further below in this chapter.

In many industrialized nations, historical connections entail that there exist interdependencies between the MNC and its location (Dunning, 2000). MNCs can be important sources of technological and innovative input and, in the context of RISs, they can be central actors in connecting global flows of knowledge and innovation to local knowledge networks. Naturally, the knowledge formation process of the MNC is one that encompasses deeply integrated and networking actors that could be both geographically proximate and far away (Dicken, 2004). A reason for this is the increasingly distributed knowledge base that many firms rely on. The fact that industrial renewal takes place in between and beyond existing sectors and regions makes these knowledge formation processes highly important to take into account. The complexity of today's advanced engineering based on synthetic knowledge formation (Laestadius, 1998; 2000; Asheim and Gertler, 2005), drawing from different knowledge bases, makes it unlikely that global competitiveness can be created without significant knowledge transfer between different localities.

As the MNC, by definition, has a geographically distributed organization, it may benefit from the dynamics of scope that it derives from having activities in spatially dispersed milieus and institutional settings (Dunning, 2000). This also means that regional firms can benefit from and/or access knowledge developed elsewhere, drawing on the knowledge accessed by MNCs (Cantwell and Iammarino, 2003). In a study of Italian districts, Lazerson and Lorenzoni (1999) found that growth was spurred through the transfer of technology and skills initiated by large firms, where these firms fed the district's smaller firms with external knowledge.

Revisiting proximity

Core to creating competitive advantage is the ability to tap into pools of expertise – regardless of their location. This is valid both when discussing firms and when discussing regions. Bathelt *et al.* (2004) suggest that successful clusters are characterized by their ability to develop and sustain channels for knowledge formation that span across various hot spots around the world. External connections can be essential for providing new knowledge inputs for innovation. Knowledge formation and exchange is hence plausibly less about geographical proximity and more about enabling/achieving a "low-cost exchange" of knowledge (Bathelt *et al.*, 2004). It is thus about the efficiency with which knowledge formation processes can be established and knowledge can be exchanged and transferred. This opens up for the exploration of other dimensions of proximity that may become more important as enablers for knowledge formation. Localized learning processes and "sticky" knowledge are considered sources of competitive advantage for firms, regions and countries. However, this view of localized interactions as the primary place for complex knowledge formation processes to take place is currently being challenged. This opens up a search for a deeper understanding of proximity.

The critique against the dominating role given to geographical proximity is primarily concerned with the lack of non-local influences and actors in the analysis. For instance, in a review of empirical evidence supporting local knowledge creation in clusters, Malmberg and Power argue that:

> industrial sectors are a lot less spatially localized than geographers and regional scientists might like to wish, this is most likely due to the fact that in the area of innovation and knowledge formation it seems especially important for firms and clusters to seek out the best possible partners or to challenge those they value as being their closest rivals. It is not surprising that in the modern business world this does not – and indeed should not be expected to – happen locally.
>
> (Malmberg and Power, 2005: 426)

Malmberg and Power's (2005) review links to the literature that challenges the strong emphasis on the spatial aspect of knowledge formation, reflecting the core notion that an approach comprising both proximate and distant relationships is necessary to fully understand the process of knowledge formation. Recent literature increasingly highlights the need to move the locus of attention from the spatially confined borders of a region or cluster and take a broader approach to encompass and understand knowledge formation. Oinas (1999) argues that such processes are both local and non-local and that knowledge formation can take place regardless of whether proximate or more distant relations are involved. Rather than an either/or distinction, dimensions of local and non-local often work in combination (Lagendijk and Oinas, 2005). Coe and Bunnel (2003) state that innovation occurs simultaneously within both regional and national systems of innovation, and also is often a constituent element of an international innovation system. Instead, the need for a multidimensional approach to proximity is highlighted, and particularly the importance of broadening the perspective to include both the local and the "global" when attempting to capture and comprehend corporate processes of knowledge formation. The suggestion is that there are several scales, or multiple geographies, of knowledge formation, and that both local and global spaces of knowledge formation and innovation need to be considered (Coe and Bunnel, 2003; Amin and Cohendet, 2004). Lorenzen (2005: 401), also asserting that the use and creation of knowledge have distinct geographies, argues that theory "assumes too much when it a priori stipulates that knowledge coordination and transfer is a local phenomenon: clearly, knowledge is both used and created across regions, national borders and even continents."

According to the advocates of this alternative view on the relationship between proximity and knowledge formation, geographical proximity is not necessarily superior to the role of for instance organizational or relational proximity and professional closeness in the formation and flow of knowledge. Boschma (2005) argues that geographical proximity is not enough or even necessarily required for knowledge formation to take place. Nevertheless, he argues, geographical proximity facilitates interaction and co-operation, which, in turn, can strengthen the other dimensions of proximity. On this issue, Amin and Cohendet (2004) suggest that *global relational spaces* are key to knowledge formation. These thoughts are influenced by the discussion on *communities*, a concept that can be defined in varying ways, from a small group of craftsmen working in daily interaction (e.g. communities of practice, c.f. Wenger, 1998) to a more broadly defined group of professionals with a shared worldview, shared terminology and a common tradition of technological understanding (e.g. *technological communities*, cf. Constant, 1984).[1] This approach of *relational proximity* puts forward the idea that "the creation of organizational space and strong relational proximity could facilitate the flow of knowledge across geographical and institutional boundaries" (Lam, 2006: 8). Instead of viewing the local context as the main environment in which relational proximity is created, the proponents of a new view on spatiality argue that strong relational proximity can equally be created in specific organizational contexts or within groups of professionals. The connections between actors can be close in a cognitive or relational sense, although in a geographical sense they rely on a spatially "stretched" connectivity. This need not indicate that such spatially stretched ties are in any way weaker, less intense or committed than local ties (Amin and Cohendet, 2004).

This view on knowledge formation processes, much based on observations of the changing geography of innovation, also has implications for our understanding of regional development. Coe *et al.* (2004) assert that there has been a too strong focus in the regional innovation systems literature on local institutional structures in understanding regional development, although recent developments have begun to deal with the fact that there may be more to the story than the local dynamism. It is obviously important to understand the non-local dimensions of regional development also. This is in line with studies performed by Moodysson *et al.* (2008) on a cluster initiative in Southern Sweden, where they found that an exclusive focus on support for local knowledge formation might actually hamper the innovativeness of firms, particularly in the long run.

The importance of extra-regional knowledge links

The discussions above on the importance of extra-regional links to complement regional knowledge assets implies that the role of policy may be to devise strategies to manage the complex ways in which regional and extra-regional dimensions interplay. Hence, in order to build competitive regions, the role of policy is to explore ways to create mechanisms that support the connection between the global and the regional knowledge assets. As Coe *et al.* argue:

> . . . the fortunes of regions are shaped not only by what is going on within them, but also through wider sets of relations of control and dependency, of competition and markets. These relations may be with other regions within the same national territory, but increasingly occur at the international scale.
>
> (Coe *et al.*, 2004: 469)

Also, Gertler and Wolfe (2006) argue that a knowledge-based theory of a cluster must recognize that relatively few clusters are completely self-sufficient in terms of the knowledge base from which they draw. Consequently, this should also be reflected in regional innovation systems' policy initiatives. A single-handed focus on promoting the regional level is seldom sufficient. According to Bathelt *et al.* (2004: 48–49), it is "especially the development of global pipelines which requires institutional and infrastructure support," whereas recent (cluster) policies "are so predisposed toward local networking that the importance of external, translocal communication is overlooked." Lagendijk and Lorentzen (2007: 457) argue that economic performance relies more on "localized capacities to build 'global' connections, complemented with an adequate local resource base, than on local networking and clustering."

Further, different types of proximity are likely to matter more or less depending on the specific situation. Although theory has come a far way by opening up for a multi-level or multi-scale approach to geography and knowledge formation, there appears to be a need to understand what is regional and what is not when it comes to knowledge formation. For instance, Malmberg and Power (2006) suggest that there is evidence that local labor market dynamics and individual social interaction contribute to knowledge formation in firms and clusters. Results by Lagendijk and Lorenzten (2007) are on the same track, but for somewhat different reasons. They also argue that policy should focus less on building local networks, but not because local networks pretty much happen any way but because they are of limited importance. Lagendijk and Lorentzen (2007) have found that geographical proximity is important as an enabler of specific localization effects underpinning the development of a physical infrastructure, labor market specialization and business support. However, their findings do not support the idea of an intense localized interaction explaining regional development. Therefore, they suggest that policy should focus less on promoting the formation of regional and local knowledge networks, and put more resources into fostering entrepreneurship and learning in the context of global knowledge networks. According to Lagendijk and Lorentzen (2007: 8), "nurturing local talent and providing good infrastructures [. . .], as well as stimulating global knowledge searching and benchmarking [. . .] is what regional growth policy should be about." Also, further work remains on developing our understanding of who the local and non-local actors are, and their role on the regional level (as discussed by Coe *et al.*, 2004), as well as how the type of industry and the underlying knowledge base will impact on the role of the region (Asheim *et al.*, 2007). To this can be added a time dimension to the discussion – the impact of different dimensions of proximity, depending on the stage in an industry's life-cycle, where geographical proximity may be more important in the early stages of an industry's development (Boschma, 2005).

Methodology

This chapter reports on a research project that was designed and initiated in 2004, with the aim to understand the organization of corporate knowledge formation, on the one hand, and the organization of knowledge formation within a regional innovation systems policy initiative, on the other. Methodologically, a case study approach was adopted to analyze the relationship between corporate knowledge, on the one hand, and regional innovation measures to enhance regional competitiveness, on the other. Two case studies were conducted: (1) a case study of the regional innovation systems policy initiative; and (2) a case study of the knowledge formation process of a multinational firm within this regional innovation systems policy setting. As Yin (1993) writes, a case study approach is desirable when the boundaries between the phenomenon under study (in this case, knowledge formation) and the context (the regional innovation systems policy initiative, on the one hand, and the MNC, on the other) are intertwined. The adoption of a case study approach provided us with the opportunity to gain holistic insights into the regional development process and the global knowledge formation of the MNC.

The data in both case studies consist of primary sources collected through semi-structured interviews (lasting 1–2 hours) – a total of 29 interviews have been carried with regional stakeholders representing regional industry, regional academia and regional development authority, as well as staff from two of the MNC's R&D units. This material is supported and complemented by secondary sources in the form of documents and written records, such as the application documents containing the plan of action. Knowledge formation is, in this chapter, interpreted as knowledge formed by new combinations or recombinations of knowledge. These are typically the result of some sort of interaction – between individuals but also between different knowledge domains or disciplines. The aim of the interviews has been to identify such patterns of interaction in the two cases.

The first case study – the regional innovation systems policy initiative – was conducted between November 2004 and March 2005. An in-depth description and the challenges of regionalization of innovation policy have been reported in various papers (Gustavsson and Laestadius, 2006; 2007; Nuur *et al.*, 2009). The aim of conducting this case study was to understand the mechanisms that were put in place to support the evolution of a regional innovation system in the field of robotics and map out the actors involved in the development processes. Respondents in this case study were the Robot Valley process manager along with other representatives from the Robot Valley initiative, municipality representatives from the three counties, representatives from involved companies (both large and small companies) and professors from the two regional universities.

The second case study, the MNC case, referred herein to as the focal company, was conducted in 2006–7. The MNC is one of the central industry actors in Robot Valley and participates with both financial means and human resources in several of the projects within the policy initiative. Interviews were conducted with research managers and project managers at the company's research and development facilities in Västerås, Sweden, as well as the more recently established research unit in Shanghai, China.

A regional innovation systems policy initiative and MNC knowledge formation

In this section, we will first look at the policy initiative Robot Valley, which is intended to foster regional industrial development and growth. Second, we will look at the focal company of the paper and how it organizes its knowledge formation processes.

The policy case

In focus, here is the Swedish policy initiative Robot Valley, which is one initiative within the Swedish policy program, VINNVÄXT (Regional Growth through Dynamic Innovation Systems). This program aims at promoting regional growth through dynamic regional innovation systems. Robot Valley is, together with two other regions, the first generation of VINNVÄXT initiatives launched in 2003 (for details on this case, cf. Gustavsson and Laestadius, 2006; 2007). The purpose of the initiative is to create and sustain international competitiveness in robotics-related industrial activities. According to the application documents and plan of action, the vision of the Robot Valley is to make the Mälardalen region in Central Sweden a global leader in the manufacturing, research and development of robot-based automation. The objective of Robot Valley, according to the same documents, is to strengthen and develop existing technological knowledge, to support the generation of new products and new ventures, and to create a strong infrastructure that supports R&D activities in robotics. In quantitative terms, the goal is to create 30 new products and 30 new enterprises by 2013.

The Robot Valley initiative encompasses three Swedish counties in Central Sweden. Each county is represented by officials from the municipal, county council and administrative board. These counties are also home to several global firms connected to the automation industry. Strongly influenced by the triple helix theories on joint collaboration between academia, industry and state (Leydesdorff and Etzkowitz, 1998; Leydesdorff, 2005; Etzkowitz, 2008), the academic representation is also based on the collaboration of the university/university colleges of Mälardalen University in Västerås and Eskilstuna, and Örebro University.

Based on existing competence, Robot Valley has identified three application areas that are to be promoted. These relate to industrial and field robotics, as well as health care robotics, all of which have been identified as having a strong growth potential in terms of the generation of new products and technologies. Industrial robotics is a mature area, with a strong tradition in the region, and where our focal company – the MNC – is one of the largest industrial actors. Within this area, at the time of the interviews, there was expected rise in demand among SMEs in the region for increased automation. Field robotics is a relatively new industrial area, which is also seen to have growth potential including the development of, for instance, autonomous loading and mining equipment. Within this technological segment, Robot Valley has two global companies: Atlas Copco in Örebro and Volvo Construction Equipment in Eskilstuna. Also, health care is a relatively new area for robot applications in general and an area considered to have great potential, especially considering that a large part of the world is facing a future with an increasing number of elderly people. However, within this area, both industrial and academic competence within the region is weak.

As a first generation VINNVÄXT initiative, Robot Valley has been granted financing worth 10 million SEK annually (approximately €1.1 million) for up to 10 years. A condition for receiving this funding is that regional actors contribute with a corresponding amount. At the regional level, the stakeholders are the three university/university colleges of Mälardalen University in Västerås and Eskilstuna, and Örebro University, several regional firms involved in the production of robotics and related equipment, and the business development units of the counties.

The policy initiative focuses on creating competitive research and innovation environments within dynamic regional innovation systems. In this context, several regional collaborative projects with the aim of developing new products within the three areas of robotics (industrial robotics, field robotics and health care robotics) were initiated. All three areas have been identified as having strong growth potential in terms of generating new products and technologies that are also in line with the activities of the three global firms. As mentioned, industrial robotics, which relates to our focal company, is a mature area with a strong tradition in the region. Within this area, the policy initiative hopes that the rapidly increasing demand among SMEs in the region for automation will lead to growth. In addition, a number of projects directed at the regional SMEs have been initiated. For example, a very successful sub-project in industrial robotics, "Robotics for SMEs," which aims at helping SMEs to use/introduce automation solutions through adapting existing robotics technology and developing new robotics applications tailored for SMEs. At the time of the interviews, more then 100 companies have been analyzed within the project and about half of them have introduced robots or other automation solutions as a result. The project has also resulted in job opportunities for students who carried out the studies at the firms. Another example is that Robot Valley has participated in the creation of a new company, Flex Pack Robotics that makes customized packaging solutions for the pharmaceutical industry. This is a new application for the pharmaceutical industry, and, in the longer term, can hopefully be launched in other industries such as the food industry.

There are certainly strong dynamics in the region where, on the producer side – apart from the large companies – there is also a large number of robotics-related smaller firms in the region. Particularly within industrial robotics, there is an established foundation of producers, system integrators and users in the region. However, the strong industrial foundation in the region for robotics and process automation is not matched by a similar strength in the academic infrastructure regarding robotics. One measure of this is to look at which universities the three large industry actors collaborate with in Sweden. All the three companies state that Chalmers University of Technology in Gothenburg, the Faculty of Engineering at Lund University, the Institute of Technology at Linköping University and the Royal Institute of Technology in Stockholm are important partners in research, whereas the collaboration with the two regional universities has previously been rather limited. One possible reason for this could be that both universities in Mälardalen – Örebro University and Mälardalen University College – are rather young and are still in the process of building up a strong research base. This also shows that many important Swedish research groups with connection to robotics and to the robot-related companies are located outside the "Valley."

The MNC case

As indicated earlier in this chapter, the focal company in this study is a key actor in the Robot Valley initiative through its automation division. This MNC was among the pioneering companies in industrial automation in the 1970s, and it is today one of the world's largest companies within automated solutions. In Europe, the primary competitor is the German company Kuka, and in the Asian markets the largest competitors are the Japanese companies Fanuc and Motoman.

The company offers robots, services and manufacturing solutions for use in assembly, finishing and machine tending. Key markets include the automotive and manufacturing industries, in addition to applications in foundry, packaging and material handling. The division develops standardized manufacturing cells for machine tending, welding, cutting, painting and finishing, and provides packaged systems to automobile manufacturers for press automation, paint process automation and power train assembly.

Key locations for both manufacturing and R&D are Västerås in Sweden, Bryne in Norway and Shanghai in China. Approximately 600 employees work in R&D globally, of which 250 are located in Västerås. The R&D center in Shanghai was established in 2004, and engages 50 people in R&D (in 2007). Shanghai is also, since 2006, the location of the company's headquarters, which was relocated from Västerås.

Undoubtedly, globalization implies that firms have to look beyond their home locations in finding customers and knowledge sources to innovate and become globally competitive. The primary market for the focal company has traditionally been Europe, and this is still the main region for the company's sales. In 2006, Europe accounted for 58 percent of sales, the Americas 25 percent and Asia 17 percent. However, the increase in orders on group level (the focal company is one of five divisions within the group) as a total from 2005 to 2006 for different countries and regions shows clearly that it is the emerging markets that are the fastest growing. China increased by 26 percent, India by 39 percent, the Middle East by 54 percent, and Central and Eastern Europe by 49 percent. At the same time, the US increased by only 18 percent and Western Europe even less, by 14 percent. We can assume that the general picture on group level sufficiently well reflects the order distribution of our focal company. The share of total employees also shows that emerging markets are most rapidly growing in numbers of people: Asia increased from 18 percent to 21 percent, which equals an increase of 4,000 people, whereas Europe actually reduced its total share of employees from 58 percent to 56 percent, equaling an increase of 650 people. The Americas, the Middle East and Africa were unchanged.

Knowledge about markets is acquired globally as well as locally. Due to the industry structure of Central Sweden, with a high proportion of SMEs currently under pressures of an increasing competition, automation is an attractive option for these companies. Currently, robots are both expensive and not developed for smaller scale operations. Equally, a growing SME industry and a growing car industry in China, for example, calls for global presence. In this context, access to skilled people from adjacent industries and from academia is important – in the old as well as the new locations that the focal company chooses.

At the time of the interviews, the company was experiencing a slight economic downturn, which is mainly caused by the dip in the American and European automotive industries, which historically have been the largest customers of industrial robots. Therefore, the company needs to look into new markets for expansion (such as the growing Asian markets) but also to search for new applications in current markets (such as automation solutions developed for SMEs, introduction of industrial robots in new types of industries, etc.).

The company had previously not been very global in terms of its R&D activities. Much of these activities have been concentrated to the Swedish site, and specific applications developed in the other sites. However, the decision to move the company's headquarters to Shanghai in 2006 reflects both market trends and the strategic importance of the company's business in China. The R&D unit that, since 2004, is located in Shanghai is currently growing rapidly and attracting highly educated people. According to our interviews, the Chinese unit has a higher number of PhDs than its Swedish counterpart. The company is viewed as a highly attractive company to work for[2] and the Shanghai R&D center employs a higher proportion of PhDs than in Västerås. The implications of this is that Swedish MNCs are looking beyond Sweden for knowledge necessary for innovations, and that globalization has contributed to the footlooseness of not only production sites, but also the location of research and development.

According to the company, the organization of knowledge formation is focused on getting the fastest and most demanding market/customer input. The corporate aim is thus to locate its innovative activities as close as possible to markets/customers in order to gain from the inputs and/or information flow back to its knowledge centers. A growing car industry in China is one major reason for local R&D, but another important reason is to target a growing SME industry with a currently low level of automation.

The decision to locate production, as well as R&D activities, in China is a consequence of its growing industry and the great potential of the Chinese market. In particular, the rapidly growing automotive industry in China makes the Chinese market highly interesting. One of the interviewees reflected on this:

> Automation grows as the industry grows. Even though today labor costs are still cheap if compared to Europe or the US, but that has increased. In a few years, the labor costs will no longer be cheap. In China today, the production costs may be, say, US$2/hour and, in the US, US$7–8/hour – but this margin is decreasing year by year. That is why also companies in China are looking for automation solutions.
>
> (R&D Manager, Shanghai)

The vision is that China is to become the long-term number one market for the company, and 80–90 percent of revenues are to be derived from locally produced products and services. China is also to become an important base for global sourcing of components and semi-manufactured goods.

The implementation of automation solutions is driven also by factors other than costs. If manufacturers in China want to export products, automation is a way to guarantee quality standards as stated by one of the interviewees:

> . . . if you are to manufacture world-class cars, you have to automate. There is no human being who can achieve that manufacturing exactitude. And, in order to sell cars in Europe, they have to comply with certain demands, and that requires automation.
>
> (Manager R&D Strategy, Västerås)

At the time of collecting the data, the market had yet not gained sufficient momentum, and as such the main economic driver for a distributed research and development organization was local sourcing. However, the fact that the company has local activities is essential for competitiveness, primarily with the Japanese companies, in getting shares of the growing Chinese market. The potential with a globally distributed organization is the possibility to beat the competition to the market. For instance, by locating in China, the time to market is drastically reduced, as shipping products by boat from Europe takes about five weeks, and this gives the company a stronger competitive position relative to its Japanese competitors, who formerly held a great advantage in closeness to the market. Also, the distributed organization is considered advantageous for the potential to roll out global solutions faster.

In addition, the company plans to start up R&D activities in India. In this case, it is more evident that knowledge and expertise is the main focus. On group level, there is already a rather large activity in India – mainly within software development and test and validation – for some years now (the R&D laboratory in Bangalore opened in 1998), so it is mainly the technological competence that is interesting in that case. In fact, the research laboratory in Bangalore is the single largest R&D location, with more than 500 people.

Hence, from our focal company's perspective, there are at least three reasons that underlie the rationale for locating activities abroad: (1) economic; (2) market access (both in terms of improving time to market and the learning possibilities); and (3) expertise (engineering – primarily software). Yet, despite the growing importance and role of foreign R&D sites, Sweden is still the major site for both production and R&D. In 2007, approximately 8,500 robots were produced in Sweden, around 500 in Norway and around 1,000 in China.

The MNC and the regional innovation systems policy initiative

Given a situation with a small domestic market, knowledge formation processes that have a global reach are a necessity and natural part of MNC activities. This has not always been the case, however: two decades ago, many global firms were strongly "non-globalized" as regards their R&D activities (Patel and Pavitt, 1991). As a part of the process analyzed in, for example, Castells (2000), firms have learned not only to relocate but also to operate at least some of these functions globally. Still, some aspects of the region continue to matter. Thus, as regards the role of Robot Valley for our focal company, this must be analyzed in relation to the global market where it is present. The company needs to compensate for the decline in the global automotive industry, which hitherto has been a dominant purchaser of robots. One means for that is to expand into SMEs, which all over the world increase their robotization efforts. Robots for SMEs, however, typically need to be smaller, cheaper, easier to program and built for rapid change of tasks. Within the Robot Valley initiative, the very successful project "Robotics for SMEs" targets this particular area. For our focal company, one of the greatest values with the Robot Valley initiative is the access to a local network of end customers – primarily SMEs, but also larger companies and a network of systems integrators. Many of these end customers in the region of Mälardalen are highly sophisticated and can provide valuable input to the development work. This directly relates to the company's needs to find new solutions and application areas for its existing products in larger sections of the industry, and particularly products and services that also suit SME producers. Thus, the project enables the company to increase knowledge about its products and applications to a new group of customers, and it also provides the company with the possibility to capture new market needs that can be transformed into technical solutions. One company official expresses the role of the Robot Valley for the company as follows:

> I think that the Robot Valley contributes as a think tank. Via the Robot Valley and the network that exists there we capture new trends and new needs that we perhaps cannot capture on our own. So the think tank is an important purpose.
>
> (Manager Process and Project Office, Västerås)

The engagement in Robot Valley is, however, only one of the strings on the company's guitar. The establishment in Shanghai has a similar aim. In fact, it seems that the R&D focus in Shanghai is, to a large extent, on SMEs. The young researchers recruited there are explicitly assumed to transform their computer game skills into easy-to-hand robots attractive for SMEs in China and the rest of the world.

As discussed above, the large industrial actors in the region primarily collaborate with universities outside the region of Mälardalen. Our focal company collaborates with about 70 universities worldwide, including, for example, the Massachusetts Institute of Technology (MIT). As regards research collaborations in Sweden within robot automation, the company has far-reaching collaborations with primarily Chalmers University of Technology in Gothenburg, the Faculty of Engineering at Lund University, the Institute of Technology at Linköping University; all located outside the Robot Valley region. Collaborations with the two regional universities in the Robot Valley were basically non-existent before the initiative and, therefore, the policy initiative has definitely led to increased contacts with these two universities; and, for instance, the "Robotics for SMEs" project is coordinated by the Mälardalen University College and engages both university staff and students, but the project leader is an ABB employee.

To conclude this section, we can see that the policy initiative has had a great, positive effect on creating new networks within the region – networks that have importance both for regional actors such as the local universities and for the companies in the region, such as our focal company. Yet, the case reveals a need to strengthen the region in its research capacity. In order to build a strong region, it is crucial to connect it to existing or emerging centers of expertise – nationally as well as internationally. The need to look outside the region for knowledge and expertise is corroborated by the existing research networks that the company has, and also in the stated need for Robot Valley to build research networks outside the region. One reason why policy is not achieving this is the financial set-up of the policy program. By conditioning the financial support from VINNOVA with the requirement that local actors should contribute with the equivalent amount of money, a strong incentive for locating projects and initiatives within the own region is created. However, policy incentives to search for expertise outside the region are currently lacking. The need for extra-regional links is acknowledged among the actors in the policy initiative. However, how to deal with this issue is still not clear:

> Globalization is a really difficult question in relation to VINNOVA and to this regional financing. What we have looked at in the board is this with internationalization. To build research networks – to start in that end, in addition to the work that goes on here at Örebro University and Mälardalen University [. . .] But if you look at the other two pillars, with society and primarily industry – companies and end customers – then [. . .] I am not sure that the Robot Valley can drive that outside the region. I think one should start with research. And that is a tough enough challenge, actually.
>
> (Manager Process and Project Office, Västerås)

Concluding discussions

The aim of this chapter was to contrast the organization of knowledge formation processes of an MNC with a regional policy initiative aimed at contributing to the development of a regional innovation system. We have shed some light and given insights into the paradoxical situation that may arise between globalization and regionalization as regards knowledge formation processes in industry and technology, i.e. R&D, innovation and design processes, knowledge formation processes and the creation of industrial capabilities in a wide sense. Regional interactions and networks can be of great importance in certain cases, whereas non-regional connections may be more important in others. In many cases, one does not exclude the other, but both regional and non-regional interactions matter – in different proportions and in different ways, depending on the specific situation and the type of knowledge sought in the collaboration of course. What we want to stress in this chapter is that one dimension should not overshadow the other dimension.

We argue that the cases reveal two different strategies as regards the organization of knowledge formation and the role of regional and non-regional interactions. Whereas the policy initiative reveals an approach to knowledge formation as an essentially geographically sticky process, the MNC is not necessarily dependent on geographically proximate relations and interactions. This is illustrated by the strong regional focus, on the one hand, from the policy initiative in terms of promoting new collaborations, and, on the other, an essentially non-geographical focus illustrated by the MNC where collaborations are sought regionally as well as nationally and globally, depending on the purpose of a specific collaboration. What this indicates is that the MNC is dealing with the dual development of regionalization and globalization. However, the strategies adopted by the studied policy initiative remain highly regional in terms of measures aimed at stimulating knowledge formation, which neglects the dimension of globalization. The policy initiative is – despite the objective of building a global industry – highly regionally constrained. For instance, the features of the research map (i.e. where the leading research in Sweden and also the rest of the world is located) should be reflected in – and have consequences for – how regional innovation systems policy initiatives of the type that Robot Valley represents are organized.

We do not argue that Robot Valley has failed in its undertaking – on the contrary, the project has been quite successful in achieving a shared vision and a regional arena for co-operation, as stipulated both in the VINNVÄXT call and the application made to receive the financial support. It has also led to efforts at the two university colleges to seriously build up competencies in robotics, as well as measures to increase the collaboration between the university colleges and the regional SMEs (Nuur *et al.*, 2009). Instead, we argue that the initiative risks not capturing vital dynamics that are important in the realization of a viable regional innovation system. Regional excellence is today rarely – if ever – created in isolation – but is promoted by external linkages. Few regions, if any, can claim to have all relevant knowledge accessible within their own geographic area. Therefore, in order to tap from the available world-leading knowledge and expertise, links to other regions are often an integral part of regional development. Yet, it may be claimed that the policy studied in this chapter initiative is not dealing sufficiently with this issue. It is important, therefore, that tools are developed for policy to deal with this strategic coupling of regional assets and global networks. Here, work remains to be done to develop stronger policy mechanisms that support and create incentives for extra-regional links.

What should be important for this type of policy initiative is to ensure that essential parts of the activities will be located in Sweden. For instance, the MNC's market is primarily non-domestic, and this makes it rather natural that significant parts of the activities are located outside of Sweden. Thus, a primary goal for policy-makers should be to ensure that efforts to enhance the competitiveness of Swedish industry and Swedish regions focus on making these industries and regions integral parts of larger industrial systems – national as well as international. Based on this chapter, it appears as if policy needs to more clearly incorporate action plans that acknowledge and allow for the interdependence between regional and non-regional activities and it is essential that these extra-regional connections are actively encouraged and facilitated. If the regional focus is too strongly emphasized – in particular, within a small economy like the Swedish – it may create lock-ins rather than contribute to innovative, knowledge creating regions. The financial setup as discussed in this chapter is one example of how the initiative creates regional lock-in effects. To avoid such a situation, a two-folded approach is desirable: on the one hand, policy needs to focus on strengthening the domestic (national and/or regional) innovation system, while, on the other hand, policy should have the aim of enlarging the arena for collaboration (Lundin and Schwaag Serger, 2007). In this case, as robotics is a highly globalized industry, consequently the arena for collaboration can potentially be global in its scope rather than regional. One means is to apply a network approach that focuses on developing certain essential and externally connected nodes within the region/nation. If we look at the MNC in this regional setting, there is no question that it views knowledge formation as a process involving both regional and extra-regional links. The regional innovation systems policy initiative, on the other hand, is mainly focused on the former, i.e. a regionalization of knowledge formation. In conclusion, the two concurrent forces at work today – regionalization and globalization – should also be reflected in regional innovation systems policy initiatives when measures to promote knowledge formation are formulated.

Notes

1. We may point out here that the concepts of technological communities and technological systems (Dosi, 1984; Carlsson and Stankiewicz, 1991), and even Erik Dahmén's (1950) development blocs have no a priori spatial dimension connected to them; and for instance empirical studies of the development of the Swedish forest industry technological system shows strong global links in terms of industrial knowledge formation (Laestadius, 2000).
2. The company ranked "top 10 best employers" in the 2003 "Employee Opinion Survey" conducted jointly by Watson Wyatt Worldwide, Fortune China and Sohu, which analyzed 65 Chinese and foreign-invested companies in China.

References

Amin, A. and Cohendet, P. (2004) *Architectures of Knowledge: Firms, capabilities and communities*. Oxford: Oxford University Press.

Archiburgi, D. and Lundvall, B.-Å. (2001) *The Globalizing Learning Economy*. Oxford: Oxford University Press.

Asheim, B.T. and Coenen, L. (2005) "Knowledge bases and regional innovation systems: Comparing Nordic clusters." *Research Policy* 34(8), 1173–1190.

Asheim, B.T. and Coenen, L. (2006) "Contextualizing regional innovation systems in a globalising learning economy: On knowledge bases and institutional frameworks." *The Journal of Technology Transfer* 31(1), 163–173

Asheim, B.T. and Gertler, M. (2005) "The geography of innovation: Regional innovation systems." In J. Fagerberg *et al.* (eds) *The Oxford Handbook of Innovation*. Oxford: Oxford University Press.

Asheim, B.T., Coenen, L., Moodysson, J. and Vang, J. (2007) "Constructing knowledge-based regional advantage: Implications for regional innovation policy." *International Journal of Entrepreneurship and Innovation Management* 7(2–5), 140–155.

Bathelt, H., Malmberg, A. and Maskell, P. (2004) "Clusters and knowledge: Local buzz, global pipelines and the process of knowledge creation." *Progress in Human Geography* 28(1), 31–56.

Boschma, R.A. (2005) "Proximity and innovation: A critical assessment." *Regional Studies* 39(1), 61–74.

Cantwell, J. and Iammarino, S. (2003) *Multinational Corporations and European Regional Systems of Innovation*. Routledge: London .

Carlsson, B. and Stankiewicz, R. (1991) "On the nature, function and composition of technological systems." *Journal of Evolutionary Economics* 1(2), 93–118.

Castells, M. (2000) *The Rise of the Network Society*. 2nd edn. Oxford: Blackwell.

Coe, N.E. and Bunnell, T.G. (2003). "Spatializing knowledge communities: Towards a conceptualization of transnational innovation networks." *Global Networks* 3, 437–456.

Coe, N.M., Hess, M., Yeung, H.W.C., Dicken, P. and Henderson, J. (2004) "Globalizing regional development: A global production networks perspective." *Transactions of the Institute of British Geographers* 29(4), 468–484.

Constant, E. (1984) "Communities and hierarchies: Structure in the practice of science and technology." In R. Laudan (ed.) *The Nature of Technological Knowledge: Are models of scientific change relevant?* Dordrecht: Reidel.

Cooke, P. (2005) "Rational drug design, the knowledge value chain and bioscience megacentres." *Cambridge Journal of Economics* 29(3), 325–341.

Dahmén, E. (1950) *Svensk industriell företagarverksamhet*. Stockholm: Industriens Utredningsinstitut.

Dicken, P. (2004) "Geographers and 'globalization': (Yet) another missed boat?" *Transactions of the Institute of British Geographers* 29, 5–26.

Dosi, G. (1984) *Technical Change and Industrial Transformation*. London: Macmillan.

Dreher, A. (2006) "Does globalization affect growth? Evidence from a new index of globalization." *Applied Economics* 38(10), 1091–1110.

Dunning, J.H. (2000) *Regions, and the Knowledge Economy: The issues stated*. Oxford: Oxford University Press.

Epstein, G. (2003) "The role and control of multinational corporations in the world economy." In J. Michie (ed.) *The Handbook of Globalisation*. Northampton, MA: Edward Elgar.

Etzkowitz, H. (2008) *The Triple Helix: University–industry–government innovation in action*. New York: Routledge.

Freeman, C. (1987) *Technology Policy and Economic Performance: Lessons from Japan*. London: Pinter.

Friedman, T. (2005) *The World is Flat: A brief history of the globalized world in the 21st century*. London: Penguin, Allen Lane.

Gertler, M. and Wolfe, D. (2006) "Spaces of knowledge flows." In: B. Asheim, P. Cooke and R. Martin (eds) *Clusters and Regional Development: Critical reflections and explorations*. London: Routledge.

Gustavsson, L. and Laestadius, S. (2006) *Innovationssystem som politisk praktik- en studie av Vinnväxtprojektet Robotdalen*. Stockholm: KTH Working paper.

Gustavsson, L. and Laestadius, S. (2007) "Innovationssystem som politisk praktik- en studie av VINNVÄXTprojektet Robotdalen." In S. Laestadius, Nuur, C. and Ylinenpää, H. (eds) *Regional växtkraft i en global ekonomi: Det svenska vinnväxtprogrammet*. Stockholm: Santérus.

Isaksen, A. (2001) "Building regional innovations system: Is endogenous industrial development possible in the global economy?" *Canadian Journal of Regional Science* 24(1), 101–120.

ITPS (2007) *Svenskt näringsliv i en globaliserad värld*. A report from the Swedish Institute for Growth Policy Studies. A2007:004.

Laage-Hellman, J., Rickne, A. and Stenborg, E. (2007) *Biomedical Firms in Western Sweden: A study of a regional innovation system from a biomedical firm's perspective*. RIDE/IMIT Working Paper No 84426-024.

Laestadius, S. (1998) "Technology level, knowledge formation, and industrial competence in paper manufacturing." In G. Eliasson and C. Green (eds) *Micro-foundations of Economic Growth: A Schumpeterian perspective.* Ann Arbor, MI: University of Michigan Press.

Laestadius, S. (2000) "Biotechnology and the potential for a radical shift of technology in forest industry," *Technology Analysis and Strategic Management*, 12(2), 193–212.

Lagendijk, A. and Lorentzen, A. (2007) "Proximity, knowledge and innovation in peripheral regions: On the intersection between geographical and organizational proximity." *European Planning Studies* 15(4), 457–466.

Lagendijk, A. and Oinas, P. (2005) "Proximity, external relations, and local economic development." In A. Lagendijk and P. Oinas (eds) *Proximity, Distance and Diversity*. Aldershot: Ashgate.

Lam, A. (2006) "The Tacit Knowledge Problem in MNC's: A comparative analysis of Japanese and US offshore knowledge incubators." *The Journal of International Business Studies.*

Lazerson, M. and Lorenzoni, G. (1999) "The firms that feed industrial districts: A return to the Italian source." *Industrial and Corporate Change* 8(2), 235–266.

Leydesdorff, L. (2005) "The triple helix model and the study of knowledge-based innovation systems." *International Journal of Contemporary Sociology* 42(1), 1–16.

Leydesdorff, L. and Etzkowitz, H. (1998) "The triple helix as a model for innovation studies." *Science and Public Policy* 25(3), 195–203.

Lorenzen, M. (2005) "Knowledge and geography." *Industry and Innovation* 12(4), 399–407.

Lundin, N. and Schwaag-Serger, S. (2007) "Globalization of R&D and China: Empirical observations and policy implications." Swedish Institute for Growth Policy Studies and Lund University, Working Paper R2007:013.

Lundvall, B-Å. (1992) *National Systems of Innovations: Toward a theory of innovation and interactive learning*. London: Pinter.

Malmberg, A. (1998) "Den gäckande agglomerationsfördelen: teoretiska principer och empiriska paradoxer." *Svensk Geografisk Årsbok* 74, 110–124.

Malmberg, A. (2002) *Klusterdynamik och regional näringslivsutveckling: begrepp, diskussioner och forskningsöversikt.* A report for ITPS.

Malmberg, A. and Maskell, P. (2005) *Localized Learning Revisited.* DRUID Working Paper No. 05–19.

Malmberg, A. and Power, D. (2005) "(How) do (firms in) clusters create knowledge?" *Industry and Innovation* 12(4), 409–431.

Malmberg, A. and Power, D. (2006) "True clusters: A severe case of conceptual headache." In: B. Asheim, P. Cooke and R. Martin (eds) *Clusters and Regional Development: Critical reflections and explorations*. London: Routledge.

Marshall, A. (1890/1920) *Principles of Economics: An introductory volume.* Canada: McMaster University.

Martin, R. and Sunley, B. (2003) "Deconstructing clusters: Chaotic concept or policy panacea?" *Journal of Economic Geography* 3, 5–35.

Moodysson, J., Coenen L. and Asheim, B.T. (2008) "Explaining spatial patterns of innovation: Analytical and synthetic modes of knowledge creation in the Medicon Valley Life Science Cluster." *Environment and Planning A* 40, 1040–1056.

Nuur, C. and Laestadius, S. (2010) "Development in peripheral regions: Case studies in Sweden." *European Urban and Regional Studies* 17(2).

Nuur, C., Gustavsson, L. and Laestadius, S. (2009) "Promoting regional innovation systems in a global context." *Industry and Innovation* 16(1), 123–139.

Oinas, P. (1999) "Activity-specificity in organisational learning: Implications for analysing the role of proximity." *GeoJournal* 49(4), 363–372.

Patel, P. and Pavitt, K. (1991) "Large firms in the production of the world's technology: An important case of 'non-globalization'." *Journal of International Business Studies*, 22, 1–21.

Porter, M. (1990) *The Competitive Advantage of Nations*. London: Macmillan.

Porter, M. (1998) "Clusters and the new economics of competition." *Harvard Business Review* November–December.

Porter, M. (2000) "Locations, clusters and company strategy." In: G. Clark, M. Feldmann and M. Gertler (eds) *Oxford Handbook of Economic Geography*. Oxford: Oxford University Press.

Saxenian, A. (1994) *Regional Advantage: Culture and competition in Silicon Valley and Route 128*. Cambridge, MA: Harvard University Press.

Statistics Sweden (2008) www.scb.se/Statistik/HA/HA0201/2008M02D/HA0201_2008M02D_SM_HA22SM0801.pdf (downloaded September 26, 2008).

Storper, M. (2000) "Globalization and knowledge flows: An industrial geographer's perspective." In H.J. Dunning (ed.) *Regions, Globalization and the Knowledge-Based Economy*. Oxford: Oxford University Press.

Sölvell, Ö. and Birkinshaw, J. (2000) "Multinational enterprises and the knowledge economy: Leveraging global practices." In H.J. Dunning (ed.) *Regions, Globalization and the Knowledge-Based Economy*. Oxford: Oxford University Press.

Waits, M. (2000) "The added value of the industry cluster approach to economic analysis, strategy development, and service delivery." *Economic Development Quarterly* 14(1), 35–50.

Wenger, E. (1998) *Communities of Practice: Learning, meaning, and identity*. Cambridge: Cambridge University Press.

Yin, R. K. (1993) *Applications of Case Study Research*. Newbury Park, CA: Sage.

7 Regional policy as change management

Theoretical discussion and empirical illustrations

Martin Andersson and Börje Johansson

Introduction

There are two generic features of the long-term evolution of regions. The first is persistence in structures, which is revealed by the stable pattern of city-size distributions and spatial hierarchies (Beckmann, 1958; Gabaix, 1999; Gabaix and Ioannides, 2004). The second is durability and persistence in change processes, which manifest itself in that evolution and change tend to follow a predictable path governed by initial conditions (Glaeser *et al.*, 1995). We focus, in this chapter, on what the latter type of change processes imply for regional policy. Our contribution addresses the general policy issue of the focus and potential for regional policy (Johansson *et al.*, 2002). The chapter is conceptual and our arguments are based on theoretical discussions substantiated with empirical examples and illustrations.

Specifically, we address challenges and potentials for regional policy for situations in which durable change processes are present. We refer to such processes as *fundamental change processes*.[1] An important characteristic of fundamental change processes is their persistence, such that they repeat themselves over time. In this way, they influence the long-term evolution of regions and are often associated with structural change. Examples of fundamental change processes at the macro-level include the steady increase of the importance of knowledge-intensive services in advanced economies over the last decades (Schettkat and Yocarini, 2006), as well as the increase in the share of population living in urbanized areas that has taken place since at least the 1920s (UNFPA, 2007). The important role of services of contemporary economies and the large fraction of peoples livening in urban regions are results of persistent change processes that have operated over long periods of time. At the regional level, fundamental change processes pertain to phenomena such as in- and out-migration, entry and exit of firms, changes in the types of products and services firms in the regions produce, as well as changes in the type of inputs they require, and inflow of labor with different education profiles, skills and competences.

We argue that a major task of regional policy is to identify and respond to fundamental change processes. In a given period of time, such processes are 'gifts from the past' that define the context for regional policy. For certain regions, they constitute a positive evolution trajectory that policy may aim to stimulate. For other regions such trajectories imply a negative trend that policy often tries to counteract or transform to something more desirable. Some regions may, for instance, experience a gradual net outflow of population over time, whereas others experience a net inflow. Irrespectively, fundamental change processes form an important context to which policy must respond. In this regard we may view policy as 'management of change'.

The chapter focuses on two basic questions. The first question is about understanding the processes that policies attempt to strengthen, counteract and transform, while the second is about the identification of policies that have a potential to influence change patterns in a warranted way.[2]

1. What are the characteristics and underlying forces of fundamental regional change processes?
2. What is the scope for regional policy to alter and adjust to these change processes, and what aspects of the processes should policy focus on?

A major argument of the paper is that fundamental change processes should be understood as development paths with persistent mechanisms of change, driven by slowly changing (regional) characteristics and self-organizing feedback adjustments. This associates to the notion of cumulative causation that is a prominent feature of the work by, for example, Myrdal (1957) and Kaldor (1970, 1981), as well as of modern models of agglomeration economies and the evolution of spatial structures (Krugman, 1991; Fujita *et al.*, 1999; Fujita and Thisse, 2002). It also adheres to the recent literature on evolutionary economic geography, which emphasizes the role of cumulative processes in the patterns of regional change (Boschma and Frenken, 2006; Martin and Sunley, 2006; Frenken and Boschma, 2007).

The discussion of the scope and focus of policy is based on a distinction between slow and fast processes, which adjust on different timescales. Policy, viewed as the management of change, must recognize and deal with both types of processes while giving priority to the slowly changing factors. Variables that change on a slow timescale can be characterized as parameters that govern the evolution of fast processes. We make a strong case that parameters of a change process represent those things on which policy decisions should concentrate. Consequently, regional policy should be directed towards altering parameters that govern slowly changing regional characteristics. By doing so, policy may influence the evolution trajectory of regions. We use the concept of infrastructure as a comprehensive term for regional characteristics that develop on a slow timescale (cf. Johansson and Wigren, 1996; Johansson, 1998). Infrastructure pertains to various sorts of endowments, including the knowledge embodied in the labor force and firms and other organizations located in the region, including durable contact links with the rest of the world. For medium-term policy decisions, tangible and intangible infrastructures are parameters governing the change processes that we consider in this paper.

The remainder of the paper is organized in the following fashion: first, we outline the 'paradigm' for analysis of fundamental change processes and demonstrate the durable and persistent character of regional development and innovation phenomena with a set of empirical examples over long-term population change, new firm formation and innovation activity across regions in Sweden. In the next section, we analyze properties of the dynamic systems derived from the 'paradigm' outlined in the previous section, and focus on how they can change and reach bifurcation points. Changes in the evolution path of the systems and bifurcation points are associated to policy decisions. A fourth section concludes.

Fundamental regional change processes

A 'paradigm model' of fundamental change processes

How can we understand fundamental change processes? The basic argument here is that they depend on slowly changing characteristics and invariant response mechanisms. The latter implies that a process feeds itself in such a way that the state in current periods depends on the state in previous periods.

Figure 7.1 shows the paradigm model for fundamental change processes, with regional business renewal in the form of entry and exit processes as an example. It illustrates that the change processes at hand depend on two main factors. The first is the state of slowly changing regional characteristics conducive for entry and exit processes. Slowly changing characteristics of a region refer, in particular, to regional endowments in the form of climate, natural resources, infrastructure capital, skills of the labor force, social capital and many similar things. It is a trivial conclusion that existing endowments provide regions with different opportunities for business renewal and development. At any moment in time, endowments are 'gifts from the past' that influence the development of each region's economy.[3] Glaeser and Gottlieb (2009) show, for instance, how the elasticity housing supply – an important factor for regions' potential to adjust to population flows – depend on 'fixed' attributes of the land in different regions. Moreover, Cheshire and Magrini (2006) illustrate how a previous dependence on coalfields and ports influences current development of regions.

The second factor is self-reinforcing adjustment mechanisms, which imply that the phenomena in the current period are partly a response to the same phenomena in previous periods. An example is that entrepreneurial activity in a region in current periods can generate demonstration effects that stimulate entrepreneurial activity in the future (Andersson and Koster, 2011). When such response (or feedback) mechanisms are invariant, they imply cumulative change processes, eventually combined with conservation principles. Conservation principles prevent explosive dynamics.[4]

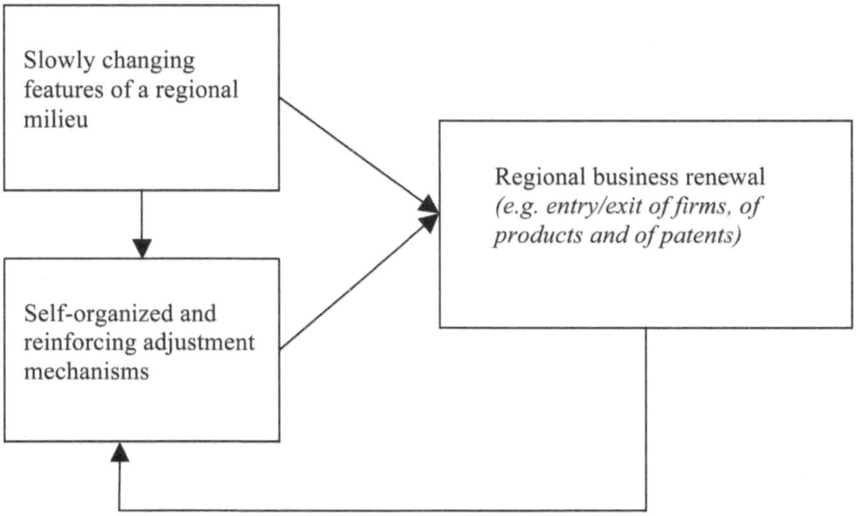

Figure 7.1 Paradigm model for regional change processes (Andersson and Koster, 2011).

It is evident that a change process driven by slowly changing (regional) characteristics and invariant self-organized feedback mechanisms will remain persistent in its way to develop over quite some time.[5] It is equally evident that such fundamental change processes imply evolution paths that may lead to new structures of economies over long-term horizons. Examples of such paths of evolution include:

- The growth of the share of manufacturing products in Sweden's export sales, where we can observe 10 percent in 1920 and more than 50 percent in 1990.
- The growth of the share of service sector production in the Swedish economy, with around 20 percent in 1920 and above 60 percent in 1990.
- The increase in the share of population living in urbanized areas, from about 25 percent in the 1920s to around 85 percent in 1990.
- The growth of knowledge-intensive producer services.

Having explained the 'paradigm' model for fundamental change processes, we now turn to a set of examples of the general model in Figure 7.1 in a regional context. We describe two examples conceptually and provide empirical illustrations. The two conceptual examples focus on exemplifying the response mechanism emphasized in the paradigm model. We end the section with two further empirical illustrations of the relevance of response mechanism and slowly changing regional characteristics.

Example 1: Long-term growth in regional size

An important aspect of the paradigm model in Figure 7.1 is invariant response mechanisms, and we shall here illustrate how cumulative processes pertaining to the long-term growth in size of a region can be based on such invariant feedback phenomena.[6]

Consider firms that supply distance-sensitive products (goods and services). The exit, entry and expansion of such firms in a functional region can be described as a response to the demand for customers located in the same region. The self-reinforcing feedback is set into operation when the expansion of the pertinent activities attract new customers to the region or stimulate the output of customer firms to grow. Figure 7.2 illustrates such a generic change process.

The response process illustrated in Figure 7.2 consists of two coupled adjustments. Consider, first, a situation with regional growth. Then the first adjustment is based on a stimulus for new activities to establish themselves and expand, generated by the size of regional demand at each moment in time. The second response is an increase of labor input, which is stimulated by the growth of economic activities. The described growth trajectory obtains when the size of the regional market is sufficiently large, implying that the exit activities are smaller than the entry.

A second trajectory obtains when the regional market potential is too small. Also in this case there will be entry and exit of firms as well as expansion and contraction of firms selling distance sensitive output. However, in the second trajectory, exit and decline will dominate over entry and expansion.

To illustrate the above phenomenon, we can apply a simple differential equation showing the net change of distance-sensitive output per time unit,

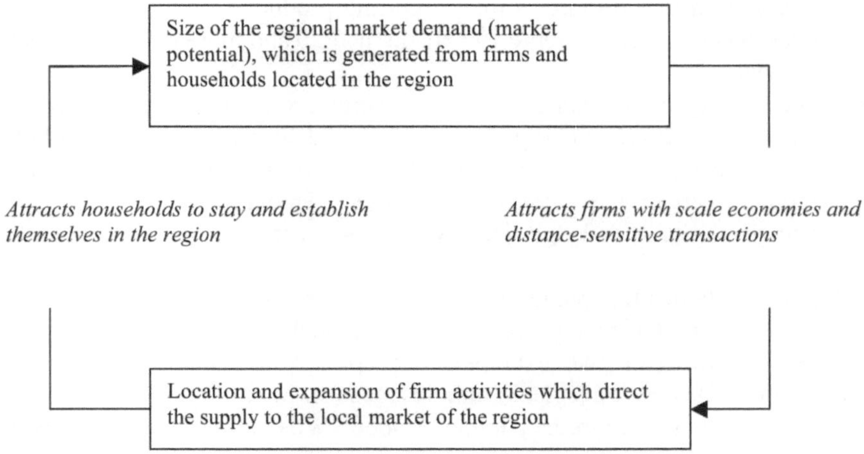

Figure 7.2 Cumulative change based on adjustment in the internal market of a region.

denoted by $x = dx/d$. With the help of the following parameterized equation, we can depict both expansion and contraction:

(1) for $x = dx/dt = \alpha + \beta M$ for $\alpha < 0$

where M reflects the size of the regional demand. The output from the pertinent activities will decline as M is small enough to satisfy $\alpha + \beta M < 0$.

Empirical illustration of example 1: Population growth across Swedish regions

In the previous section, we introduced a simple change equation to depict the cumulative growth in functional regions that follow from the home-market effect, which obtains strength in an economy with an increasing share of distance-sensitive transactions. This example was compatible with the paradigm model for fundamental change processes. What, then, is the empirical relevance of this example?

The example suggests that functional regions change in response to the size of the internal market, such that a large region attracts households to settle and remain in the region, whereas a smaller region is unable to attract and retain sufficiently many of their respective populations. Hence, the smaller regions have to experience a gradual decline in population. The empirical relevance of this statement is illustrated in Tables 7.1 and 7.2.

Table 7.1 shows that a clear majority of all municipalities were already declining in the period 1981–95, whereas less than 5 percent were declining in the metropolitan regions. For the medium-sized regions, we can observe that 70 percent of the associated municipalities were growing. Table 7.2 presents data over a longer time horizon (1970–2005). The table illustrates that 47 percent of all municipalities had a declining population in 1970–2005, whereas 53 percent had an increasing population. The latter 53 percent are typically municipalities that were large in 1970.

The results make it clear that it is not just the size of the municipalities that matters, but more so the size of the region to which it belongs. The reason for this is that the functional region represents the local market for each of the

Table 7.1 Population change in municipalities belonging to three types of regions, 1981–95

Municipalities in	*Negative*	*0–5 percent*	*5–15 percent*	*Above 15 percent*	*Total*
Metropolitan regions	4	9	38	49	100
Medium-sized regions	30	28	30	12	100
Small regions	63	21	16	0	100

Source: Statistics Sweden (SCB), own calculations.

Table 7.2 Municipalities and population change, 1970–2005

Population growth (%)	*Number of municipalities*	*Share (%)*
Less than –20	42	15
–20 to –10	37	13
–10 to 0	53	19
0 to 10	52	18
10 to 20	27	9
More than 20	73	26
Total	284	100

Source: Statistics Sweden (SCB), own calculations.

municipalities in the functional region. This conclusion is strengthened by Figure 7.3, which illustrates the population development in three types of municipalities during the 1990s.

Type I municipalities are defined as the largest central municipality, in terms of population, in each functional region. Type II municipalities are defined as non-central municipalities in large functional regions, again in population terms. Large, functional regions are defined as those with a population size of at least 100,000. Finally, type III municipalities are defined as non-central municipalities in those functional regions not considered as large, i.e. with a population below 100,000.

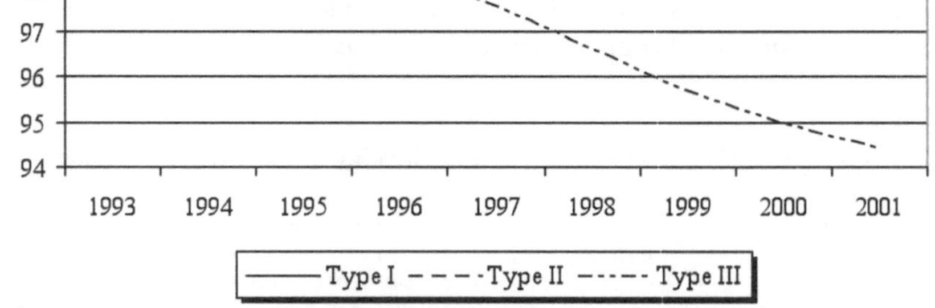

Figure 7.3 Population development, 1993–2001, in three types of municipalities in Sweden (based on Andersson and Klaesson, 2009).

Note: Total population of type I municipalities, 1993 = 4,597,511; total population of type II municipalities, 1993 = 3,288,816; total population of type III municipalities, 1993 = 858,782.

A similar message is conveyed in Figure 7.4, which shows the share of shrinking functional regions in 1994–2006 in different size classes. Of the functional regions with an initial population size less than 100,000, 93 percent were shrinking in terms of population between 1994 and 2006. The same figure for functional regions with a population between 100,000 and 500,000 amounts to 78 percent. In the group of functional regions with a population of more than 500 000, the share of shrinking municipalities was only 15 percent. Hence, the pattern is: small becomes smaller, whereas large becomes larger.

Example 2: The frequency of innovation activity in regions

Our second example pertains to how a regional 'innovation milieu' may develop in a cumulative process, where innovation activity by firms stimulates the development of a milieu, which, in turn, stimulates further innovation activity.

Consider firms supplying to external markets. These firms are influenced by demand as it develops in other markets outside the own region. An important part of this external demand develops in export markets. Existing firms, as well as potential start-ups, have to transform information about the export markets into innovation ideas, comprising 'know-how' about how to find customers in each such markets and knowledge about the preferences of the pertinent customers. Such information and knowledge is often generated in the regional milieu where a firm is located. A potential exporter's conceptualization of such an opportunity is an 'export innovation idea', which,

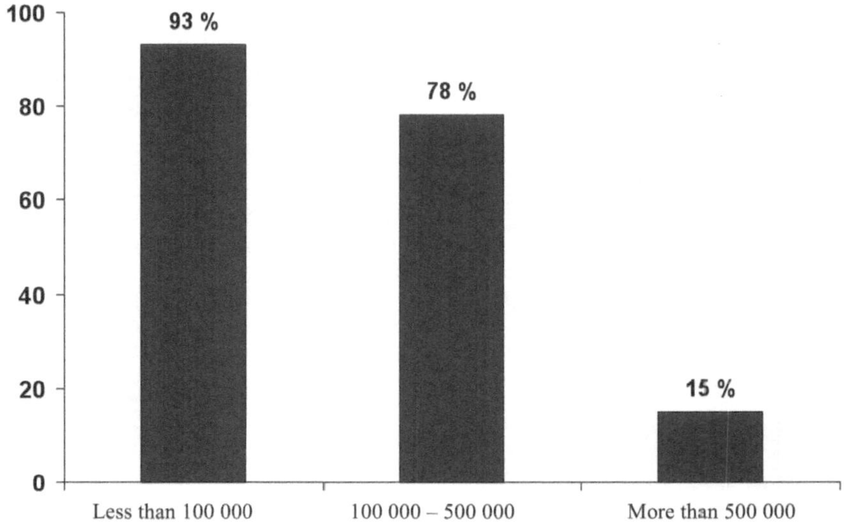

Figure 7.4 Share of shrinking municipalities for three size classes of regions, 1994–2006.

according to empirical observations, can be described as a stochastic process where innovation ideas arrive to exporters with different frequency parameters in different regions.

Figure 7.5 describes how characteristics of the regional milieu influence the arrival of ideas. In the figure, the characteristics are divided into: (1) knowledge sources in the region; (2) communication opportunities in the region; and (3) the absorption capacity in the region. When the frequency of new ideas is high, the milieu improves and the export-innovation intensity can increase. This further improves the milieu characteristics in a self-reinforcing circle. Such a process is consistent with the response mechanisms discussed in the paradigm model.

In a recent study by Andersson and Johansson (2008), regional characteristics are reflected by the variables presented in Table 7.3. These variables are used as explanatory factors determining the basic parameter of a Poisson process, which depicts the arrival rate of innovation ideas.

Empirical illustration of example 2: Export dynamics in regions

Above, we maintained that export firms, existing and potential, have to transform information about the export markets into innovation ideas, comprising 'know-how' about how to find customers in each such market and knowledge about the preferences of the pertinent customers. We also claimed that the regional milieu is often an important source for such information and knowledge.

Before entering into this discussion, we first present a diagram illustrating the correspondence between the emergence of new export firms and the size of a region. A second diagram reveals the correspondence between the introduction of new export varieties and regional size. We also present results from a recent econometric analysis by Andersson and Johansson (2008), where the frequency of new export varieties and new export firms, respectively,

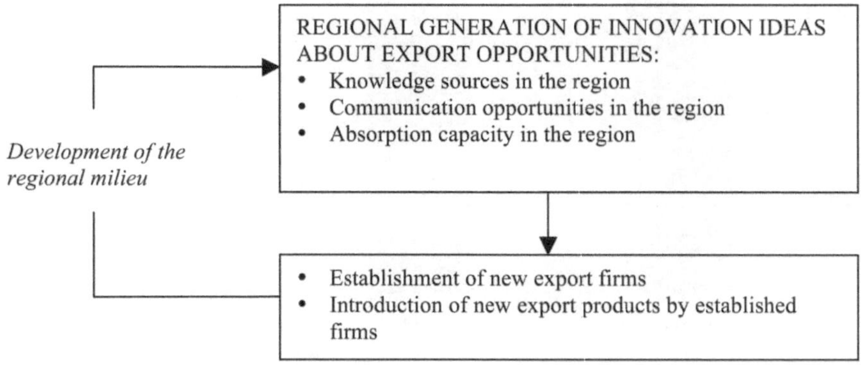

Figure 7.5 Dynamics describing the generation of export-innovation ideas.

Table 7.3 Regional characteristics influencing the arrival of export-innovation ideas

Knowledge sources in the region	• The number of existing export varieties in a given industry of a region, reflecting the localized *knowledge about the spectrum of export varieties* • The number of exporting firms in an industry of a region, reflecting the localized *knowledge about neighboring export firms* • Number of destination markets in an industry of a region, reflecting *knowledge about foreign markets*
Communication Opportunities in the region	• *Localization or cluster economies*, measured as export specialization of an industry in a region • *Urbanization or diversity economies*, measured by the size of the region • *Metropolitan contact opportunities*, measured by the distance to the nearest metropolitan region
Absorption capacity	• *Knowledge intensity*, measured as the share of the employees with university education of an industry in a region

is regressed on initial conditions of each region. The start up of new export firms, i.e. firms initiating exports, is depicted in Figure 7.6 as a function of the size of each region. The figure illustrates that the number of new export firms is clearly higher in larger regions.

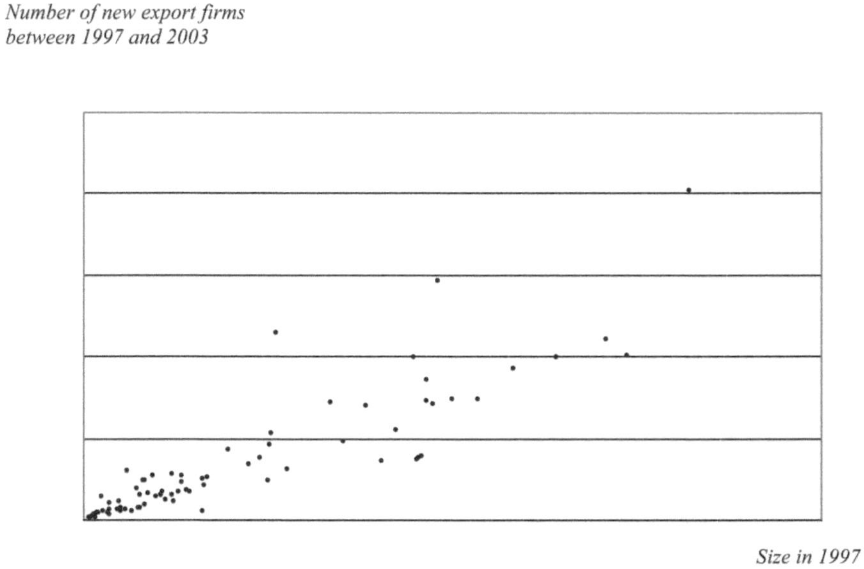

Figure 7.6 Relationship between the number of new export firms between 1997 and 2003 (vertical axis) and regional size in 1997 (horizontal axis) (Stockholm, Göteborg and Malmö excluded).

The information in the figure can be enriched with the help of econometric analyses that make use of the regional-milieu characteristics as described in Table 7.3 in the previous section. Andersson and Johansson (2008) show, in an econometric analysis, where a negative binomial Poisson model is used to estimate the likelihood of new exporting firms in a region, that the new export firms are positively influenced (in a statistically significant way) by the following factors:[7]

- the number of already established export firms in the sector;
- the number of destination markets in the sector;
- the export specialization of the sector, reflecting localization economies;
- the size of the region, reflecting urbanization economies in the sense of Jacobs;
- the distance to the closest metropolitan region; and
- the knowledge intensity of the sector, reflecting knowledge absorption capacity.

Returning to the list of regional-milieu characteristics in Table 7.3, these characteristics are also candidates to be included in an analysis of the frequency of new export varieties in different regions. Again, we can plot the number of new varieties against the size of each region, as illustrated in Figure 7.7. In addition, we can refer to the estimation results in Andersson and Johansson (2008), which show that the previous listed properties of the regional milieu

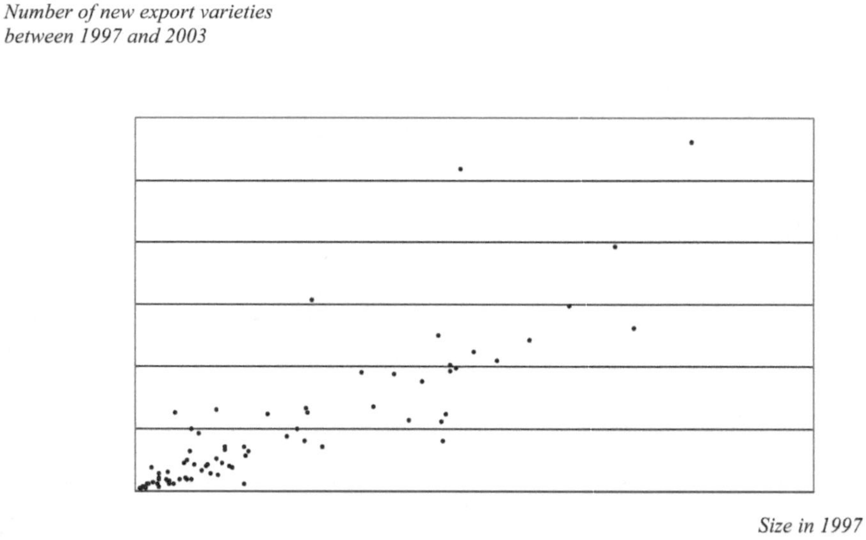

Figure 7.7 Relationship between the number of new export varieties between 1997 and 2003 (vertical axis) and regional size in 1997 (horizontal axis) (Stockholm, Göteborg and Malmö excluded).

also positively influence (in a significant way) the likelihood of new export varieties.

Empirical illustration 3: Entry of new firms as a time invariant process

We stated earlier that an example of self-reinforcing adjustment mechanisms is that entrepreneurial activity in a region in current periods can generate demonstration effects that stimulate entrepreneurial activity in the future (cf. Andersson and Koster, 2011). When such effects operate, entrepreneurial activity in regions should be persistent over time. We will now illustrate the empirical relevance of this argument.

The entry of new firms varies strongly across sectors. Since the concentration of sectors in regions differs in a systematic way, we can conclude that one may expect the entry rate to differ between regions. The decision to start a new firm is frequently assumed to be determined by: (1) profitability conditions of the firm's sector; (2) the market growth in the sector; (3) tangible and intangible capital intensity; (4) scale economies of the sector; and (5) industry concentration (Nyström, 2007).

For sectors with distance-sensitive transactions that mainly compete on regional markets, some of the described features will be region-specific. That applies to profitability, market growth and sector concentration, which all can be expected to stimulate entrants in a positive way. To this we can add the demonstration and learning effect that stems from the frequency of entry events in the region's recent history. Factors such as profitability and market expansion can be referred to as pull factors. In contradistinction to these, one may also consider push factors that 'force' an individual to start a business as the only available means to earn an income. A typical push factor is unemployment. Pull factors that can be assumed invariant regional-milieu characteristics include the concentration of sectors with low entry barriers and positive profit opportunities, and the demonstration and learning effects of a region's experience of frequent entry events in the past.

At this stage, we first consult Figures 7.8 and 7.9 from Andersson and Koster (2011), which show that there is clear pattern of persistence in new firm formation across regions, both in terms of business start-ups (Figure 7.8) and start-ups in the form of self-employment (Figure 7.9).[8] The figures illustrate the relationship between the level of current start-ups rates in Swedish municipalities and the level of start-up rates a decade earlier. The figures suggest that there are invariant regional characteristics that sustain the persistent pattern we can observe. It is astonishing that vibrant processes of entry and exit can generate observed patterns that remain approximately unchanged or invariant over long periods of time.

As the paradigm model for fundamental regional change processes suggests, Andersson and Koster (2011) maintain that there are two basic sources for persistence in start-up rates: (1) slowly changing regional characteristics that are conducive for start-ups; and (2) high levels of start-up rates can generate demonstration effects and an 'entrepreneurial' climate, such that there is a self-reinforcing element in start-up rates across regions. Clearly, these two hypotheses are not mutually exclusive.

Controlling for regional characteristics assumed to influence start-ups, Andersson and Koster (2011) find that lagged start-up rates have a significant impact on current start-up rates. Moreover, quantile regressions and analyses of transition probabilities provide empirical support for that both factors (slowly changing regional characteristics and self-reinforcing mechanisms), i.e. the paradigm model of fundamental change processes, are important for explaining the level of start-up rates across regions in Sweden.

Empirical illustration 4: Slowly changing regional characteristics

In the previous sections, we illustrated the empirical relevance of the paradigm model of fundamental change processes across regions with regard to: (1) population growth; (2) new export firms and export varieties; and (3) start-ups. The observed invariance and path dependence in these dynamic phenomena can be explained by slowly changing regional characteristics, in combination with self-reinforcing mechanisms.

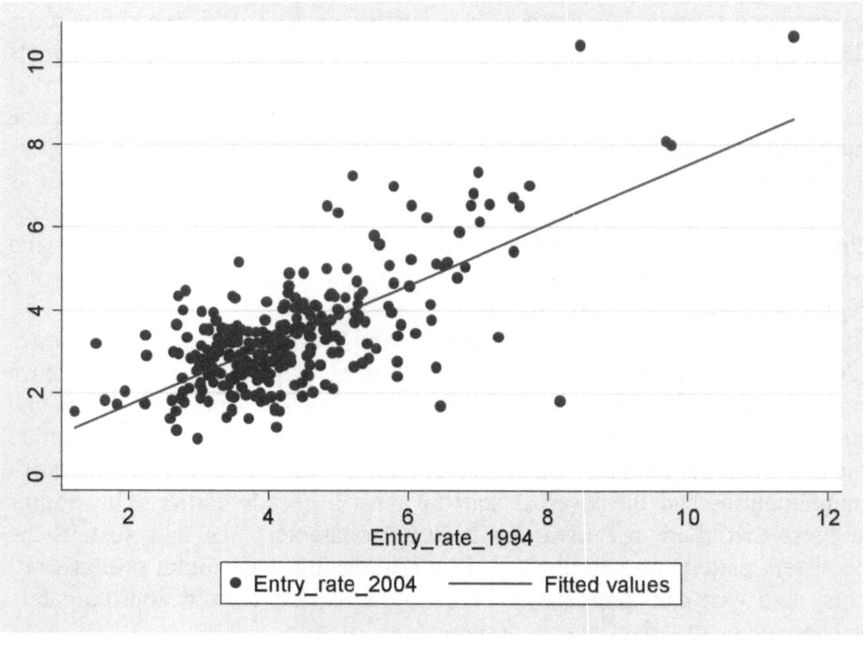

Figure 7.8 Relationship between business start-ups rates in t (2004) and t-10 (1994) across Swedish municipalities.

Source: Andersson and Koster (2011)

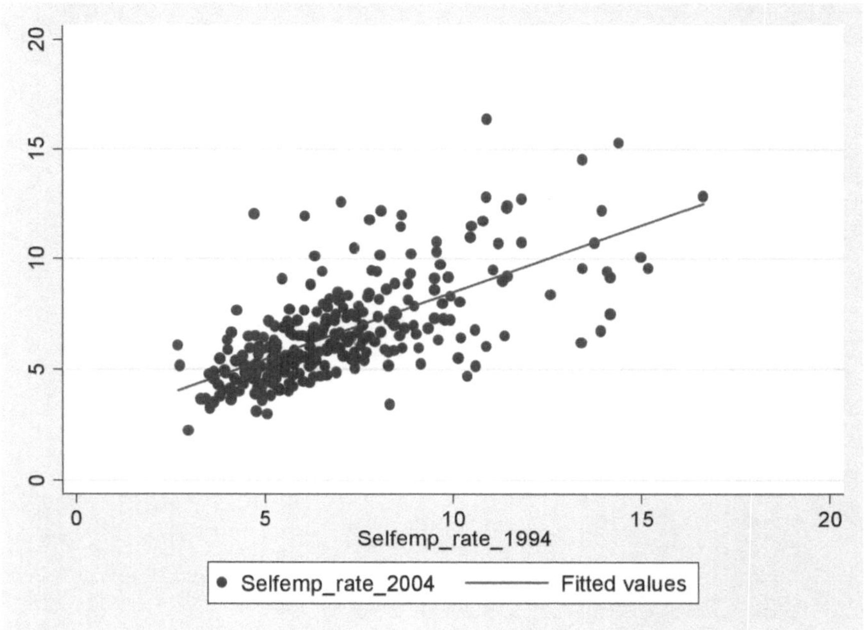

Figure 7.9 Relationship between self-employment start-ups rates t (1994) and t-10 (2004) across Swedish municipalities.

Source: Andersson and Koster (2011)

Our final empirical illustration focuses on the invariance in regional characteristics assumed to influence the conditions for business renewal. In particular, we will illustrate invariance in structures among municipalities in Sweden, with respect to the following set of regional properties:

- education level of the labor force;
- employment share; and
- share of services in the regional economy.

Figures 7.10–7.12 illustrate invariance in the structure of these characteristics across Swedish municipalities over a decade (1994–2004). In each figure, the horizontal axis measures the level of the variable in 1994, whereas the vertical axis measures the level of the variable in 2004 for each municipality.

The education level of the workforce is here measured as the share of the employees with a long university education (at least three years). It can be interpreted as the ability to absorb and assimilate new techniques, new knowledge, innovations and ideas (cf. Cohen and Levinthal, 1990). As such, the education level of the workforce in a region represents an important condition for business renewal processes. The employment share of a region can be interpreted as the inverse of a push factor, in the sense that low

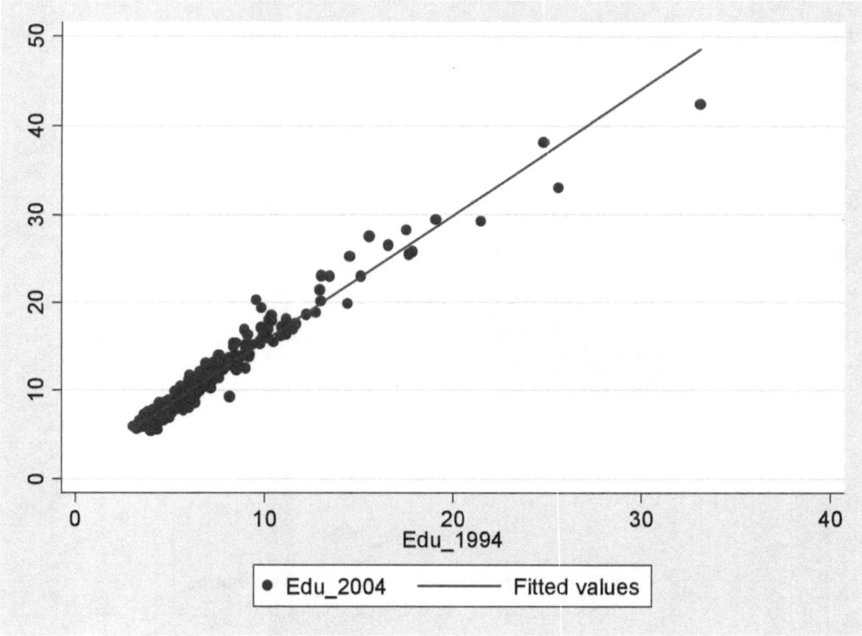

Figure 7.10 Relationship between education level of the workforce, 1994 and 2004, across municipalities.

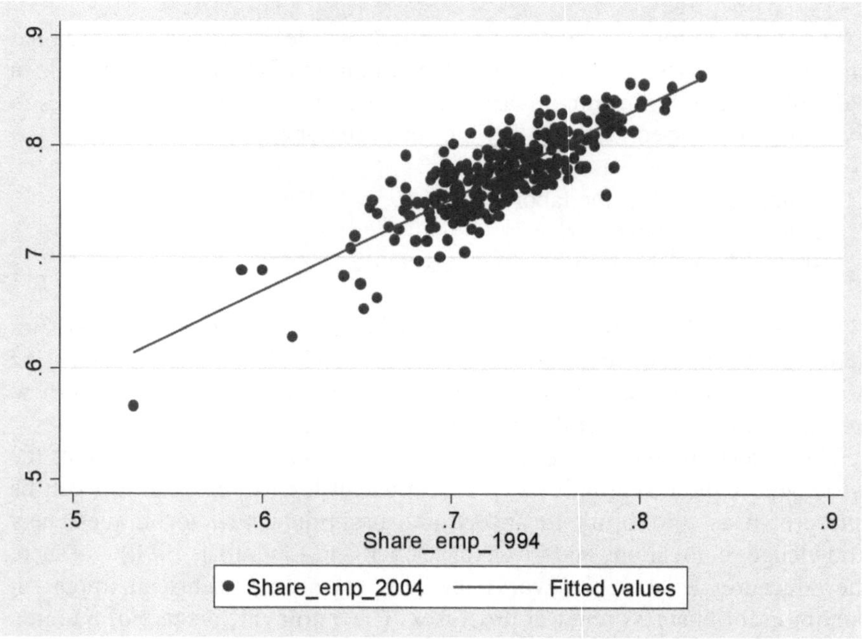

Figure 7.11 Relationship between the employment share in 1994 and 2004 across Swedish municipalities.

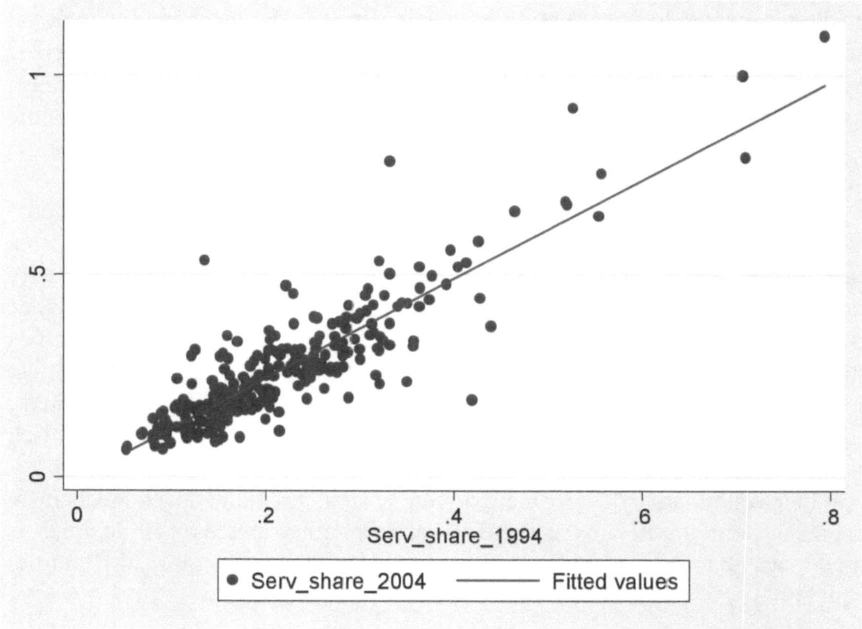

Figure 7.12 Relationship between the share of services of municipalities in 1994 and 2004.

employment shares imply that a large fraction of the workforce is not employed. Suppose that this factor influences start-ups and is invariant in the sense that the relative size of unemployment rates remains unchanged across regions over time. This would make the start-up rate of regions persistent over time.

Moreover, considering that entry barriers in service sectors are relatively low and that private services are the fastest-growing part of the Swedish economy during the past 10–12 years, the share of service activities in a region should influence the frequency of start-ups and business renewal in regions. The fastest growing segment of services is the group of knowledge-intensive producer services (Klaesson and Johansson, 2008), which are often maintained to be vital in the upgrading of regional industry.

Behavior of dynamic systems

The previous sections present the paradigm model for fundamental regional change processes and provide examples of the empirical relevance of invariant regional change processes in terms of population growth, export dynamics and start-up rates. Examples of slowly changing regional characteristics were also presented.

Given the observed persistence in the behavior of indicators of regional business renewal and innovation, we may ask: what is the scope for regional policy? Can policy alter the observed invariant change processes characterized by path-dependence? To provide a basis for a discussion of the above questions, we analyze the stability and evolution trajectories of simple non-linear dynamic systems that incorporate the essential characteristics of the paradigm model of change processes outlined in Figure 7.1.

Consider a system for which we can identify invariance over time. Such a lack of change can be associated with two alternative phenomena. The first case obtains when the seemingly invariant variable is just changing on a very slow timescale. The other phenomenon that can be associated with invariance is a system with feedback properties causing the system to behave in a stationary way. In this latter case, the response mechanism functions like a homeostasis mechanism that brings the system back to its stationary equilibrium whenever it is disturbed. In this way, we may think of the feedback as the major case for stationary processes.

When a cumulative development driven by an invariant feedback mechanism is present, this implies that a focused variable either increases or declines as time goes by. Such phenomena can be described by simple differential equations:

$$(2)\quad x = \alpha(N - x)x$$

With this equation, x will expand as soon as x attains a value greater than 0 and it will continue to grow in response to the size of x itself, as long as $x < N$. The change process has two equilibria. The first is $x = 0$ and the second is $x = N$. It can describe an economic activity that is dependent on a resource, the size of which is signified by N. As long as the resource constraint is not binding, the process will develop along a sigmoid (S-shaped) curve (Figure 7.13).

The curve in Figure 7.13 is a generic form of the evolution of shares, such as the share of a specific product in a product group or the share of a sector in an economy. The essential feature of this model is that the system is conservative and thus does not explode.

With a system with more than one change variable, one can sometimes order the variables with regard to their speed of adjustment. In a two-variable system, one variable can be classified as fast relative to the other. This is a basic element of Haken's (1983) 'slaving principle' in which the fast-adjusting variable is slaved by the evolution of the slower one. In accordance with the paradigm model outlined above in this chapter, we may think of business renewal processes being slaved (or governed) by the education profile, competences and skills of the regional labor force which change slowly. This implies that the type of firms that can develop in a short time period adjusts quickly to a temporary equilibrium solution given the state of slow-adjusting regional attributes. As an example, consider the following coupled dynamics:

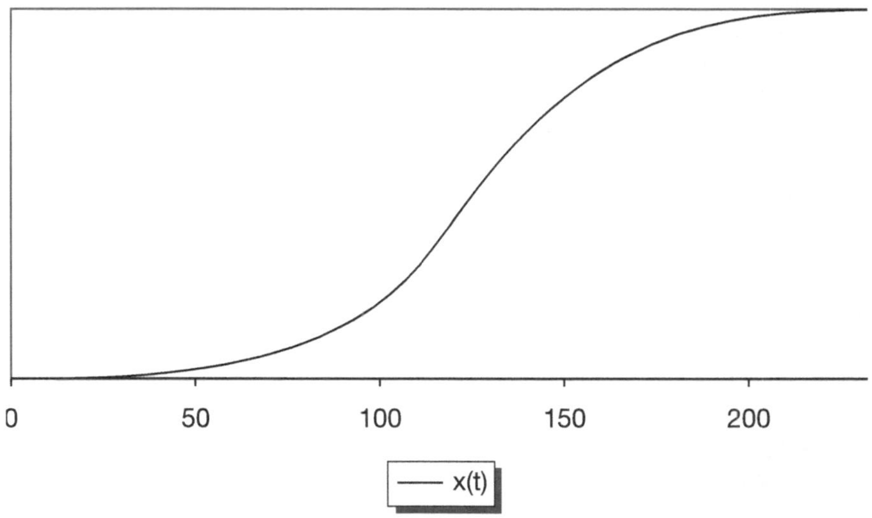

Figure 7.13 The sigmoid shape of the development path of *x* in Equation (2), figure based on N = 500 and α = 0.0001.

(3) $x = \alpha x - xy + f(t)$

(4) $y = \beta y + x^2 + g(t)$

where *x* represents a slow variable and *y* a fast one, and where the functions *f* and *g* describe influences from a surrounding system, such that this influence evolves as time goes by. Suppose now that the external influences are absent so that $f(t) = g(t) = 0$. The change of *x* is described by $x = dx / dt$ and takes place on a slow timescale when α is positive and small, while β is much lager than α. When this is the case, we can introduce the approximation $y = dy / dt = 0$, which yields $y = x^2 / \beta$. This implies that *y* is slaved by *x*, and we can insert this approximation into (3) and thereby reduce the coupled system into just one equation. Thus, as long as the approximation is valid and as long as $f(t) = 0$, we have:

(5) $x = \alpha x - x^3 / \beta$

When $\alpha > 0$ there are two stable equilibria at $x \pm \alpha\beta$. However, when $\alpha < 0$, $x = 0$ is a stable equilibrium. If we make the *y*-variable a constant then Equation (3) simplifies into:

(6) $x = \alpha + \beta x - xy$

where α, β and *y* are positive fixed parameters and $y > \beta$. This system comes to rest as a stationary equilibrium when $x = \alpha/(y - \beta)$. In this case, we obtain a system that will safeguard a stationary solution, i.e. it is stable with regard to disturbances.

Consider again Equation (5) and assume that we add still another change process, this time describing the evolution of α. If this process starts for some negative value of α and expands gradually over time, then a bifurcation point obtains when it reaches the critical value 0 (see Figure 7.14). This illustrates how the nature of a developing system can change in character for small alterations of parameter values. It also puts into perspective the prediction capability for a system with parameter invariances. At the very bifurcation point, the model does not advise about which of the two alternative development paths the system will settle on.

Our last example of dynamic systems is used to illustrate that a variable may be changing at a slow pace during certain phases of a development process and then suddenly adjust to much faster change (Johansson, 1993). To make this example concrete, we may consider that an urban region consists of several different urban areas. Each such urban area will have its own local urban infrastructure and other support systems that generally develop on a slow

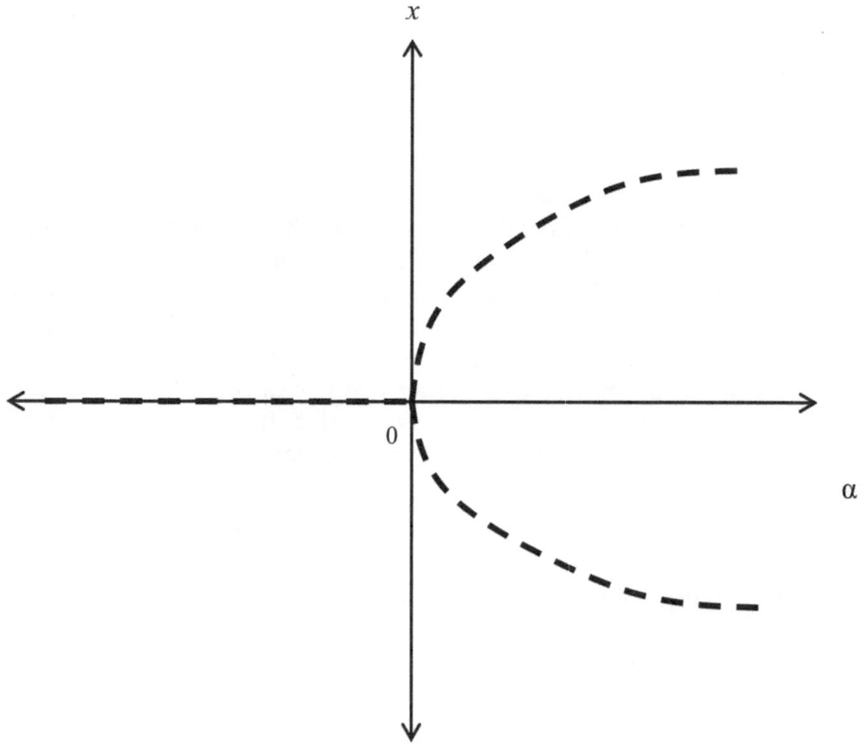

Figure 7.14 Bifurcation consequence of changing values of a slow response parameter (α).

timescale. The response to infrastructure development, in terms of location of firms in the area, may display a modest or even slow response. At a sudden point in time, the infrastructure may reach such a quality and capacity that the attraction of new firms will give rise to a fast economic development that eventually settles at a new path of slow change, now with a declining direction. This is often observed in historical analyses of business location in sub-areas of an urban region.

This type of dynamics is typically labeled 'relaxation dynamics' and is illustrated in Figure 7.15. The horizontal axis measures the gradual increase of slowly changing variables, i.e. with infrastructural properties and other similar milieu-factors. The vertical axis measures the intensity of the economic activity in question, e.g. the number of firms in an urban area. As can be seen, there are four phases: (1) gradual increase in economic activity; followed by (2) a fast expansion of activities; and (3) a gradual decline in activities; ending with (4) a fast contraction. In our example the nature of the dynamics may be moderated by location and infrastructure policies.

A classic example of relaxation dynamics, as illustrated in Figure 7.15, is a newly opened restaurant where the quality of the meals served improves over time. At the first relaxation point, there is a rapid increase in customers attracted by the qualities of the restaurant. As this happens, crowding and congestion features sets into play due to the tensions of more customers than

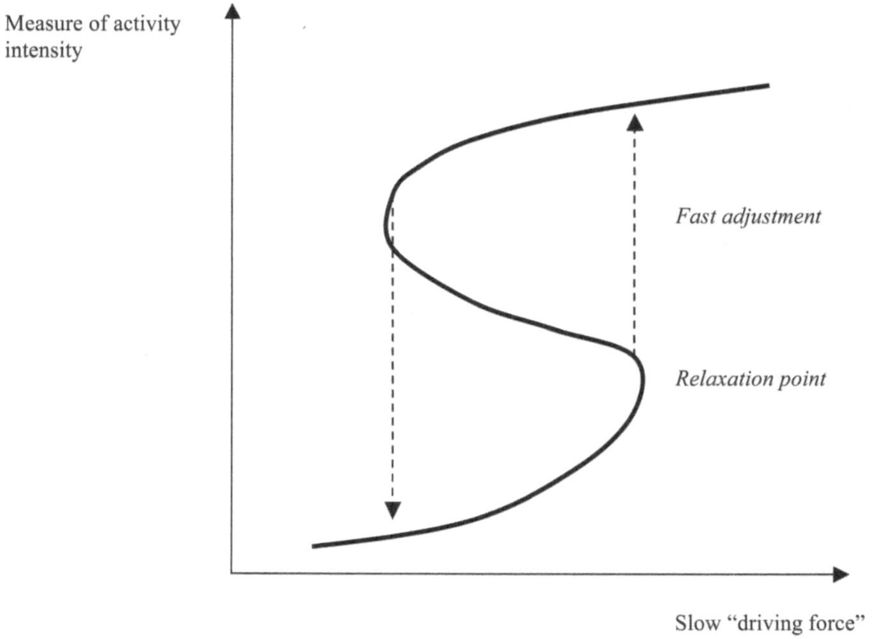

Figure 7.15 Illustration of relaxation dynamics.

the capacity of the restaurant can handle. As a consequence, the quality starts to worsen. At a new relaxation point, customers are rapidly turning their back to the restaurant. For a district of an urban region, one may think of cluster evolution as another example. At the first relaxation point, many firms are strongly attracted by the milieu created by the already clustered firms. Congestion and tensions, e.g. manifested in high land prices and competition for labor, or just capacity shortages, eventually imply that the cluster is not capable of renewing its business activities. Hence, firms find it less advantageous to stay in cluster, or just lose their capacity to survive.

In summary, we have reviewed four basic models of aggregate change processes comprising the essential features of the paradigm model of change processes initially outlined in this chapter. A common characteristic of the discussed prototype models is feedback phenomena. Once growth (decline) is initiated, continued growth (decline) is stimulated by the growth in previous periods. A major reason for this is invariant response mechanisms phenomena. We also showed how slowly changing variables govern the evolution of variables that adjust on faster timescales.

What are the policy lessons from the model exercise? There are two basic lessons. The first is that policy must be capable of recognizing the nature of the different parameters that govern slow and fast change processes. Equations (3)–(6) illustrate that there are parameters whose levels are critical for the direction of change. Importantly, changes in those parameters that make processes slow can imply that such processes increase their tempo of change. As this happens, the system reaches a bifurcation point that may impact the process to jump or shift to a new evolution path (Figure 7.14). We consider these types of factors as defining the basic area for regional policy. Policies that correspond to influencing the determinants of slow processes are a major field for regional development policy.

The second lesson points to the fact that regional policies must be persistent, in the sense illustrated by the phenomenon in Figure 7.15. What this means, in practice, is that a policy altering the slow variables, such that a new evolution path can be accessed or adopted, may have to build up resources for several decades to sustain the entrance into new evolution paths. This can, for example, pertain to shifts to new persistent tax policy, R&D and education policy, and systems for facilitating entry and exit processes. Another lesson from Figure 7.15 is that, if the relaxation dynamics in the figure represent one of several industries in a region, then every industry that has attained a large share of the regional activity eventually comes to a point where its importance is quickly degraded. As a consequence, a region that has managed to adjust to a new evolution path in which a sector gradually increases its share of the economy, a structural change process will eventually enter a new situation in which the very same industry is phased out. Thus, the structural change process may be repeated for new industries.

Summary and conclusions

We have discussed the challenges and potentials for regional policy in situations where durable and persistent change processes are present. These are labeled fundamental change processes and are claimed to form an important context to which policy must respond. Fundamental change processes exert a strong influence on the conditions for long-term renewal and growth of regions, and are often associated with evolution paths that imply structural change. We maintained that policy under these conditions may be described as 'management of change'.

It is, of course, important that policy decisions are based on a solid understanding of the processes that govern the evolution of regions. We outlined a paradigm model of fundamental change processes, which suggests that they should be understood as development paths with persistent mechanisms of change, driven by slowly changing (regional) characteristics and self-organizing feedback adjustments. This model abstracts from the specific circumstances that apply to each individual micro-case. It is designed to understand and predict the behavior of systems with many agents.

To illustrate the empirical relevance of the paradigm model, we investigated and identified regimes of change processes in the system of Swedish regions. We showed that change processes, in the form of population growth, export dynamics and new firm formation, feature a high degree of invariance over time. As such, they are characterized by path dependence and sensitivity to initial conditions.

By analyzing the behavior of dynamic systems in concordance with the paradigm model, we also illustrated how they can be subject to self-generated sudden change and shifts in change processes, either stimulated endogenously or by exogenous phenomena. The latter implies a potential role for policy.

Our main conclusions for policy are as follows:

1. Fundamental change processes constitute a major challenge for policy. A prerequisite for policy is to identify and understand the nature of any pertinent change processes.
2. Parameters that alter these change processes represent those things on which policy decisions should concentrate, and there are parameters whose levels are critical for the direction of change. Policy has thus a potential to influence the evolution of a fundamental change process by primarily influencing the parameters of slow processes. This, of course, requires a capability to recognize the nature of the different parameters that govern slow and fast change processes.
3. For medium-term decisions, infrastructure – tangible and intangible – may often be thought of as a parameter governing slow processes.
4. Policy may need to build up resources for several decades to sustain the entrance into a new evolution path. Such a strategy can, for instance,

pertain to shifts to new persistent R&D and education policies, and systems for facilitating entry and exit processes.

The first conclusion is based on the argument that fundamental change processes influence the long-term evolution of regions and are often associated with structural change. Many regions in Sweden, for instance, experience a net outflow of individuals each year, i.e. they are depopulated. This depopulation may be understood as a fundamental change process of these regions, which clearly exert a major influence on their growth potential. At the same time, certain regions manage to attract firms and individuals, and maintain innovation activity over long periods of time. The typical example is vibrant cluster environments such as Silicon Valley in California and large city regions. In this case, fundamental change processes may operate in the same direction as the objectives of policy. Nonetheless, in both cases, fundamental change processes set the context for policy; they are 'gifts from the past'. To be able to manage this context, irrespective of whether policy wishes to counteract or reinforce the change processes, policy-makers must identify and understand the nature of the pertinent change processes.

The second and third conclusions emphasize that policy indeed has a potential to influence change processes, and that this can be done by influencing the parameters of slow processes. Infrastructure is a typical parameter of slow processes. Infrastructure can be tangible (e.g. roads and buildings) and intangible (e.g. knowledge and competences of regional workforce, public and private R&D). Investments in such slowly changing features of a region are a way in which change processes may be altered, because these slowly changing characteristics are parameters of the change processes.

The fourth conclusion states that substantial resources and complementary investments may be necessary to sustain a shift in a region's evolution path. As an example, a well-known regional policy (motivated by the arguments presented here) is to establish regional universities in order to boost the regional innovation potential and the education level of the regional workforce (Florax, 1992). Such a strategy often needs to be complemented with investments designed to making the region attractive as place of residence and work for educated people. Otherwise, there is a risk that the region becomes an 'exporter' of knowledge labor.

We end our contribution with a set of examples that provide a more concrete understanding of the conceptual conclusions above, and illustrate that policy may matter. Let us start with an example of the Strömstad and Åre municipalities, which, during the past 15 years or so, have transformed from stagnating peripheral locations to areas with a sharply increasing share of tourist exports and other service sectors. In both cases, the local policy has adjusted in response to exogenous changes in conditions. Partly in line with the general transformation of the Swedish economy towards increasing importance of service activities, these conditions have increased the demand for tourism, retail and other services. Policy-makers in the Åre and Strömstad municipalities managed to support a new evolution path initiated by exogenous phenomena. The enhanced attractiveness of both regions is revealed by the expanding prices of property in the regions. Our next example illustrates the importance of policy-makers' capacity to hold onto long-term efforts in transforming a region into something new. In the middle of the 1970s, the shipyards on the west coast in Sweden were closed down in sequence. In Gothenburg, three different shipbuilders discontinued their production. They were all located in an area of the city that made contact with central business district and had waterfront opportunities. Some 25 years later, most of these areas were completely rebuilt and designed from vibrating knowledge and producer service activities, culture and exclusive housing. This process was accompanied by the transformation of the Gothenburg region from a previously industrial economy towards a knowledge economy. Even if this is part of a general trend in reshaping city regions, the change in Gothenburg required a great deal of persistence and patience in the regional policy-making. Of course, the policy-making was based on an understanding of the overall fundamental change processes at hand. Our final, and perhaps most eminent, example is the remarkable growth of the Umeå region during the past 40 years, in which an initially very small town has developed to an international research center and producer of university-educated individuals for export to other parts of Sweden. In this case, national resources were part of the enduring efforts of change. However, the example illustrates that policy decisions, in this case the establishment of a university, can push a region onto new trajectories.

Notes

1. Regional policy is here broadly defined and refers to policies aimed at stimulating regional growth.
2. The second question is related to a recent analysis by Dolfsma and Leydesdorff (2009), who analyze theoretically the conditions for breakout from technological trajectories.
3. The role of initial conditions is also often discussed in the literature on regional clusters and innovation systems (see e.g. Afuah and Utterback, 2000).
4. A general example of a conservation principle is congestion effects.
5. Invariant response mechanisms and cumulative expansion/contraction are closely associated with the popular notion of path-dependence. As noted by Antonelli (1997: 643–644), 'path-dependence defines the set of dynamic processes where small events have long lasting consequences that economic action in each moment can modify only to a limited extent'. Path dependence implies that once a process, for some reason, is initiated, it tend to persist over time. Invariant response mechanisms may cause such persistence.
6. We discuss regions in terms of *functional regions*. A functional region can be delineated in such a way that firms and households have a common market within the boundaries of the region. Such a region will also be an arena for interaction between firms in exchange of products, information and knowledge. The most important feature of the concept of functional regions is that they allow for frequent face-to-face contacts between individuals and firms.

7. It should be stressed that regional factors are measured for the year 1997, while the entrance of new export firms is recorded as firms exporting in 2003, but not exporting in 1997.
8. Fritsch and Mueller (2007) find a similar pattern for regions in Germany.

References

Afuah, A.N. and J.M. Utterback (2000), 'Sources of Innovative Environments: A Technological Evolution Perspective', in Z.J. Acs (ed.), *Regional Innovation, Knowledge and Global Change*, Pinter Publishers, London .

Andersson, M. and B. Johansson (2008), 'Innovation Ideas and Regional Characteristics: Product Innovation and Export Entrepreneurship by Firms in Swedish Regions', *Growth and Change*, 39(2), 193–224.

Andersson, M. and J. Klaesson (2009), 'Regional Interaction and Economic Diversity: Exploring the Role of Geographically Overlapping Markets for a Municipality's Diversity in Retail and Durables', in C. Karlsson, B. Johansson and R.R. Stough (eds), *Innovation, Agglomeration and Regional Competition*, Edward Elgar, Cheltenham, UK.

Andersson, M. and S. Koster (2011), 'Sources of Persistence in Regional Start-Up Rates: Evidence from Sweden', *Journal of Economic Geography*, 11, 179–201.

Antonelli, C. (1997), 'The Economics of Path-Dependence in Industrial Organization', *International Journal of Industrial Organization*, 15, 643–675.

Beckmann, M.J. (1958), 'City Hierarchies and the Distribution of City Size', *Economic Development and Cultural Change*, 6, 243–248.

Boschma R. and K. Frenken (2006), 'Why is Economic Geography not an Evolutionary Science?' *Journal of Economic Geography*, 6, 273–302.

Cheshire, P. and S. Magrini (2006), 'Population Growth in European Cities: Weather Matters, but Only Nationally', *Regional Studies*, 40, 23–37.

Cohen, W.M. and D.A Levinthal (1990), 'Absorptive Capacity: A New Perspective on Learning and Innovation', *Administrative Science Quarterly,* 35, 128–152.

Dolfsma, W. and L. Leydesdorff (2009), 'Lock-In and Break-Out from Technological Trajectories: Modelling and Policy Implications', *Technological Forecasting and Social Change*, 76(7), 932–941.

Florax, R. (1992), *The University: A Regional Booster?: Economic Impacts of Knowledge Infrastructure*, Aldershot, Avebury, UK.

Frenken K. and R.A. Boschma (2007), 'A Theoretical Framework for Evolutionary Economic Geography: Industrial Dynamics and Urban Growth as a Branching Process', *Journal of Economic Geography*, 7, 1–15.

Fritsch M. and P. Mueller (2007), 'The Persistence of Regional New Business Formation Activity Over Time: Assessing the Potential of Policy Promotion Programs', *Journal of Evolutionary Economics*, 17, 299–315.

Fujita, M. and J.-F. Thisse (2002), *Economics of Agglomeration: Cities, Industrial Location and Regional Growth*, Cambridge University Press, Cambridge.

Fujita, M., P. Krugman and A. Venables (1999), *The Spatial Economy*, The MIT Press, Cambridge.

Gabaix, X. (1999), 'Zipf's Law for Cities: An Explanation', *Quarterly Journal of Economics*, 114, 739–767.

Gabaix, X. and Y. Ioannides (2004), 'Evolution of City Size Distributions', in V. Henderson and J.-F. Thisse (eds), *Handbook of Regional and Urban Economics 4*, Elsevier North-Holland, Amsterdam.

Glaeser, E. and J. Gottlieb (2009), 'The Wealth of Cities: Agglomeration Economies and Spatial Equilibrium in the United States', *NBER WP 14373*.

Glaeser, E., J.A. Scheinkman and A. Shleifer (1995), 'Economic Growth in a Cross-Section of Cities', *Journal of Monetary Economics*, 36(1), 117–143.

Haken, H. (1983), *Synergetics: An Introduction*, Springer-Verlag, Berlin.

Johansson, B. (1993), 'Economic Evolution and Urban Infrastructure Dynamics', in Å.E. Andersson *et al.* (eds), *The Cosmo-Creative Society*, Springer-Verlag, Berlin.

Johansson, B. (1998), 'Infrastructure and Economic Milieu: Swedish Contributions 1960–1995', in L. Lundqvist, L.-G. Mattsson and T.J. Kim (eds), *Network Infrastructure and the Urban Environment*, Springer-Verlag, Berlin.

Johansson, B. and R. Wigren (1996), 'Production Milieu and Competitive Advantage', in D. Batten, and C. Karlsson (eds), *Infrastructure and the Complexity of Economic Development*, Springer Verlag, Berlin.

Johansson, B., C. Karlsson and R.R. Stough (eds) (2002), *Regional Policies and Comparative Advantage*, Edward Elgar, Cheltenham, UK.

Kaldor, N. (1979), 'The Case for Regional Policies', *Scottish Journal of Political Economy*, 17, 337–348.

Kaldor, N. (1981), 'The Role of Increasing Returns, Technical Progress and Cumulative Causation in the Theory of International Trade and Economic Growth', *Economie Applique*, 34: 593–617.

Klaesson, J. and B. Johansson (2008), 'Agglomeration Dynamics of Business Services', CESIS WP 153.

Krugman, P. (1991), 'Increasing Returns and Economic Geography', *Journal of Political Economy*, 99, 483–499.

Martin, R. and P. Sunley (2006), 'Path Dependence and Regional Economic Evolution', *Journal of Economic Geography*, 6, 395–437.

Myrdal, G. (1957), *Economic Theory and Under-Developed Regions*. Gerald Duckworth and Co., London.

Nyström, K. (2007), 'An Industry Disaggregated Analysis of the Determinants of Regional Entry and Exit', *Annals of Regional Science*, 41, 877–896.

Schettkat, R. and L. Yocarini (2006), 'The Shift to Services Employment: A Review of the Literature', *Structural Change and Economic Dynamics*, 17(2), 127–147.

UNFPA (2007), 'State of World Population 2007, Unleashing the Potential of Urban Growth', UN, available at: www.unfpa.org/swp/2007/english/introduction.html.

8 Constructing an innovation policy agency

The case of the Swedish Governmental Agency for Innovation Systems

Bo Persson

Introduction

The traditional role of the state in science and technology policy has been challenged in recent decades.[1] A new set of ideas, sometimes referred to as the innovation paradigm, has increasingly influenced the policy development in this area. According to these ideas, the role of the state should no longer be to support only R&D, but the whole dynamic process, or system, in which innovations develop, diffuse and are utilized (Edquist, 1997; Fagerberg *et al.*, 2005). The supporters of this paradigm argue that innovation policy should focus on promoting innovation within a given institutional context (like traditional science and technology policy) and on changing the institutional context in order to promote innovation (Biegelbauer and Borrás, 2003: 9). The focus of state activities in innovation policy should also be to facilitate collaborations and entrepreneurship among actors from different sectors, not only focusing on the role of universities or research institutes, but also on industry and different governmental actors (see e.g. Etzkowitz and Leydesdorff, 1997). Furthermore, the perspective suggests the need to integrate or coordinate different policy areas and lead them in the same direction, or at least persuade different policy actors to consider the innovation perspective within decision-making (Lundvall and Borrás, 2005; Pelkonen, 2008).

These new expectations and ideas present challenges in terms of the development of new state capacities and legitimacy of the state. Traditional research funding organizations, primarily based on relatively stable relations with the research community (research councils), or alignment with specific sectorial interests or policy areas (mission-oriented agencies), would seem less adequate for a more proactive and flexible innovation policy. Scholars in the field of science policy studies have argued that these new expectations imply a less stable situation for research funding organizations, requiring an increasing focus on repeatedly developing new capacities and acquiring legitimacy among stakeholders (van der Meulen, 2003). This raises the question of how different policy systems have handled this "innovation policy challenge," and the possible tensions involved in changing existing research funding structures.

This chapter analyzes this issue by looking more closely at the establishment of the Swedish Governmental Agency for Innovation Systems (VINNOVA) in 2001. The new agency is explicitly related to a system view of how innovation processes should be understood (Jacob, 2006; Eklund, 2007), and based, to a greater extent than earlier technology and research polices, on an intention to promote coordination and co-operation between different policy areas. The idea of the new agency is a distinct departure from traditional research councils and mission-oriented funding agencies, which support research through peer-review procedures and fund R&D in specific sectors, respectively. The profile of the new agency is instead to take "a different view," through focus on the whole system of innovation, using unconventional working methods for steering and mobilizing support.

The overall purpose of this chapter is to analyze how an agency of this kind, seemingly reflecting a more idea-oriented, strategic and proactive role than traditional bureaucracies, was established, and how it has developed its role and steering strategies and acquired legitimacy and support for them. How have new policy ideas been integrated in a political process and "translated" into organizational outputs? I argue that the establishment of the new agency can be understood as a dynamic process where new ideas interact with existing and developing institutional contexts. The birth of the new agency was a complex process where policy-makers translated new ideas into suitable solutions, a process also shaped by the institutional legacy and political traditions of the Swedish political system. The establishment of a new public agency also meant that the institutional context changed; a new "policy actor" was created with a need for survival and legitimacy. The legitimacy of the agency is not based on legal power, like a traditional bureaucracy, or on the professional legitimacy of academia as in the case of a research council, but rather on a capacity to "deliver." New innovation policy ideas play the role of giving substance not only to policy positions and programs, but also to legitimacy and support from multiply networks.

The case is an example of a new trend where the role of state agencies seems to be changing; sometimes described as a process from "government to governance." The limitation of state capacity and the emergence of new policy problems also seem to have led to the development of different governmental roles and policy instruments (Pierre and Peters, 2000; Jessop, 2002). According to Pierre and Peters, the capacity of the state to act is based less on "constitutional" aspects, i.e. what used to be the "ultimate source of state power" (Pierre and Peters, 2000: 81–3). Instead, the capacity to act is based on what it has to offer, and resources and abilities to "read and exploit unique contexts" (Pierre and Peters, 2000: 83). Several scholars analyzing Swedish public policy have also pointed out that this development tends to involve a stronger focus on persuasiveness and the imposing of new values on the government apparatus or the society at large. Instead of regulating behavior, state agencies more often create common visions and spread ideas (Lindvall and Rothstein, 2006; Stenlås, 2006). This means that ideas and scientific knowledge tend to become more important in the activities of governmental agencies, not least due to agencies' embeddedness in international epistemic communities, sharing underlying assumptions about how problems should be perceived and handled (Haas, 1992). The chapter is also a contribution to research about the roles and function of research funding organizations (Braun, 1993; Braun and Guston, 2003; van der Meulen, 2003). While many earlier studies have focused on how new challenges are handled by existing governmental agencies (primarily research councils), my study focuses on the establishment of a new innovation-oriented agency.

The chapter may be characterized as a case study, analyzing one particular case of a phenomenon in its context. The focus is thus not on the historical episode itself, but on a specific, defined aspect of it (George and Bennet, 2004: 18). The analysis is primarily based on written material gathered from the VINNOVA archives; primarily documentary material, such as memoranda from the VINNOVA board, and other public material such as government bills and annual reports. The minutes of the board are especially useful, since they include detailed notes of discussions and decisions. Furthermore, I have also conducted nine interviews with key informants (seven with representatives of the management and staff of VINNOVA, one with the former Minister of Industry and Minister of Education and research, and one with a senior representative of the Ministry of Industry). The objectives of these interviews have primarily been complementary, to fill gaps and fill a function in a process of data-triangulation. Four of the interviews with program managers at VINNOVA have been granted anonymity and are only indirectly referred to here. Finally, the study also relies on secondary material, not least to provide the context of events, including research literature and the government-initiated evaluation of VINNOVA performed in 2008.

This chapter is structured as follows. The next section outlines the theoretical starting points of the analysis, focusing on theses about the changing role and legitimacy of state research funding agencies, and the role of new ideas in this process. The following section presents the results of the case study. As a background, I begin with a sketch of the historical development of the state apparatus of Swedish technology policy; that is, the institutional legacy inherited by the innovation policy agency. I then analyze the processes through which VINNOVA was inaugurated and how it designed and developed its mission. I also provide some examples of how the agency has dealt with its mission through different initiatives and forms of implementation, e.g. in the field of regional policy. Finally, I present a theoretically informed analysis of what the case tells us about the tension between dealing with the challenge of developing new capacities (based on new ideas of steering), on the one hand, and the need to uphold legitimacy and support in public research funding bodies, on the other.

The changing role of research funding agencies

Theories of governmental agencies are primarily focused on the relationship between politics and bureaucracy, a relationship that is considered to be characterized by strong tensions, especially in a democratic political system (see e.g. Meier and O'Toole, 2006). On the one hand, bureaucracy can be seen as a government tool with which to accomplish goals and exercise coercion. On the other hand, the expertise and stability of a bureaucratic organization tend to make it hard to steer and control. In this chapter, the focus is on a specific kind of governmental agency, the research funding agency, which may be characterized as an "intermediary agency" mediating between different "worlds," mainly politics and science, but in the context of this case is analyzed as an intermediary among a multitude of actors.

While research funding bodies share many characteristics with other types of governmental agencies, scholars in the field of science policy studies have argued, on the basis of principal-agent theory, that the policy context of these agencies, research policy, is characterized by its markedly strong imbalance, asymmetry, between political (the principals) and scientific actors (the agents). In the terminology of the principal-agent literature, such asymmetries can be expressed as marked problems of adverse selection and moral hazard. The adverse selection problem means that the politicians usually lack the knowledge to pick the most competent agent; the moral hazard problem that the politicians usually have problems evaluating the results of the work of the scientists (thus a monitoring problem). Although this model is simplistic, it points to important circumstances in the relationship between politics and research, which relates to the conditions of research funding bodies. In principal-agent literature, research-funding agencies have specifically important roles in managing or mediating this relationship (Braun and Guston, 2003). Dietmar Braun's discussion of the issue treats funding bodies as the intermediary institutions in a "triadic structure" (Braun, 1993). Thus, the funding body is, in this perspective, both an agent to the government (or the politician), and a principal in relation to the researchers (Braun and Guston, 2003).

However, as shown by many studies, the role of the intermediary agency varies, partly depending on the strength of political demands and the position of the scientific community, as well as on the capacity and strategic capability of the intermediary agency to balance the relationship. Hence, the "configurations" among principal, intermediary and agent vary. Barend van der Meulen suggests that the specific form of different configurations depends, for example, on the interests of different actors and on the different structural characteristics: the possibilities of principals finding other agents, and vice versa, and the existence of alternative mediation structures (van der Meulen, 2003). The different types of configuration can be illustrated by different "ideal-type intermediaries" of the twentieth century. The traditional research council, for example, may be seen as an expression of the Bush model – a model based on trust between politics and science and self-steering of science. The councils identify themselves strongly with the scientific community, and the monitoring is organized by peer-review processes. The mission-oriented agency may be seen as an intermediary, more driven by the interest of policy objectives, although, as shown by Braun, still strongly dependent on the interest and support of the scientific community (Braun, 1993; Benner and Sandström, 2000).

What are the implications of new innovation policy ideas and challenges of governance for the role and strategies of research funding agencies? As mentioned earlier, the new discourse on science and innovation, implying more interaction with users and steering through networks, implies new roles for intermediaries. One implication is that the configuration of actors may be considered more complex, involving several actors and possible positions for the funding agency. If, for example, a new actor (e.g. a private company) enters the configuration, this actor can function as an extra principal (e.g. funding research at the agency), and be allied to the agency (e.g. the researchers) (van der Meulen, 2003: 326). However, the challenges of the new discourse also include aspects where the principal-agent model seems to be insufficient. The principal-agent perspectives' "clear-cut" divide seems to underrate the importance of context and the co-production of knowledge and ideas. The development of the "innovation agency" indicates a development towards a "broader goal boundary organization," not only acting towards the research community (Klerkx and Leeuwis, 2008). The more idea-oriented organization also resembles what the organization theorist Henry Mintzberg has called a "missionary organization" (Mintzberg, 1979). This type of organization is characterized by a common ideology and professional norms, rather than governed by universal rules or stable relations to certain interests. It is furthermore "change-oriented" rather that predictable, and actively focused on "altering the norms and behavior of outsiders rather than responding to them" (Rothstein, 1996: 50–51).

My understanding of the process of establishment of the new agency is, furthermore, inspired by a historical institutionalist perspective of public policy change (Thelen and Steinmo, 1992). In this case, it implies that the establishment of a new idea-oriented agency is not clear-cut, but shaped in a dynamic process, not only by new ideas and challenges, but also the interest of actors and the institutional context. In the words of Desmond King, "ideas must be translated into languages and slogans appropriate for political decision-making" (King, 1992: 242). Ideas not only provide concrete prescriptions for policy action; they are also used by policy-makers and other elites to create legitimacy for different policy solutions (Campbell, 1998). The development of new state agencies is, furthermore, a process influenced by new challenges and ideas, path dependencies and institutional legacies. These concepts, central to historical institutional analysis, are all trying to express the ideas that new policies are often reactions to, or consequences of, old policies (Hall and Taylor, 1996). Bo Rothstein (1992), for example, has shown that politicians, when they formulate proposals and reforms, largely take the existing administrative capacity of the state as their starting point. Therefore, this capacity is in itself an important explanation of changes in politics and society.

In line with this reasoning, I would also argue that the establishment of new agencies may have "path creating" consequences, thus also putting policy into new trajectories, not least in policy areas where the political interest in steering is often limited or varied. The institutionalization of new ideas in this case is closely linked to the logic of governmental agencies, which are not only instruments of the state; they also eventually become "actors in their own right," seeking to enhance an institutional interest, and to secure survival and legitimacy among relevant actors in their environment (see e.g. Wilson, 1989). New ideas can play different roles in this process. For example, an administrative agency can "bolster its claim to resources or jurisdiction" over particular policy areas, by being seen to draw heavily on expert knowledge and new ideas, as well as to lend authority to agencies' particular policy positions, especially in highly contested policy areas (Boswell, 2009: 7). As pointed out earlier, ideas also seem increasingly important for the construction of identities and missions for public agencies, pointing to the importance of ideas in the construction of meaning (see e.g. March and Olsen, 1995).

Let us summarize: research funding agencies are characterized by their intermediary position not only between politics and science, but also increasingly between industry and other actors. The handling of the intermediary position can be assumed to become more complex, creating opportunities, but also tensions, and they operate more like intermediaries and spiders in the web than regulators. The entrance of new ideas in research policy presents a challenge to the more "simple" structure and logic of research funding bodies. The use of new ideas cannot be understood from an instrumental perspective alone. I argue that ideas interact with interests and institutions in a dynamic process. New agencies are created on the basis of old ones operating in fields that are reinvented, but build upon established issue networks and definitions of problems, and the creation of specific policies and practices over time become closely linked to the institutional interest of the new agency. Now, let us move on to the analysis of VINNOVA.

The making of an innovation agency: The case of VINNOVA

The aim of this section is to analyze the establishment of VINNOVA, and how it has developed its role and mission in the light of new ideas. After a short background, this process is described in three different phases of the establishment identified as important for the agency: the initiation of the agency (how did the decision-makers design the agency?), the creation of a mission and identity (how did the management develop its mission?), and finally the practice and implementation phase (how did the organization handle its mission?). Although these phases overlap, in my mind they still demonstrate how ideas and interests interact, although they may be based on, as will be shown, somewhat different logic.

The setting: Swedish agencies for research funding

The current structure of Swedish research funding agencies stretches back at least to the 1940s, when several research councils were established, some more focused on basic-oriented research and others on more mission-oriented research. As in many other countries, this was a formative moment in the establishment of two different modes of research funding. First, the research council model, characterized by a rather "passive" role, allocating research funding by means of peer-review procedures, mainly through a bottom-up process where researchers send in applications; a model mainly used for allocation of funding to basic research. Second, the mission-oriented type of agency, characterized by a more active role in supporting research of relevance for specific political objectives; a model used for allocating research funding in a more selective manner. This model was expanded in the 1970s into a more interventionist model of governance (Stevrin, 1978).

VINNOVA is the inheritor of a legacy from more mission-oriented agencies, mainly supporting technical R&D. The first of these was the Swedish Technical Research Council (TFR) in 1942. This was, in fact, the first governmental agency of this type mainly supporting R&D financially, but also assuming an overall responsibility for the development of technical research, taking new research initiatives and spreading research results. The Swedish National Board for Technical Development (STU), established in 1968, took over the role of TFR and other organizations created in the 1960s. The role of STU was broader than that of TFR, with a more emphasized industry policy role and more active interventionist role. In the early 1990s, STU merged with the Agency for Industrial Development (Industriverket) and the Energy Agency to form a broad industry policy organization, the Swedish National Board for Technical and Industrial development (NUTEK). Both STU and NUTEK were profiled as networking agencies, focusing in particular on mobilizing different actors, primarily the research community, for goals of industrial renewal (Weinberger, 1997).

The two major types of research funding bodies (research councils and mission-oriented agencies) were complemented by a third type in the 1990s: semi-public foundations for strategic research.[2] Of primary importance for technology policy was the creation of the Swedish Foundation for Strategic Research (SSF), with the goal of supporting "strategic research" (Benner, 2001). The role of SSF has partly overlapped that of earlier STU and NUTEK, but has been focused to a greater extent on fostering "a new spirit of excellence and competitiveness," mainly supporting larger research programs within areas of strategic interest to Swedish industry. While their "innovative" role in the Swedish system of research funding is unclear, their existence has clearly made the landscape of research funding bodies more complex (Benner and Sörlin, 2007).

Initiating a new agency: Expectations, ideas and interests

What, then, were the ideas and motives behind the construction of a new innovation agency in the late 1990s? The creation of VINNOVA was primarily the result of a struggle over the direction of Swedish research policy that culminated in the late 1990s. In 1998, a governmental commission (called Research 2000) proposed the abolishment of governmental sectorial or applied research agencies and a stronger focus on basic research. The general principle was that research funding should either be controlled by the research community (by universities themselves, or by research councils controlled by elected members of the research community) or by governmental agencies responsible for specific sectors and funding R&D for their own needs and activities. Specialized sectorial research funding agencies such as NUTEK or the Swedish Council for Building Research, which funded research and acted as representatives for the needs of industry or societal sectors, were to be abolished. This third category constituted a "blurred" type of governance, where the criteria for funding research, according to the investigation, were unclear (SOU, 1998).

The proposal, however, met strong resistance from many actors, not least many industrial organizations, including industry-oriented unions, the Industry Committee, consisting of representatives of industry and industry-oriented unions, and sectorial agencies such as NUTEK. The proposal also seemed too radical and focused too much on a university perspective for the Social Democratic government. The lack of support from the "industry policy coalition" also made the proposal almost impossible to implement politically. In the bill on the Research 2000 proposal, the government made it quite clear that the state also had to take responsibility for applied research of the kind supported by NUTEK (Benner, 2001). To come up with a more "realistic" proposal, the government therefore set up a working group under the auspices of the Ministry of Education. The working group suggested a new Research Council, merging the existing ones for basic research into a larger organization with more "muscle," and proposed creating two new mission-oriented research funding councils, consisting of parts of earlier sectorial research agencies and units. The new mission-oriented agencies were, unlike their predecessors, organized along much the same lines as the research councils for basic research, thus reflecting a strong influence by the research community (e.g. by using peer-review procedures in decisions on research grants). However, the investigation also proposed a new R&D agency for more needs-oriented research, mainly related to growth and industrial policy, but also connected to, for example, transport and infrastructure. The important point was that this organization was to have a different organizational structure to research councils, with a strong emphasis on societal relevance (Ds, 1999).

The proposal for a new R&D agency was developed in close collaboration with another investigation on the organizational structure of the industrial policy implementation organization. This investigation had been set up as a "parallel inquiry" to the working group analyzing the research funding system from an industry policy perspective. Formally, the mission of this investigation was to analyze how the activities of industrial policy agencies of relevance to the Swedish innovation system could become more effective and contribute more to economic growth and the development of society. It also suggested a new "R&D agency," but specified the ideas behind the agency by putting a stronger focus on its role in an innovation policy perspective, driven by explicit ideas about innovation and the development of industry and society. Interestingly, the proposals were largely inspired by the innovation and research policy of Finland, at the time considered a great innovation policy success (Flodström, 1999; Eklund, 2007).

Finally, the proposals for a new agency for R&D were presented in two governmental bills to parliament in the spring of 2000. In the bill dealing with the whole research funding organization, the new R&D agency was supposed to consist of parts of three sectorial agencies: the R&D-parts of NUTEK, The Council for Working Life Research and an agency funding transport and communication research. The objective was to create one strong governmental agency to support research of industrial and societal relevance. The motives for establishing the new agency were mostly of an administrative kind: an agency with stronger financial "muscle" would make it easier to focus resources and to set priorities, and merging three agencies would lead to fewer administrative costs. However, it was also pointed out that the growing need for innovation-based economic growth had made co-operation between sectors and disciplines increasingly important; something that the new agency would make it easier to accomplish (Prop. 1999/2000:81). The creation of the agency must also be seen in the light of a reorganization of the whole central administration for industrial and growth policy, as proposed in a bill formulated by the Ministry of Industry earlier that spring (Prop. 1999/2000:71).

The focus of the new R&D agency was further specified in the more comprehensive research bill that was presented in the autumn of 2000. The government proposed that the new agency should be named The Swedish Governmental Agency for Innovation Systems, VINNOVA. The agency's role was more or less described as it had been in the earlier decisions, but in this bill more emphasis was put on the role of innovations, innovation systems and innovation policy. These terms were also defined and discussed, even though it was not explicitly stated that the government had formulated an innovation policy, distinct from research policy (Prop. 2000/01:3). The naming of the new agency had been dealt with during the spring by an organization committee, and according to the member of the committee responsible for the R&D agency the coming director of VINNOVA, Per Eriksson, the intention was to follow the ideas of the investigation led by Anders Flodström, and to develop them in a dialogue with members of NUTEK and the research community (Eriksson, 2008). According to another central actor in this process, several names were considered, but the choice of innovation system was in line with "fashion" more than anything else (Sandberg, 2007). As pointed out by Magnus Eklund, the innovation system concept had also been used during the whole process of investigation, described above, to gather a coalition of agencies, industry and unions around the applied technical research and innovation policy. Eklund interprets this as an attempt to transfer legitimacy from the OECD, which had adopted the innovation system concept in the 1990s, to the project on saving sectorial research funding (Eklund, 2007: 143–5).

Seemingly, the expectations of the new agency were quite high. It seemed clear that the political leadership expected the creation of something new; the more specific objectives of the agency were to strengthen the knowledge bases of growth areas in the Swedish economy, promote the development of strong research and innovation environments, work for more effective use of R&D, and act as an expert agency contributing to the development of an effective innovation policy in Sweden. The new agency was mainly described in the context of industrial development and economic growth. Even if its role was still supposed to be to support research of relevance for society and industry, the focus in this bill was, to a larger extent, on the latter, e.g. through concentration on certain technology areas and collaboration with industry in sectorial R&D-programs (Prop. 2000/01:3). The expectations seem to have developed in close interplay with several actors, especially around a coalition of industry policy actors. The position of the political leadership seems, in reality, to have been less clearly expressed; for example, the interest of the leading politicians within research policy at this time was directed more towards the role of basic research and the new merged research council (Sandberg, 2007; Östros, 2007).

Establishing an organization: Forging a mission and identity

How, then, did the leadership of the new agency handle this rather broad mandate, and unclear expectations? In an early stage of the process, it was quite clear that the intention of the leadership of the agency was to create something new, separating the organization from its predecessors and other research funding actors in the Swedish system. The main idea was to position VINNOVA as an agency focused not on classical sectors or specific actors, but on collaboration between different types of actors and a concentration on effects, innovation and growth (VINNOVA, 2002). The annual report of 2002 stated that the agency was supposed to be an actor working in a modern and trend-shaping mode. The agency was also described as an "expansive agency" that wanted to be visionary, active and future oriented (VINNOVA, 2003a). Interestingly, the management and the board went on a study visit to the Finnish Technology agency (TEKES) in its first year. The board members and management were highly impressed, and thought the Finnish agency a good example for VINNOVA to follow (Minutes of the board, April 2001).

Although the use of social science knowledge in formulating policy had also been present in the "old NUTEK" and STU (see above), innovation concepts were more explicitly used in the new organization. In its strategy of 2003, VINNOVA pointed out that its way of working was based on "an integrated system of innovation perspective," and that its working management practice was expressly inspired by the triple helix concept, meaning a strong focus on the importance of the interaction of university, industry and public sector (VINNOVA, 2003b). The R&D programs of VINNOVA became more streamlined than before; they were designed to follow a certain rationale or logic in that they were all motivated by and presented within a narrative based on systems of innovation, triple helix, and cluster. The logic of the agency followed the recommendations of researchers within the field of innovation systems, emphasizing the identification of important actors in the innovation system and development of new programs to remove obstacles and develop possibilities (Jacob, 2006). By relating to these concepts, the agency also related the organizational idea to an international knowledge community.

Nonetheless, how these concepts should be interpreted was not always obvious to the staff of the agency. The staff was partly new, but a large part of it had been taken over from the predecessors of VINNOVA, with different traditions and routines when it came to the management of R&D programs (Minutes of the board; Interviews with Program managers at VINNOVA). Several initiatives were therefore taken to make the identity of the agency more visible internally. The document "This is VINNOVA, what we want and where we want to go" (my translation) was produced at the beginning of 2003. According to the foreword, the purpose of the document was to "work as a compass" to the staff of the agency, in internal as well as in external relations. The document described, in relative detail, the visions and goals of the agency, as well as central concepts and starting points for what was called the "idea basis of VINNOVA." It also provided an interpretation of the history of the agency, and can be seen as an effort not only to shape the identity of the agency, but also to spread certain beliefs about the world (prescribing the avoidance of concepts such as "applied research," as well as claiming that it is not ethical to commercialize research, etc.) (VINNOVA, 2003c).

VINNOVA's leadership has actively put a lot of effort and resources into building a strong network and making "allies" in the political system (Eriksson, 2008). As pointed out earlier, the neo-corporatist arrangements in public administration were abolished in the early 1990s. The point of the earlier arrangements had been, at least partly, to acquire a strong legitimacy for steering, through representation of those whose interests were affected in the implementing organization (Rothstein, 1992). Although the influence of the VINNOVA board, as with most other governmental agencies, has become more restricted, it has been important in achieving legitimacy for VINNOVA's organizational mission, according to interviews and written sources. To collaborate closely with other actors can be interpreted as a natural consequence of the organizational idea and implementation strategy of the agency. The emphasis on triple helix as a starting point for R&D programs has led to a strategy of "network steering" and to a need for the inclusion of different interests and competences to participate in discussion about the design of strategies and R&D programs, reviews of applications, etc. In particular, the agency has tried to create support for its organizational idea in the business community and among industrial unions. The industrial committee, for example, has been an important "discussing partner" for the agency, but the board of VINNOVA has also been important in channeling contacts between VINNOVA and society (Minutes of the board, April 2006; Eriksson, 2008).

The importance of communication issues was already discussed during the preparatory work of the organization committee. For example, one of the first tasks of the new agency was to establish a department for communication that would guide the work on forming a communication policy for the new agency. The General Director emphasized that the objective of the new policy would not only be traditional government information:

> The overall objective of the communication task for VINNOVA is much greater than only to introduce the agency to the environment and an interested public. A major challenge is to achieve a wide acceptance of the opinion that research (within the responsibility of VINNOVA) is an investment that yields profit.
>
> (Minutes of the board, December 2001, my translation)

The planning of the new department made it clear that the starting point would be to "establish the brand" of VINNOVA "and to profile the agency as a central actor in supporting innovation systems and need-oriented research (Minutes of the board, December 2001). A memo about the external communication of VINOVA also stressed that external and internal communication had to be coordinated and that "all of the staff at VINNOVA must realize that they play an important role in external communication" (Minutes of the board, October 2001). During the first year of the new agency, the board often discussed whether VINNOVA should be more visible in the environment and more active in, for example, publishing articles in newspapers. One expression of the importance that the agency placed on external communication was the initiation of an interview survey to measure how VINNOVA was conceived by relevant groups in the environment of the agency (it was carried out for the first time in 2001; see Minutes of the board, September 2001).

In parallel and interplay with the establishment and activities of VINNOVA, the government took several decisions that were eventually packaged as a "new innovation policy." In the governmental bill *R&D and co-operation in the innovation system*, presented in the fall of 2001, the concepts of "innovation policy" and "innovation system" formed the framework for the discussion. The bill, drafted in the Ministry of Industry, formulated the basis for a Swedish innovation policy as "a tool to enable an effective innovation system" mainly by instruments of coordination and collaboration among different actors (Prop. 2001/02:2). Although the focus on innovation policy in the bill was partly a way of "packaging" rather than a strong political statement, the government – surely influenced largely by the rhetoric around the creation of VINNOVA – seems to have been inclined to formulate a more fully articulated innovation policy at the turn of the millennium. Later on, the government took other initiatives to strengthen an innovation perspective, e.g. through the formulation of an innovation policy strategy (Ds, 2004). The main concrete results of this "new innovation policy offensive" were the negotiations with representatives of sectors, which were initiated by the Swedish government in 2004 to strengthen R&D and industrial renewal in "key sectors" of the Swedish economy (Östros, 2007). Obviously, these initiatives were strongly triggered by the upcoming election; the formulation of an innovation policy thus partly functioned as a strategy for the Social Democrats to show that they were capable of dealing with these issues in the future (Benner, 2008: 334–6).

The stronger formulation of innovation policy on a political level was naturally embraced and supported by VINNOVA; this development was very much in line with the mission of the new agency (Eriksson, 2008). During its first years, the agency was very active in arguing for a more comprehensive innovation policy, with many documents pointing to the weak capacity for coordination in the Swedish government (see e.g. VINNOVA, 2003b). The agency also saw its role as that of spreading knowledge about the systems of innovation perspective, partly as a consequence of policy activities connected to the OECD and the EU. In some VINNOVA documents the acceptance of the concept in the Swedish innovation and research policy network was described in a deterministic manner, as just a matter of time (see e.g. Granat Thorslund *et al.*, 2005). In its annual report of 2006, the agency also pointed out that it had been successful in playing a major part in innovation policy, and had managed to get other actors to act in a similar fashion (VINNOVA, 2007). At the same time, the minutes of the board showed that the leadership of VINNOVA felt that the position of the new agency and its mission were threatened, and encouraged members of the board (representing different interests) to work more actively towards strengthening the network of the agency (see e.g. Minutes of the board, February 2005).

The new center-right government, which came into power after the election in 2006, had been critical of VINNOVA while in opposition, and had even proposed abolishing the agency. Now in power, the center-right "alliance" has shown a stronger interest in VINNOVA and innovation policy issues, and not implemented any major changes (Benner, 2008: 336). The management of VINNOVA also put a lot of effort into presenting the idea of VINNOVA to the opposition before the election (Minutes of the board, April 2006). The legitimacy of the mission of VINNOVA was questioned in connection with a government-initiated evaluation of the research funding organization in spring 2008. The evaluation expressed severe criticism against several of the new agencies, and proposed a merger of the larger ones into a large research council. According to the evaluators, VINNOVA was successful in getting support for its organizational idea (they described VINNOVA as an "idea-oriented, relation-concentrated center for Swedish innovation policy with relatively strong resources"), but was less successful in implementing and evaluating its programs in accordance with the formulated steering strategy. New programs were not designed on the basis of system-oriented analysis, not systematically evaluated, and the evaluations were not used in the "rational" way prescribed by VINNOVA's objectives. VINNOVA, like the other research funding agencies, was also criticized for not putting more effort into collaborating with other agencies, and in that way enhancing the prospects for a more effective use of resources (SOU, 2008: 89–92).

VINNOVA reacted strongly against the conclusions of the evaluation. In its formal response, the agency rejected the merger-proposal, stating that most of the critical remarks on the working method of VINNOVA were based on irrelevant information. It then made an alternative proposal where the main idea was to merge the existing funding bodies into two organizations: one for basic research and one for innovation and need-oriented research (VINNOVA, 2008). Several other policy actors also put forward alternative proposals similar to VINNOVA's, which was partly due to a joint investigation by VINNOVA in collaboration with, among others, the Royal Academy of Engineering Science (IVA), one of the supporters of the agency (see the discussion in Benner, 2008). The overall interpretation is hence that support for VINNOVA, at least from business and industry, seemed strong. However, attention was directed to the problem of evaluation; the organization model of VINNOVA promised not only good research, but also results in the form of measurable effects. Even though VINNOVA, prompted by the Ministry, put considerable effort into strengthening its capacity in this area, it was a notoriously difficult task (VINNOVA, 2006).

To summarize, the leadership of VINNOVA devoted a lot of effort to constructing a mission based on the concepts of innovation systems and triple helix. It is quite clear that this reflects a continuation (learning) of an idea legacy from its predecessors, and a way to give the new agency a profile, distinctive from other agencies, thereby strengthening the possibility of holding an originally rather broad agency together and enhancing organizational survival.

Implementing an idea-oriented mission

How has VINNOVA implemented its objectives in practice? According to the principal-agent inspired literature on research funding bodies, intermediary agencies such as VINNOVA are strongly dependent on "agents," i.e. primarily researchers and different industrial actors. How was this structural condition dealt with in relation to new ideas and challenges of the innovation paradigm? The main impression is that VINNOVA has been active in developing new initiatives to interpret its role, at the same time as strategically adapting to different policies and demands from the government. Apparently, successful initiatives have been "Research and grow" (Forska och väx), funding R&D in small- and medium-sized companies, and programs related to specific regional, industrial and labor policy needs (see below). VINNOVA has also functioned as a "fixer" for the government, e.g. in the crisis in the vehicle industry and in problems concerning IT and risk capital (Eriksson, 2008). The size of the agency has also increased from about 150 employees and a budget of €100 million at the beginning of the century (VINNOVA, 2002) to about 200 employees and a budget of €220 million ten years later (VINNOVA, 2010).

A typical example of the willingness of innovation and entrepreneurship in the new agency was the VINNVÄXT initiative (vividly described and analyzed in other chapters of this book). When the agency was established, one of the first steps of the leadership was to establish a new department for "actor-oriented initiatives," focusing not on specific sectors or technology areas, but different types of actors. One of the first initiatives within the unit, inspired by the General Director, Per Eriksson, was a program focusing on regional systems of innovation, the "Regional growth for dynamic innovation systems program" (VINNVÄXT), which was launched in 2001 as a way to implement regional development policy by supporting regional partnerships selected according to a "competition" procedure. VINNVÄXT can be characterized as typical of the mobilizing and collaboration state, as described in the discussion of governance. The program was set up to encourage "self organization" and provided visions and conceptual tools for the actors involved. Furthermore, the program provided an alternative to a more traditional redistributive regional policy (Laestadius and Ylienenpää, 2007).

The VINNVÄXT program was explicitly inspired by ideas of innovation systems and triple helix relations, focusing on supporting collaborations among different types of societal actors and using ambitious systems of monitoring and evaluation (Christensen *et al.*, 2006; Laestadius and Ylinenpää, 2007). The program was, to a large extent, the product of a learning process; it was inspired by social science knowledge of regional innovation systems and clusters, the experiences of its General Director, programs from other countries and VINNOVA's predecessor NUTEK Technology (Christensen *et al.*, 2006). However, it is also obvious that the VINNVÄXT program was a way for the new agency to "receive the ownership" of the issue of innovation systems (to quote a VINNOVA initiated report about the program). The policy innovation VINNVÄXT was obviously a program that included both "puzzling and powering" (Heclo, 1974: 305).

The agency has been active in stretching the boundaries of its mission. One expression of this has been a strong willingness to collaborate with other agencies and broaden its role to that of a more "network-oriented" agency. The objective has partly been to influence other actors in the system and achieve "more value for the money." For example, in its research strategy from 2003, VINNOVA stated that with increased funding it could influence other actors by collaboration and co-funding and thereby function as an "agent of change" (VINNOVA, 2003b: 32). The agency has also shown a strong willingness to expand in new directions. For example, it has been active in supporting new areas such as academic entrepreneurship (see Benner, 2008) and areas in other policy domains, such as the regional policy-oriented VINNVÄXT-initiative described above. The somewhat expansive and entrepreneurial style of the agency has even led to some tensions with actors in neighboring policy areas, e.g. conflicts over turf and fuzzy borders of responsibilities. The former General Director Per Eriksson has stated, for example, that VINNVÄXT and have led to tensions and that the agency has made "enemies" (Eriksson, 2008).

Regardless of the many new initiatives, it is also important to emphasize that a large share of the activities of VINNOVA – as discussed above – continued along the same path as its predecessors, the "old" NUTEK and STU. The support to strong research environments, such as VINN centers of excellence, was more or less new versions of earlier programs such as the material consortium and the competence centers program, although somewhat differently motivated and "framed" (Hjorth, 2006). The focus on co-funding, usually 50 percent demanded from industry, or in a few cases the public sector, was a general rule, with exceptions, that had already been established during the NUTEK period (see above). In the early 1990s, co-funding arrangements were introduced, primarily as a way to manage a lower budget by sharing the burden with industry, but later on, in the context of the new innovation policy, these arrangements were legitimized by the need for collaboration within an effective innovation system, or the importance of triple helix relations (Persson, 2001). The difference is often claimed to be that VINNOVA, in accordance with a triple helix view on the innovation process, also includes funding from the public sector, but that this is probably partly a consequence of the fact that VINNOVA, unlike NUTEK, is also responsible for "softer" areas such as part of the working life research sector.

How, then, has the rather innovative character of the agency been integrated with more specific demands and relations on the levels of single programs? According to interviews with program managers at VINNOVA, the use of innovation systems analysis has become more complicated than originally expected, which seems to have led to a more pragmatic view on initiatives for new programs. The program managers need to deal with complex technology areas, with varying types of actor configurations. In particular, it seems that the usefulness of the new concept for the actual design of specific instruments has been difficult for the VINNOVA management to communicate to actors addressed by or involved in VINNOVA programs (Minutes of the board, June 2006). According to the interviews, the concepts have been most important internally, to make sense of initiatives and create a logic for the program. The decision-making rationale of basing the design of new programs on an analysis of the relevant system of innovation has not always been applied systematically (see e.g. the analysis in SOU, 2008). Obviously, the development varies depending on technology area, sector, etc., and it might take more time to implement new ideas in a new organization, but the above observations still indicate that the new concepts have not provided the policy-makers with a "clear and specific course of policy action," in the wording of Campbell (1998: 385). The interviews seem to suggest that the agency has partly reconsidered the somewhat streamlined use of the concepts in the design of R&D programs (Marklund, 2007; Eriksson, 2008).

As was pointed out earlier, the objective of VINNOVA has not only been to steer research programs, but also to take the responsibility for spreading ideas, a "vision," about the importance of a system of innovation view. This objective has been expressed in at least two ways. First, through certain more high-profile programs (apart from the more traditional R&D programs) such as VINNVÄXT, where ideas of TH and IS have been spread through, for example, courses, conferences and seminars connected to the program. The objective has been to create a common understanding, or at least a common platform of knowledge. Second, knowledge and research results have been spread through communication on a more general level, through reports, conferences, etc., often based on system of innovation and triple helix perspectives.

Conclusions

This chapter has tried to show how the Swedish policy system has dealt with the challenge of what has sometimes been called "a new innovation paradigm," through the establishment of a new governmental agency for innovation systems, VINNOVA. It has also shown how the process of initiating and forging a new agency in this field has been shaped by a complex process where more narrow interests of actors and organizations, new ideas, and the institutional context play roles. Interestingly, the creation of the agency was more an expression of a will to establish a "strong" and more forceful agency for technical research policy, rather than a clear expression of an innovation policy paradigm. The proposal of an "innovation agency" was one that was supported by a broad but temporary coalition of experts, bureaucrats and interest groups and came, so to speak, at the right moment in time. The innovation policy strategy of the Social Democratic government at the beginning of the new millennium also played an important role in strengthening VINNOVA's position.

The forging of the new agency can be understood as a blend of two processes: the willingness of the management to create something new, translate new ideas into programs, and "survive" in what was conceived as a rather unfriendly political environment. This is expressed in the forming of an identity of the new agency – a process that can be interpreted as a mix of idea-oriented and organizational survival. New concepts, such as systems of innovation and triple helix, have been frequently used to "frame" or give structure to arguments and proposals connected to innovation. These concepts thus seem to have been useful as tools to mobilize actors for a common project, not least within the "industrial policy network," pointing to a strategic use of ideas. The agency has, over time, become both an arena for and a representative of the new paradigm in the Swedish political context; picking up new ideas and concepts originated in an epistemic community of experts and bureaucrats. One typical example of the willingness of innovation and entrepreneurship in the new agency was a program focused on regional innovation. However, the development of the activities of the agency is, to a large extent, based on the institutional legacy of the policy area, and a more pragmatic view on research funding on the program manager level. As has been shown, VINNOVA has indeed added new instruments to the existing "toolbox," and old programs have been reframed to fit into a new context. The capacities of the state have thus been changed incrementally, through processes of layering and redirection of old arrangements. Furthermore, the establishment of VINNOVA, and the connotations of the naming of the agency, were important in triggering the subsequent development of a more coherent governmental innovation policy (at least efforts in that direction). VINNOVA can be characterized as a rather typical "modern" agency as described in the debate on governance; an agency largely focused on taking new initiatives, enabling actors' networking and "policy learning." It is also striking that the agency, to a large extent, has taken its role to be the representative or advocate of the new ideas (broadly speaking), separating it from more traditional research funding agencies, and seemingly making it an expression of the more visionary-oriented character of Swedish governmental agencies noted by other researchers.

The establishment of this innovation agency is an interesting experiment, but not without tensions and problems. It seems like the new idea paradigm has had consequences for the functioning of this agency, and been important in finding new solutions and translating new ideas into a specific national context. Obviously, new ideas have also played a role in obtaining legitimacy and support from a wider community of actors. Still, the new agency's role is also characterized by instability; to uphold support for this idea means unstable coalitions with constantly varying types of actors, depending on, at least formally, the needs identified through the analysis of the innovation system. Compared with the research council model, where the simple objective most often is "only" to deliver high quality research, the mission of an innovation agency also creates high expectations (from the political principals) to deliver more concrete results or effects, which, in this area, is extremely difficult to show. An expansive and entrepreneurial style can also lead to tensions with actors in neighboring policy areas, e.g. conflicts over turf and fuzzy borders of responsibilities. There is, thus, a tension between, on the one hand, the demands for flexibility and entrepreneurship and, on the other, for legitimacy and support to guarantee survival. This probably explains the importance given to building strong relations with an industry policy coalition and nurturing stable support for technical research, the safest constituency for an agency such as this.

Notes

1. The author is grateful for comments on earlier drafts of this chapter by Mats Benner, Henry Etzkowitz, Leif Hommen, Staffan Laestadius, Annika Rickne and Håkan Ylinenpää.
2. The reform was made possible due to the center-right-wing government's abolishment of the "wage earners." These funds were created a decade earlier by the Social Democratic government to increase the wage earners' influence over industrial production.

References

Benner, M. (2001) *Kontrovers och konsensus. Vetenskap och politik i svenskt 1990-tal*, Stockholm/Nora: SISTER/Nya Doxa.

Benner, M. (2008) *Kunskapsnation i kris? Politik, pengar och makt i svensk forskning*, Nora: Nya Doxa.

Benner, M. and Sandström, U. (2000) 'Institutionalizing the triple helix: Research funding and norms in the academic system', *Research Policy*, 29: 291–301.

Benner, M. and Sörlin, S. (2007) 'Shaping strategic research: Power, resources, and interests in Swedish research policy', *Minerva*, 45: 31–48.

Biegelbauer, P. and Borrás, S. (2003) 'Introduction', in P. S. Biegelbauer and S. Borrás (eds) *Innovation Policies in Europe and the US*, Aldershot: Ashgate.

Boswell, C. (2009) *The Political Uses of Expert Knowledge: Immigration Policy and Social Research*, Cambridge: Cambridge University Press.

Braun, D. (1993) 'Who governs funding agencies? The principal agent relationship in research policy making', *Journal of Public Policy*, 13: 135–62.

Braun, D. and Guston, D. (2003) 'Principal-agent theory and research policy: An introduction', *Science and Public Policy*, 30: 302–8.

Campbell, J. L. (1998) 'Institutional analysis and the role of ideas in political economy', *Theory and Society*, 27: 377–409.

Christensen, L., Hallencreutz, D. and Lundequist, P. (2006) 'Policylärande i och av VINNVÄXT', Stockholm: VINNOVA. Available: www.vinnova.se/upload/Policy l%c3%a4rande%20i%20och%20av%20VINNV%c3%84XT%20060516.pdf (accessed 15 August 2011).

Ds 1999:68 *Att finansiera forskning och utveckling: rapport från Arbetsgruppen för fortsatt beredning av myndighetsstruktur för forskningsfinansiering*, Government report, Stockholm: Ministry of Education.

Ds 2004:63 *Innovativa Sverige: en strategi för tillväxt genom förnyelse*, Government report, Stockholm: Ministry of Industry, Employment and Communication/ Ministry of Education.

Edquist, C. (1997) *Systems of Innovation: Technologies, Institutions, and Organizations*, London: Pinter.

Eklund, M. (2007) *Adoption of the Innovation System Concept in Sweden*, Uppsala, Sweden: Uppsala University, Department of Economic History.

Eriksson, Per, General Director of VINNOVA. Interview, Lund, Sweden, 19 June 2008.

Etzkowitz, H. and Leydesdorff, L. (eds) (1997) *Universities and the Global Knowledge Economy: A Triple Helix of University–Industry–Government Relations*, London: Pinter.

Fagerberg, J., Movery, D. C. and Nelson, R. R. (2005) *The Oxford Handbook of Innovation*, Oxford: Oxford University Press.

Flodström, A. (1999) *Utredning om vissa myndigheter*, Stockholm: Ministry of Industry.

George, A. L. and Bennett, A. (2004) *Case Studies and Theory Development in the Social Sciences*, Cambridge, MA: MIT Press.

Granat Thorslund, J., Elg, L. and Sandgren, P. (2005) *En lärande innovationspolitik – samordning och samverkan?* Stockholm: VINNOVA (VINNOVA analys 2005:07).

Haas, P. (1992) 'Introduction: Epistemic communities and international policy coordination', *International Organization*, 46: 1–35.

Hall, P. and Taylor, R. (1996) 'Political science and the three new institutionalisms', *Political Studies*, 1–23.

Heclo, H. (1974) *Modern Social Politics in Britain and Sweden: From Relief to Income Maintenance*, New Haven, CT: Yale University Press.

Hjorth, Staffan, programme manager at VINNOVA. Interview, Stockholm, 30 November 2006.

Jacob, M. (2006) 'Utilization of social science knowledge in science policy: Systems of innovation, triple helix and VINNOVA', *Social Science Information*, 45: 431–62.

Jessop, B. (2002) *The Future of the Capitalist State*, Cambridge: Polity Press.

King, D. S. (1992) 'The establishment of work-welfare programs in the United States and Britain: Politics ideas and institutions', in S. Steinmo, K. Thelen and F. Longstreth (eds) *Structuring Politics: Historical institutionalism in Comparative Analysis*, New York: Cambridge University Press.

Klerkx, L. and Leeuwis, C. (2008) 'Delegation of authority in research funding to networks: Experiences with a multiple goal boundary organization', *Science and Public Policy*, 35: 183–96.

Laestadius, S. and Ylinenpää, H. (2007) 'Att utveckla VINNVÄXTprogrammet – erfarenheter och policyimplikationer', in S. Laestadius, C. Nuur and H. Ylinenpää (eds) *Regional växtkraft i en global ekonomi*, Stockholm: Santérus förlag.

Lindvall, J. and Rothstein, B. (2006) 'Sweden: The fall of the strong state', *Scandinavian Political Studies*, 29: 47–63.

Lundvall, B.-Å. and Borrás, S. (2005) 'Science, technology, and innovation policy', in J. Fagerberg, D. C. Movey and R. R. Nelson (eds) *The Oxford Handbook of Innovation*, New York: Oxford University Press.

March, J. G. and Olsen, J. P. (1995) *Democratic Governance*, New York: Free Press.

Marklund, Göran, Director of the Strategy Development Division at VINNOVA. Interview, Stockholm, 20 August 2007.

Meier, K. and O'Toole, L. J. (2006) *Bureaucracy in a Democratic State: A Governance Perspective*, Baltimore, MD: John Hopkins.

Mintzberg, H. (1979) *The Structuring of Organizations*, Englewood Cliffs, NJ: Prentice-Hall.

Minutes of the board 2001–6, Stockholm: VINNOVA.

Östros, Thomas, Member of the Swedish Parliament and former Minister of Education (1998–2004) and Minister Industry (2004–6). Interview, Stockholm, 24 January 2007.

Pelkonen, A. (2008) *The Finnish Competition State and Entrepreneurial Policies in the Helsinki Region*, Helsinki: University of Helsinki.

Persson, B. (2001) *Motsträviga myndigheter: Sektorsforskning och politisk styrning under 1980-talet*, Linköping/Stockholm: Tema Teknik och social förändring/SISTER.

Pierre, J. and Peters, B. G. (2000) *Governance, Politics and the State*, Basingstoke: Palgrave Macmillan.

Prop. 1999/2000:71 *Vissa organisationsfrågor inom näringspolitiken*, Government bill.

Prop. 1999/2000:81 *Forskning för framtiden: en ny organisation för forskningsfinansiering*, Government bill.

Prop. 2000/01:3 *Forskning och förnyelse*, Government bill.

Prop. 2001/02:2 *FoU och samverkan i innovationssystemet*, Government bill.

Rothstein, B. (1992) *Den korporativa staten*, Stockholm: Norstedts.

Rothstein, B. (1996) *The Social Democratic State: The Swedish Model and the Bureaucratic Problem of Social Reforms*, Pittsburgh, PA: University of Pittsburgh Press.

Sandberg, Olof, senior official at the Ministry of Industry. Interview, Stockholm, 24 January 2007.

SOU (1998) *Forskningspolitik*, Government report, Stockholm: Ministry of Education.

SOU (2008) *Forskningsfinansiering: kvalitet och relevans*, Government report, Stockholm: Ministry of Education.

Stenlås, N. (2006) 'Konkurrensstaten tågar in: epokskifte eller mest bara prat?' in T. Lundqvist (ed.) *Den kreativa staten: Framtidspolitiska tendenser*, Stockholm: Institutet för Framtidsstudier.

Stevrin, P. (1978) *Den samhällsstyrda forskningen: en samhällsorganisatorisk studie av den sektoriella forskningspolitikens framväxt och tillämpning i Sverige*, Stockholm: Liber förlag.

Thelen, K. and Steinmo, S. (1992) 'Historical institutionalism in comparative politics', in S. Steinmo, K. Thelen and F. Longstreth (eds) *Structuring Politics: Historical Institutionalism in Comparative Analysis*, New York: Cambridge University Press.

van der Meulen, B. J. R. (2003) 'New roles and strategies of a research council: Intermediation of the principal-agent relationship', *Science and Public Policy*, 30: 323–36.

Weinberger, H. (1997) *Nätverksentreprenören: En historia om teknisk forskning från den Malmska utredningen till Styrelsen för teknisk utveckling*, Stockholm: Kungliga tekniska högskolan, Avdelningen för teknik-och vetenskapshistoria.

Wilson, J. Q. (1989) *Bureaucracy: What Government Agencies Do and Why They Do It*, New York: Basic Books.

VINNOVA (2002) *Årsredovisning 2001*, Stockholm: VINNOVA.

VINNOVA (2003a) *Årsredovisning 2002*, Stockholm: VINNOVA.

VINNOVA (2003b) *VINNOVA:s forskningsstrategi. Strategi för hållbar tillväxt*, Stockholm: VINNOVA.

VINNOVA (2003c) 'Detta är VINNOVA, vad vi står för och vart vi vill', unpublished report, Stockholm: VINNOVA.

VINNOVA (2006) 'Strategi för ambitionshöjning av VINNOVAs effektanalyser', unpublished report, Stockholm: VINNOVA.

VINNOVA (2007) *Årsredovisning 2006*, Stockholm: VINNOVA.

VINNOVA (2008) 'VINNOVAs yttrande över Forskningsfinansiering: kvalitet och relevans (SOU 2008:30)', Stockholm: VINNOVA. Available: www.vinnova.se/sv/Om VINNOVA/Remissvar/2008/Betankandet-Forskningsfinansiering---Kvalitet-och-relevans-SOU-200830/ (accessed 15 August 2011).

VINNOVA (2010) *Årsredovisning 2010*, Stockholm: VINNOVA.

9 Policy scripts and practice[1]

Fredrik Lavén, Björn Remneland-Wikhamn and Torbjörn Stjernberg

Introduction

Much attention is paid to theories and models for innovation, such as *clusters* (Porter, 1998), *innovation systems* (Freeman, 1987; Nelson, 1993), *networks* (Simmie, 1997), *communities* (Wenger, 1998) and *triple helix* interaction (Etzkowitz, 2002). These theories and models emphasize how innovation occurs in inter-organizational and collaborative networks. From a political point of view, such theories and models have become important instruments used to stimulate regional economic growth as policy-makers combine and translate innovation theories into innovation policies aimed at encouraging innovative, inter-organizational arrangements. There have been numerous governmental programs[2] intended to support local and regional competitiveness by fostering knowledge sharing and cross-disciplinary collaboration among various organizations. The hi-tech clusters in Silicon Valley are often held up as role models for innovation (Saxenian, 1990).

In this attention on policies as stimulants to inter-organizational innovation, theoretically based polices may be seen as innovation recipes. Lavén (2008) refers to such recipes as action scripts of written or otherwise symbolic instructions for organizing innovation. This chapter explores what happens when innovation policy meets practice. The analysis describes the similarities and differences in the organizing of two innovation initiatives. This analysis is of particular interest since it shows the unexpected consequences (that are not necessarily negative) that may occur when policy is not followed as intended. The chapter accentuates the need for flexible policies that allow both policy guidance and local adaptation.

Purpose of the study

Because of our interest in politicians' and policy-makers' support of regional development through innovation programs, we decided to investigate what happens when policy meets practice. Traditional planning perspectives often depart from the assumption that, as long as an activity has been properly planned, its implementation will be unproblematic (Suchman, 1987/2007). However, plans, orders, artifacts, and other objects are translated as they travel in time and space (Latour, 1986; Czarniawska and Joerges, 1996; Czarniawska and Sevón, 1996). This translation means that that such entities are in the hands of people who may, as Latour states (1986: 266–267), "act in different ways, letting the token drop, or modifying it, or deflecting it, or betraying it, or adding to it, or appropriating it." Similarly, Van de Ven (1986) notes that innovations, defined as new ideas, may be carried, developed, reacted to or modified by people, depending on their skills, frames of reference and energy levels. In this chapter we explore what happens as the innovation polices are acted upon. Our research question is: *How are policies translated into practice in organizing innovation initiatives?*

To answer this question, we examined the VINNVÄXT policy program for regional innovation systems, initiated by the Swedish agency for innovation systems (VINNOVA), as it relates to two innovation initiatives in Western Sweden (see Chapter 1 in this book for an overview of VINNOVA and the VINNVÄXT program). Our inquiry builds on a previous study (Lavén, 2008) of how innovation policies in the VINNVÄXT program and in the similar program – "National program for innovation systems and clusters" (VISANU) – were translated into practice in an initiative focused on collaborative development of microwave technology in Western Sweden. In this chapter, we compare the findings of that study with our findings from an innovation initiative in the biomedical field, also in Western Sweden.

We next discuss innovation policies and their relationship to practice and introduce the concepts "scripts" and "editing" in order to apply our perspective on policy as high level plans and practices. We use the VINNVÄXT program since it typifies the Swedish effort to develop regional innovation systems and since the two innovation initiatives of this study applied to that program for funding, one successfully, the other not. The method section that follows explains how we collected our empirical data. In the findings section, we compare the two innovation initiatives and analyze their relationship to the policy scripts. In conclusion, we discuss our findings and suggest some implications for policy-makers.

On innovation policies, scripts and practice

A policy is a set of decision-making principles, often described as high level plans for achieving general goals. For example, Fischer (2003: 60) defines a policy as a "political agreement on a course of action (or inaction) designed to resolve or mitigate problems on the political agenda." Policy-making may be understood as *agenda-setting* where the intent is to steer action in specific directions. Governments and agencies often formulate policies intended to promote economic development through innovation programs.

An example of such a program is VINNVÄXT, launched by VINNOVA in 2003, that was aimed at the development of regional innovation systems initiatives in Sweden. The VINNVÄXT program was based on theories of innovation systems (Freeman, 1987; Nelson, 1993) and the triple helix model

(Etzkowitz, 2002), all of which emphasize the importance of interaction in innovation settings, in particular the interaction among academia, industry and the government. In its creation of this program, VINNVOA imitated the OECD's national innovation systems policy recommendations and innovation policy efforts in Finland. VINNVÄXT's purpose statement is key to understanding its high level plan, agenda-setting, and its plan for governance of innovation activities (VINNVÄXT program description, 2002: 4):

> To promote sustainable growth in the regions based on international competitive ability by successively developing or further developing the functioning, dynamics and effectiveness of innovation systems in functional regions at an international level. A prerequisite for the program is the active participation of players from the business community, research organizations, politics, and public administration.

The program was set up as a competition for funding initiatives with innovation potential. The VINNVÄXT policy program also explicitly *prescribed* that a networked triple helix organization structure, with representatives from academia, industry and government, was necessary for such initiatives. VINNOVA's language supported these intentions: "key factors for regional growth" that include a focused, yet multi-disciplinary "shared vision," "high growth potentials," and "wide commercialization opportunities" were central to the policy documents. In addition, VINNOVA stated that an initiative's history of relationships would evidence "lingual and social closeness," as well as strong communities and an innovative climate. The implication is that such relationships, in which "future growth areas" were encouraged, require both a successful *past* and a promising *future*.

Policies as scripts for organizing

One may view VINNOVA's policy program as carrying a set of *scripts* for action. A script may be inscribed in, for example, a textual document or other object that gives explicit or implicit instructions for guiding or controlling action (Woolgar, 1991; Akrich, 1992; Latour, 1992; Joerges and Czarniawska, 1998). A script may thus specify the norms for action, impose authoritarian standards, and generally indicate what should be done rather than what has been done. Similarly, policies are often based on normative ideas about what is "right" and what is "working." Fischer (2003) argues that traditional policy research often views policies as *rational instruments* designed to achieve some desired result and ready for use in practice. We argue that such rational policy instruments may be viewed as scripts for how innovation initiatives should be developed, supported and organized. As such, the scripts become ordering devices (Suchman, 1987/2007) that are similar to plans and standards.

The VINNVÄXT program's policy scripts are directed at the funding applicants for use in the organization of their innovation initiatives. These policy scripts appear in VINNOVA's program announcement, in its numerous policy information documents, reports and websites, and in the activities it supports. As the VINNVÄXT program is organized as a competition with rules and with winners and losers (Wigren and Melin, 2007), it is not surprising that the program has policy scripts. However, we argue that there is still much to learn about how these scripts are translated into practice and how this translation affects the organization of innovation work. Do the innovation initiatives follow the scripts? Do they change or even ignore them? What consequences follow as a result of such actions?

Scripts may be edited in practice

Although a script may be consistent with Akrich and Latour's (1992) *programs-for-action* (i.e. action programs one actor wants others to pursue), actors do not always enact the scripts as the scriptwriters intended. This is also true for policy scripts. Fischer (2003), referring to Stone's (1988) work on policy paradoxes in political decision-making, suggests that policies are characterized by constant struggles over ideas. Thus, in considering policies as rational instruments solely from the sender's point of view, one takes a rather narrow perspective. Policy scripts, with their programs-for-action, may encounter *anti-programs* (Latour, 1992) when the script receivers follow or ignore the scripts. The receivers may even deconstruct a script into smaller components, perhaps with new configurations or combinations of multiple scripts that they then enact (Akrich, 1992). Such actions, as scripts are translated and changed according to local desires and beliefs (Czarniawska and Sevón, 1996), may sidetrack the sender's directions so that unintended outcomes result. How a script is enacted is critical to understanding its effects.

One way to discuss this process of translation is in terms of script *editing* (Sahlin-Andersson, 1996) where, for instance, scripts are assembled, rearranged and adapted to a standard or a purpose. Sahlin-Andersson (1996) uses the idea of editing to explain how success stories on science parks were circulated through adaptation and reformulation in new settings according to local conditions and visions of the future. Later in the chapter, we discuss editing in more detail.

Thus, if we are interested in the relationship between policy and practice, the outcome of enacting a policy cannot be taken for granted a priori; it can only be understood in relation to practice *ex post*. This means that we cannot assume that innovation policy leads to the development of organized innovation systems or clusters. Instead, when examining the translation of policy into practice, we need to analyze the arguments behind the policy and to examine how they meet practice in efforts of organizing innovation (i.e. how the policies influence action). In other words, policies must be evaluated in context.

Method for the study of innovation policy and practice

During the years 2003 to 2008, we studied a cluster initiative called Microwave Road (MWR) located in Western Sweden. At its founding, MWR applied for funding from various governmental agencies. It was an unsuccessful applicant in the VINNVÄXT 2003 funding competition but received some funding from VISANU, a program connected to VINNOVA. For our study of the initiative, we collected empirical data using an ethnographically inspired methodology consisting primarily of observations, interviews and document studies.[3] The findings indicated that the enactment of innovation theory-based scripts for inter-organizational arrangements led to a focus on structure rather than on innovation activities. The study showed that it was only when the policy scripts were translated and adapted to the local context of microwave engineering that innovation work began (Lavén, 2008).

To compare the findings furthering the MWR study to findings in another setting, we conducted a follow-up study with another innovation initiative. The second study was of an innovation initiative called Biomedical Development in West Sweden (BMV). The focus of this initiative is biomedical innovation. The two innovation initiatives had obvious similarities as far as their geographic location, organizational form, type of actors and some activities. However, they differed in terms of industry focus and funding resources; BMV was a biomedical initiative with VINNVÄXT funding and MWR was a microwave technology initiative without VINNVÄXT funding.

We collected data on BMV in interviews with three BMV representatives,[4] in meetings with researchers who documented the BMV initiative,[5] at a workshop we hosted with representatives from BMV and VINNOVA, and in the analysis of documents describing BMV and its activities. These documents included VINNVÄXT applications, brochures, website information and self-appraisals. Laage-Hellman and Rickne's (2007) chapter on the early development of BMV also provided useful data.

We then analyzed and compared the data from the two innovation initiatives. We examined their history and early development, their presentations and their activities. In short, our research design was to investigate who did what, when and where, as well as how and why. Given this design, we could also explore the two initiatives' relationships to the policy scripts from the VINNVÄXT program. Our analysis covers the five years from 2003, when the initiatives made their first VINNVÄXT applications, to early 2008.

Findings – comparing how policy meets practice in BMV and MWR

In this chapter, we structure our findings for the two initiatives according to the following topics: (1) their historic roots and early development; (2) their organizational presentations; (3) their stated purposes; (4) their activities; and (5) their organizing logics.

Historic roots and early development

Biomedicine West (BMV)

There is a long history of high-quality biomedical research in Gothenburg, Sweden. For example, the Sahlgrenska Academy at the University of Gothenburg conducts research in numerous medical fields, and AstraZeneca, with a legacy from, among others, the Nobel Prize laureate Arvid Carlsson, and Mölnlycke Healthcare are Gothenburg-based companies that conduct medical research and manufacture pharmaceutical products. However, in 2003, the Gothenburg region in Western Sweden was not internationally recognized as a prominent region with regard to research and development in the life sciences. In fact, according to Laage-Hellman and Rickne (2007), there were at least two other regional clusters in Sweden with higher recognition in biomedical research: the Medicon Valley cluster in Southern Sweden (and Denmark) and the UppsalaBio cluster in the Uppsala-Stockholm area. When VINNVÄXT was founded, a number of regional actors in the Gothenburg region saw the potential for increasing international recognition of their life sciences research and for promotion of the life sciences industry.

BMV applied for VINNVÄXT funding in 2003 after receiving planning support from VINNOVA in 2002. The BMV proposal was rejected in the first round in 2003. Although VINNOVA found the BMV proposal promising, it wanted clarification of the initiative's strategy and an increase in the initiative's support from the regional triple helix actors. Nevertheless, VINNVÄXT awarded some funding to BMV so that it could develop its application for the next competition in 2004. BMV became a "winner" in 2004 and received 60 million Swedish crowns (approximately €6 million), to be distributed over a 10-year period.

Microwave Road (MWR)

Microwave and telecom technologies have held a strong industrial and academic position in Western Sweden for a long time. The region is known for the presence of a number of large companies in the field, such as Ericsson and Saab, as well as numerous SMEs and research institutes. Chalmers University of Technology (hereafter, Chalmers) has collaborated on innovations with the microwave industry since the 1960s. In 2001, when the telecom industry was in financial crisis, a group of telecom industry researchers and managers began thinking about how the microwave technology industry could be strengthened. They argued that, since the network of companies was in place, only financial support was needed to begin developing new microwave technology and applications.

The MWR initiative submitted a funding application to VINNVÄXT for the 2003 competition. The initiative had received VINNOVA planning support to develop its application. This led to intensive enrolment work to mobilize the key actors. Nevertheless, the MWR initiative failed to win funding from VINNVÄXT in 2003. According to VINNOVA's rejection decision, MWR had not presented its growth potential specifically enough, its strategic idea was underdeveloped, and, perhaps of most interest, the public sector of its triple helix structure was too weak. However, when the MWR initiative was developed further, some funding was provided by VISANU (a program for developing innovation systems and clusters connected to VINNOVA). Although this funding was far less than the BMV initiative received in 2004, the funds received were enough to continue developing the MWR initiative.

Reflections

There are similarities as well as differences in the histories and funding applications of BMV and MWR. Both initiatives had extensive professional, academic and personal collaborative relationships that crossed organizational and disciplinary boundaries. Actors with activities concentrated in the Gothenburg region represented both initiatives. In their applications, both initiatives highlighted their research strengths and their commercial prospects. Each called attention to the significant growth potential of its industry.

Apart from their different areas of expertise and different historic relationships with governmental entities, the two initiatives were similar in one important respect in their funding applications. In following the VINNVÄXT policy scripts, both initiatives emphasized that they were organized as triple helix structures. However, while BMV's application listed actors from the three helices (i.e. Sahlgrenska Academy, biomedical companies and public health care organizations), MWR's application placed more emphasis on its network of microwave companies and their relationship with Chalmers. The MWR

Table 9.1 Historic roots and early development

Similarities	*Differences*
• Both had well-established, informal relationships across organizational borders. • Both mobilized their efforts around specific competence areas. • Both targeted public money as sources of funding. • VINNVÄXT provided both with the opportunity to fund ideas they had previously contemplated. • Both applied for VINNVÄXT 2003 funding.	• The biomedical field had a history of close interaction with the public health sector, academic researchers, and medical research companies. • The microwave industry had a weaker historic relationship with the government (only as a customer to the defense industry) but stronger links with industrial companies and technology universities. • BMV received long-term funding from VINNVÄXT, MWR did not (MWR received smaller amounts from similar sources).

application was notable for the lesser focus on public actors, which may have been a reason for its failure to obtain VINNVÄXT funding.

The VINNVÄXT competition mobilized both BMV and MWR to organize and strengthen their networks of organizations and groups of people in their efforts to further innovation. Although both initiatives might have undertaken similar activities on their own, the VINNVÄXT program gave them the opportunity to formalize these existing relationships and to promote new collaborative initiatives.

Organizational presentation

Biomedicine West (BMV)

Triple helix representatives supported BMV's VINNVÄXT applications in both 2003 and 2004. A critical incident that led to the creation of BMV and its subsequent funding applications was a workshop held in January 2002. The workshop was conducted by the consultancy firm Arthur D. Little, investigating regional business development opportunities in the Life Sciences (Laage-Hellman and Rickne, 2007). From the approximately 60 representatives from biomedical organizations at the workshop, a project group was formed to promote regional efforts in the life sciences. A few months after the workshop, when the VINNVÄXT program was announced, BMV saw an opportunity to formalize its work and to seek funding.

Since BMV decided it was inappropriate to hire a private company to prepare its funding applications, it selected as the project owner Business Region Göteborg (BRG), a business development company owned by the City of Gothenburg. To begin the work with member organizations in the biomedical field, a 'process leader' was recruited and a steering committee with triple helix representation was formed. A commonly held opinion among the interviewees is that the success of BMV's funding application in 2004 was in part due to this regional triple helix structure.

Microwave Road (MWR)

After its initial failure to win funding from VINNVÄXT, MWR continued recruiting members from the various companies and organizations working with microwave technology in Western Sweden. A self-employed engineer, with a microwave technology background, was elected "process leader" and took a leading role in the recruiting of members. With funding from the VISANU program and from Region West Götaland, in addition to membership fees, MWR grew as an inter-organizational network for developing microwave technology. The original 30-member network included regional microwave- and telecom-related companies, research institutes, Chalmers and various municipal organizations such as Business Region Göteborg (BRG), and MWR had thus created a triple helix structure. In late 2003, over 150 engineers, regional developers and academics attended a kick-off event for MWR at Ericsson's offices in Mölndal outside Gothenburg. The mood was optimistic with the expectation of a prosperous future for the microwave industry as it worked collaboratively to develop microwave technology in the region.

Reflections

Despite their differences in expertise and participants, the initiatives officially presented their organizational structures in a similar fashion. Both initiatives followed the VINNVÄXT policy scripts by selecting a process leader, by forming a triple helix structure and by actively recruiting members to their networks. These organizational structures were still essentially in place for both initiatives when we concluded our research in 2008.

In order to give legitimacy to their funding applications, particularly with respect to VINNOVA, both BMV and MWR highlighted their organizational structure and leadership. BMV appointed people to top positions with prominent status in their own organizations, and MWR appointed enthusiastic people who were close to the innovation processes in their companies or in academia. The difference in these hierarchies was that BMV focused on creating both external and internal legitimacy, while MWR focused more on creating internal legitimacy.

However, having high status directors in the BMV board was not entirely unproblematic. When power struggles ensued among the directors, the more neutral BRG had to assume a leading role. As the attention shifted from fundraising to innovation work, those BMV actors who gave legitimacy to the initiative, but had little connection with the innovation work, had to co-operate with the actors at the operational levels who would drive the process. Another problem arose because the high status directors seemed to take more interest in their own organizations than in the joint initiative, causing dissension in the boardroom.

Table 9.2 Organizational presentation

Similarities	*Differences*
• Both presented organizational network structures in order to emphasize that innovation occurs in inter-organizational settings.	• BMV was "owned" by a municipal body (BRG), while MWR was organized as an independent industry association.
• Both presented triple helix structures based on the composition of their organizational membership and boards of director.	• BMV appointed people with high legitimacy to most positions (causing some conflicts of interests). MWR appointed primarily operations and industry managers.

The purpose of the initiatives

Biomedicine West (BMV)

In the 2004 VINNVÄXT application, BMV presented its vision of the Gothenburg region as one of Europe's most innovative regions in the development and application of new findings in the biomedical field. BMV set the year 2015 as its goal for the realization of this vision:

> In 10 years time, a significant increase in the number of employees in the biomedicine area is expected in the region. We estimate that current companies will approximately double the number of their employees, based on the trend of the 1990s. Among others developments, we estimate there will be around 20 companies with more than 50 employees each. Of these 20 companies, we estimate 5 will have more than 500 employees. Furthermore, we expect to see 50 new companies with up to 15 employees. The number of foreign-owned companies in the region will double. The indirect multiplication effect from this new employment will likely be significant. In total, we believe by the year 2015 around 15,000 people will be working directly in the biomedical area. We expect the same number will be working indirectly in the area.
>
> (BMV application 2004: 6, translated from Swedish)

The BMV initiative was initially formed around two main competences in the life sciences: (1) *cardiovascular and metabolic diseases (CVM)*; and (2) *biomaterial and cell therapy (IBCT)*. Both areas were considered strong in the region, in knowledge as well as by tradition. The initiative hoped to attract enough interest and venture capital to start an advanced biomedicine education program in business development and to strengthen the commercial infrastructure in the biomedical field. In general, much of the application's rhetoric focused on the future potential of the two competences – how their development would lead to economic growth and employment in the region. Most actors involved in the process agreed on the need to strengthen the Gothenburg region in these areas.

Once the funding was in place, the application rhetoric had to be transformed into practice through the planning and execution of specific activities that were linked to the initiative's stated vision and goals. This was an ambitious undertaking. It has been suggested that the goals were unrealistic, and therefore it would be a challenge for the initiative to meet them. Additionally, there were concerns about the collaborative climate among the participating organizations in the political process of resource allocation where the 'business as usual' attitude prevailed.

Microwave Road (MWR)

In the VINNVÄXT application, MWR highlighted the necessity of developing technology platforms in order to advance the microwave industry. Compared to BMV's application, however, MWR presented the growth potential less affirmatively:

> The central concept is development of technology platforms for microwave products. The platforms should be as diversified as possible, so that the technology developed can be used in different products and by as many companies as possible. They should help fill in the gaps in competency and be an engine for developing new companies that provide a more complete infrastructure in the region. [. . .] Despite its relative maturity, microwave technology can offer outstanding growth potential. Many applications [of microwave technology] in the 5GHz area [higher frequencies] may result in a large breakthrough within 10 years.
>
> (Microwave Road VINNVÄXT application, 2003: 2, translated from Swedish)

While BMV emphasized commercial development, in contrast, even after it had been established for some years, MWR's focus remained on technology development. MWR would continue to encourage technology projects. MWR explained the purpose of the initiative on its 2006 homepage:

> Microwave Road is a regional platform for national and international collaboration concerning microwave technology for industry, public administration and universities. The purpose is to develop products based on microwave technology through collaboration.
>
> (www.microwaveroad.se, 2006)

The emphasis in this purpose statement is the general arena of relationship building in the field of microwave technology. However, as explained below, while MWR arranged meetings and seminars for its members, only a few activities seemed to lead to collaborative technology development.

Reflections

Both BMV and MWR largely followed the VINNVÄXT policy scripts by proposing to develop a strong research and industrial region with high growth opportunities. They formed inter-organizational networks, built triple helix structures aimed at making the region more flexible and attractive, increased their knowledge, etc. Yet they took somewhat different approaches in their funding applications. BMV identified two specialized biomedical competences while MWR described its microwave technology competence more generally. BMV was also clearer about its intended results – new jobs and

Table 9.3 The purpose of the initiatives

Similarities	*Differences*
• Both concentrated on individual strengths and future growth areas. • Both emphasized regional development. • Both had externally articulated goals as well as more informal ones. • Both focused on providing an arena for relationship-building *and* undertaking specific innovation projects.	• BMV expressed its strategy more precisely by focusing on the two areas of IBCT and CVM. MWR aimed at furthering collaboration for technological development. • BMV had specific targets on regional development (e.g., employment rate, number of start-up companies, etc.). • BMV emphasized commercialization of existing knowledge and technology, MWR focused technology development.

new life sciences companies. The concern was whether BMV could achieve these results. Some BMV interviewees suggested that these targets were overly optimistic wish lists with unachievable or irrelevant forecasts.

The performed activities

Biomedicine West (BMV)

Each VINNVÄXT competition winner was required to submit a report to VINNOVA after three years describing and reflecting on its accomplishments and future challenges. In 2008 BMV published a report that described the focal areas of IBCT and CVM and four research projects (Alfa, Beta, Gamma and Delta) developed in collaboration with Sahlgrenska Academy and with industry. In the CVM area, BMV supported 17 projects using the 60 million Swedish crowns from VINNVÄXT plus 26 million Swedish crowns from other agency grants and investor financing. A leader at IBCT gave a more detailed description of the activities in one of the three IBCT projects:

> In one project, later named 'Beta', we developed a method to make ultra thin cuts through intricate materials, 100 nanometers thick. These cuts make it possible to create a precise model of the material that was not possible before. It has been a huge scientific success; for instance, after 2.5 years one student completed a doctoral thesis based on work on this project. This project results are also used to develop and sell the method to companies and other institutions. Several companies have understood the value of the project.
>
> (IBCT leader, 2008)

The report also pointed to how, with help from BMV, the Sahlgrenska Academy had introduced an international Master's Program, GIBBS, that, at the conclusion of this research, had graduated 23 students majoring in the management of innovation processes in the life sciences. Linked to this graduate program, a science park and an incubator (Sahlgrenska Science Park) were established and then expanded. At present, the Sahlgrenska Science Park supports 6 companies and more than 20 projects. Moreover, BMV has engaged in profiling activities and relationship building at a number of conventions and exhibitions, has organized an international scientific conference and has arranged 13 network seminars for almost 1,000 participants.

In its 2008 evaluation of BMV, VINNOVA praised both GIBBS and Sahlgrenska Science Park for their "excellent innovation system infrastructure." Although the VINNOVA evaluators also commented that the BMV board was rather "inharmonious," with its conflicts and internal competition, in general they rated the first three years' activities at BMV as satisfactory:

> Our conclusion is that BMV has focused on areas where they are leaders in industry and research and have expanded this position in their first three years as a VINNVÄXT-initiative. According to the materials presented, BMV in most cases has reached the objectives stated in their Action Plan of 2005.
>
> (VINNOVA evaluation 2008, translated from Swedish)

Microwave Road (MWR)

MWR's initial activities largely related to developing the initiative's purpose, creating a network structure, recruiting members and seeking government support and funding. The process leader and the project leader contacted and/or visited numerous regional organizations in an effort to recruit them to the initiative. Eventually, MWR had 37 organizations as members, most of which were regional microwave companies.

Having failed in its funding request to VINNVÄXT, MWR was naturally concerned with finding ways to fund the initiative. The funds received from the VISANU program and the regional government were minimal, and the membership fees only contributed marginally to the budget. At MWR, the search for funds can be referred to as a continual *quest for capital.* As a result, MWR representatives spent considerable time trying to interpret the various funding programs' instructions and policy scripts. They also investigated opportunities for joining other innovation initiatives as a way to improve funding possibilities. Consequently, rather than working with technology development projects, much of the initial activity at MWR was dedicated to building and financing a network organization. Nevertheless, despite MWR's limited financial resources, the initiative did organize numerous network meetings, conferences and theme days on microwaves and future growth opportunities in the microwave technology industry.

The stated purpose of the MWR initiative, however, was to advance technological development. Therefore the MWR process leader and board of directors began investigating how they might fulfill this purpose. For example, they regularly discussed how to organize joint technology development initiatives. They also organized microwave meetings and events for the MWR members that focused on developmental issues such as automotive radars and "microwaves for growth." In addition, they carried out a competence inventory among the members that might be used to identify complementary resources. Yet, there was a problem with these activities – they did not produce the desired technological developments. Most activities related to organizing the network rather than to developing the innovation work. The process leader explained: "We cannot just compete [e.g. with VINNVÄXT]; we have to get going with concrete projects." As form had taken priority over content, MWR's board members, process leaders and members were frustrated.

Reflections

BMV appeared to be involved in a number of activities, in particular within the action fields of IBCT and CVM, where innovation work was at the center of attention. It is not possible to conclude how much of this work was the result of the VINNVÄXT funding, but it is clear that these resources were useful for building the infrastructure and for providing seed money for research projects.

MWR was clearly less involved in innovation work than BMV. MWR focused on forming collaborations, seeking funding, arranging informational meetings and studying innovation program scripts. Unlike BMV, MWR had no action direction for its innovation work. Instead MWR focused on its network building. Still, it is plausible to assume that there were innovation

Table 9.4 Activities

Similarities	*Differences*
• Both put much effort into attracting and engaging allies to work for the initiative. This effort was also aimed at achieving legitimacy for their initiatives' existence. • Both worked extensively with application writing in order to secure funding from public sources. • Both were concerned with creating a cross-disciplinary arena to build new collaborations.	• For BMV, more attention was placed on innovation projects, while MWR struggled in the "quest for capital." • The VINNVÄXT funding allowed BMV to focus more on translating its purpose into innovation activities. • By 2008 BMV presented more innovation activities than MWR, and had developed the innovation systems initiative into an established, legitimate actor. MWR shifted to becoming an industrial association for microwave technology.

activities in the interaction among member organizations, and between them and Chalmers, outside of MWR's scope.

The organizing logics

Biomedicine West (BMV)

The two focal areas, IBCT and CVM, each had its own coordinator with authority to develop the content and the processes. Because of their different contexts and the different agendas of the coordinators, somewhat different organizing logics emerged in the two areas: IBCT focused on multilateral science and commercial projects, while CVM focused more on supporting and managing biomedical start-up companies.

In the IBCT area, an internationally well-known professor from Sahlgrenska Academy was appointed as the scientific leader. As the result of this appointment, IBCT continued to focus on the traditional, multidisciplinary research that had earned the Academy a worldwide reputation for scientific excellence. Yet, in its application to VINNVÄXT, as a way of strengthening its proposal with relationship-building possibilities, BMV had promoted the linkage between research and commercial projects that existed in other areas. As there were a number of large and small regional companies working in the biomaterial and cell therapy area, early on IBCT spun-off a number of commercial projects to private sector companies with the intention of building business relationships with them. In a sense, this relationship already existed since industry frequently looked at IBCT researchers as potential employees. As a result, in practice, a number of applications and methodologies were generated and commercialized with IBCT industry partners. However, it was not always easy to tell whether such activities were generated inside or outside the initiative. An IBCT leader described this relationship with industry:

> My own philosophy is to try to have as much academic freedom as possible. Therefore it is a dilemma to work with only one company at a time. Bilateral collaborations often lead to pure commissioned research. I do not have any spiritual peace from that kind of work. The main task in laboratory work is to seek knowledge and to explore new areas. It is not just to supply data to industry. That [multilateral projects] is the main purpose.
>
> (IBCT leader, 2008)

Unlike the diverse and growing network in the biomaterial and cell therapy area, one multinational corporation, AstraZeneca, dominated the cardiovascular and metabolic diseases area. In its VINNVÄXT application, BMV gave AstraZeneca an important role as an industry partner; without AstraZeneca's involvement, VINNOVA might have rejected the application. In the realization of the CVM projects, however, AstraZeneca's role was not as apparent.

Instead, a process was established for evaluating and financing ideas coming from the Sahlgrenska Academy and Chalmers, in which CVM acted more as a venture capitalist that supported start-up companies. It was also significant that the CVM coordinator had previously worked with portfolio management in the Life Sciences sector and was familiar with its financial structures. For these reasons, although CVM's focus was originally cardiovascular and metabolic diseases, the research broadened. A CVM leader explained this change in focus:

> Over time, the industry people have become less active. [. . .] The big corporations already have enormous resources, and the amount they spend on a small rat experiments is higher than our yearly budget. We have tried to find common projects with them, but it is not easy. [. . .] They are also difficult to work with since they demand 100 percent ownership of the intellectual property in everything and five years of loyalty from all other participants. [. . .] Now we focus instead on the whole area of life sciences. We do not see any reason why we should lock ourselves into only cardiovascular and metabolic projects.
>
> (CVM leader, 2008)

Microwave Road (MWR)

After MWR's kick-off meeting in late 2003, the initiative continued to organize a network of microwave organizations. MWR's task was to channel government funds to collaborative, microwave technology development projects. Such projects would focus on microwave technology materials or components, as well as on the engineering and production of technical sub-systems platforms. This overall organizing logic was based partly on imitation of practices used at the research institute, IVF,[6] where companies collaborated in solving shared technological problems. An important difference however was that in the IVF projects the companies themselves contributed with funds to a much greater extent than in MWR. Following the triple helix structure, the MWR initiative application stated that actors from industry, academia and the public sector were included in the development work. Once the microwave network was established, these actors could begin initiating joint technology projects. As of 2008, only two innovation projects had been initiated: one with an automotive focus and one with a ceramics focus.

In the first innovation project, called the Automotive Group, representatives from the telecom and automotive industry met to discuss the possibilities of applying microwave technology to vehicles and to the infrastructure of public roads (e.g. automotive radars and intelligent road signs). At this meeting, some 30 microwave engineers enthusiastically presented their ideas and began plans for writing proposals and specifications. A problem arose because the automotive representatives showed little interest in the projects, and few attended the subsequent meetings. To the disappointment of several MWR members, the Automotive Group dissolved. However the MWR process leader and the board of directors, who wanted to continue the technology development work, then turned toward building on past experiences from joint projects with IVF. Their goal was to identify technology issues of interest to the MWR members.

In the second innovation project, called the Ceramics Project, funded by the Region West Götaland, MWR members worked with ceramic carrier materials for microwave circuits. This work ultimately resulted in three ceramics sub-projects. While these sub-projects had some overlap, each focused on the needs of three influential industrial members in MWR. The results of the projects included a scientific paper, ideas for devising new microwave construction processes and opportunities to test new material types. The Ceramics Project, which built on local expertise and practice, was more successful than the Automotive Group since its work was within a specific field of microwave technology (as opposed to organizational structuring). The argument may be made that the two groups followed different organizing scripts: the Automotive Group focused on structure and the Ceramics Project focused on action (Lavén, 2008).

Reflections

The organizing logics in the BMV and MWR initiatives reveal how policy scripts are put into practice. Both initiatives began with the idea of strengthening their respective industries by identifying an innovation theme, recruiting complementary regional actors for projects and eventually negotiating the industry direction in various rounds of political processes at different levels.

In the application phase as well as in the later operational work, BMV prioritized the IBCT and CVM knowledge areas and their projects differently. Although its initiative concerned Life Sciences in general, BMV concentrated on two particular knowledge areas where it had a good reputation. BMV then gradually extended the scope of its activities and recruitment of actors beyond these two areas. In short, BMV began small, teaming with reliable regional actors and focusing on areas of expertise before enlarging the initiative and teaming with others in its *program-of-action*. Moreover, the people with responsibility for the IBCT and CVM areas could influence their development, thus steering the organizing logic in somewhat different directions. IBCT, which was very research-oriented, built on an existing community, while CVM focused more on a portfolio management process that supported smaller start-ups and their commercialization. Nevertheless, BMV's early decision to target two knowledge areas was instrumental in directing the overall BMV initiative innovation work.

MWR, on the other hand, was less clear in its focus since the initiative aimed at developing the microwave technology industry generally. When the VINNVÄXT application failed, MWR seemed to focus more on organizing the network and acquiring actors for the initiative. MWR saw these activities

Table 9.5 Organizing logic in linking actions and activities

Similarities	*Differences*
• Both put the innovation policy scripts into practice, albeit somewhat differently. • Both identified and enrolled "relevant" actors and followed a triple helix structure. • Both appointed process leaders and triple helix boards of directors.	• BMV focused more on innovation projects, involving existing initiatives, and attended more to innovation work. MWR focused on organizing its network, which created inertia (the Ceramics Project was an exception). • There were differences in how the triple helix structures were enacted. The public sector, academia and industry were more closely integrated in BMV. MWR was more industry-focused. • BMV identified competences, began work on a small scale and then expanded. MWR took a more general development focus. • BMV enrolled other related initiatives and links to external infrastructure that MWR did not.

as prerequisites for the collaborative technology development projects that would follow. However, since the application had identified no particular innovation or technology development areas, MWR had no specific problems requiring solutions.

The two MWR innovation projects followed two different organizing logics. The Automotive Group was a group of actors with no specific task. They were actors engaged in the effort to link the up-stream actors in the value chain to potential customers. In this setting, the automotive industry acted much like AstraZeneca in the BMV initiative. The Ceramics Project, similar to the projects in BMV's IBCT area, focused on collaborative work in a technology and knowledge field where actors could inter-link their activities. The organizing logic of the Ceramics Project, as opposed to the "grouping logic" of the Automotive Group, was thus more action focused, similar to BMV's logic. In summary, while both BMV and MWR organized according to the triple helix structure, it appears as if MWR followed the VINNVÄXT policy scripts as instructions for organizing a network, while BMV followed them more loosely with more attention paid to strong activity fields and existing relationships.

Discussion and conclusions

The centrality of innovation policies in global efforts as stimulants to economic development raises many interesting questions. For instance, what impact do policy scripts have on innovation processes? Is it possible to plan and execute such scripts "top-down," as blueprints for organizing innovation systems, or are innovation processes loose, complex, self-organizing processes, impossible to control? Alternatively, are policy scripts rhetorical or presentational tools, de-coupled from the innovation work of initiatives? Or, is there a middle way in which policy scripts can be designed as adaptable guidelines that allow local improvisation in innovation work?

In this chapter we have shown the central role the Swedish VINNVÄXT policy scripts played in the BMV and MWR innovation initiatives. In this section we discuss the consequences and implications for policy-makers when the policy scripts are translated into practice.

The scripts' organizing effects

The BMV and MWR initiatives illustrate how they followed VINNVÄXT policy scripts where an inter-organizational network, the triple helix structure and an economic growth model were the funding criteria. Yet, as the policy scripts were enacted, certain actions took *precedence* over others (Lavén, 2008). Although BMV and MWR enacted their scripts differently, each innovation initiative viewed the scripts, not only as instructions to follow and negotiate, but also as VINNOVA's evaluation criteria. Particularly for MWR, the issue was how to organize its initiative according to the policy scripts that were based on innovation theories approved by the competition jury.

At a first glance, it appears BMV and MWR acted similarly in following the VINNVÄXT policy scripts. Both initiatives applied for public funding, established an organization with a triple helix structure, and appointed a board of directors and a process leader. They also tried to recruit actors from their respective fields to join their initiatives. Moreover, both initiatives organized meetings and worked on development projects to spur innovations and economic growth. Thus, it seemed the initiatives acted similarly, but in taking a closer look, we see they enacted the scripts differently, following different organizing logics.

Organizing logics of structure versus action

Initially, MWR focused on arranging its initiative innovation structure in accordance with the policy scripts by building on hybridized innovation theories of clusters, innovation systems and a triple helix structure. The organizing logic was based on "structural precedence" (Lavén, 2008) in which the organizational structure – a network of organizations in the field of microwave technology – was the highest priority in the application process. This structural precedence may have led to an overemphasis on the *form* of the initiative with less attention paid to its *content*. A certain kind of inertia resulted. Instead of focusing on microwave innovation work, the MWR initiative focused on enacting the innovation policy scripts correctly, organizing the network structure and inviting actors who were more important to the funding application than to the work of technological innovation and commercialization. For example, the MWR initiative included local government representatives who introduced what have been described as horizontal goals, such as gender issues. We were also invited to join the initiative, not because we could contribute to the innovation processes and not because of a prior relationship with MWR, but because our presence fitted the VINNOVA policy script that called for researchers to study and document the initiative processes. BMV also organized a network and recruited actors to join its initiative, as the policy scripts required, but the BMV initiative seems to have been able to translate ongoing research activities, especially in the IBCT area, to fit the policy scripts.

According to Latour (1986), tokens such as claims, orders and artifacts and even scripts may change as people translate and transform them according to their particular interests and projects. Sahlin-Andersson (1996) observes a similar phenomenon in science park success stories that were related through editing and reformulation, depending on local circumstances of a particular science context. Similarly, the innovation policy scripts were edited as they were translated into innovation work in the BMV initiative. In the editing, the scripts focused on the benefits of commercialization of biomedical research in Western Sweden.

Once BMV received long-term funding from VINNVÄXT, it had the freedom to develop its innovation work. At this point, BMV's interest turned from enacting the policy scripts to using its funds to initiate innovation projects. In contrast, MWR, in the absence of VINNVÄXT funding, continued to seek financial support; its focus turned to the rules of the game and therefore to policy scripts. As a result, eventually the MWR's board of directors and members altered the purpose and identity of the MWR initiative. They began to regard the initiative as a community of researchers, engineers and managers with a shared interest in developing the microwave industry. Therefore, as MWR adapted to the local conditions, the initiative became an arena for interaction by the local microwave community.

As a VINNVÄXT winner, BMV could open up new channels for close dialogues with the policy-makers. These channels led to a deeper understanding of the implicit meaning of each policy script and to an opportunity to motivate local adaptation towards VINNOVA. As a result, BMV could focus on biomedical innovation work. Having successfully represented the initiative in accordance with the policy scripts, BMV could then edit them according to an action-oriented organizing logic rather than according to a structure-oriented organizing logic. Such editing space, where there is room to maneuver between scripts and action, is central to innovation.[7]

The importance of an editing space between scripts and action

Arguably, BMV had a wider editing space than MWR since it could make looser linkages between the policy scripts and action; BMV was not forced to follow the policy scripts closely. Orton and Weick (1990) note how loose couplings may allow for responsive linkages between elements. This means that the elements are coupled, but at the same time retain their separate identity. The BMV initiative was presented and partly organized according to the policy scripts; however, it was also loosely coupled to other innovation work practices that were not governed by those scripts. There were tighter couplings (Weick, 1979) between the policy scripts and the actions in MWR's organizing process. The MWR initiative followed rational procedures and principles in its attempt to copy and implement the scripts in building a network structure, possibly contributing to the structural precedence and inertia.

The tightness or looseness of the coupling between the scripts and actions is partly a consequence of the processes of interpretation and translation at the local level. Thus the editing space may have been larger than the MWR actors thought. The policy scripts were based on VINNOVA's theoretical criteria. Without a deeper understanding of those criteria, it was possible to interpret them as instructions to be followed rather than as theories that provided a language and a model to imitate or to adapt. It is even possible that interpreting policy scripts as rigid instructions may disqualify an application.

Lessons for policy-makers

Since VINNVÄXT was a competition, BMV and MWR presented their applications as favorably as possible. On one level, it requires a certain rhetorical skill to produce politically correct texts to various stakeholders. Both BMV and MWR described themselves a triple helix organized innovation systems, appointed suitable members to its board, and recruited industry actors as members. This organizing process was unproblematic so long as the scripts were translated meaningfully in ways that allowed each initiative to focus on its *program-of-action* (cf. Latour, 1992) – namely, to further innovation work. However, there is a problem if innovation initiatives focus more on abstract criteria than on the connections that bind related actors in innovation practice. It is particularly problematic if the enactment of policy scripts directs attention away from innovation action to building structures for innovation. In such instances, *script fundamentalism* may result where actors strictly follow a set of basic principles. However, as we have illustrated, a different story may emerge if the scripts are edited.

This comparison of the BMV and MWR initiatives suggests that an excessively rigid compliance with policy scripts – by policy-makers in their evaluations and by innovators in their everyday work – may raise doubts about the effectiveness of the translation from policy to practice. Policy scripts offer guidance, that, when enacted, may enhance or restrain local practice. As such scripts are translated into action they may trigger the organizing of innovation work. However, since every situation is unique, policies must also allow for flexibility and interpretation in their use; they must provide an editing space. Each initiative will have its own history, relationships and goals. It is naïve to imagine that one approach fits all.

Therefore, policy efforts must guide, not police, innovation work, by allowing for flexibility and adaptation. It is also important to examine the theories that policy-makers use when they edit policy scripts. The receiver, as well as the sender, needs to interpret and translate the scripts so that innovation action is facilitated. As noted in the previous section, it may be necessary to relate the translations more to the practices that the innovation theories build on than to the prescribed scripts.

For example, it may be necessary to develop a more theoretical understanding of the roles of the different actor categories in the triple helix structure. It may not be enough just to have the three categories represented on the board of directors. In fact, these actors may be more effective and influential as providers of resources (e.g. an innovation environment) than as board members. Powerful actors in industry, academia and the government may therefore create an "infrastructure." The strength of such milieus, where ideas and knowledge may flow freely, depends more on the mutual trust and shared knowledge of the actors than on the shared power and status of board members in any specific innovation initiative.

Conclusions

This chapter explores the research question of how innovation policies are translated into practice in organizing innovation initiatives. Our argument is that while senders of policy scripts may provide clear instructions for how to organize innovation systems and cluster initiatives, receivers may translate these instructions into innovation action in different ways. For sense-making purposes, a dialogue between senders and receivers may reduce misunderstandings between them, as well as avoid the *script fundamentalism* as interpreted by some receivers, whatever the senders' intentions. Such a dialogue may allow the receivers to recognize which scripts are normative and compulsory and which scripts permit creativity and adaptation. The chapter shows that innovation policy scripts present a pattern for initiatives: however, the pattern reflects mostly "what" to do, while the "how" of organizing logic requires editing by receivers.

The principal implication for policy-maker senders is that they should look at policy scripts as facilitating guidelines rather than as prescriptions requiring policing. At the same time, the script receivers must also have the strength and ability to make translations to the local context. Policy scripts may only provide a general guidance for innovation work, such as in providing an innovation scaffold. However, if local innovation work is to progress, this metaphorical scaffold may have to be remodeled by the receivers. Therefore policy-makers should spend more effort on the interpretational space between themselves and their scripts, on the one hand, and between the innovation initiative representatives and their activities, on the other. Widening the editing space means creating a deeper understanding of the basis of the scripts. This is by no means an easy task; both policy-makers and innovators are involved in complex processes with uncertain outcomes. However, as this study has showed, a key policy challenge is to focus on the facilitation of innovation action and not solely on the organization of innovation structures.

Notes

1. This study was made possible by financial support from VINNOVA's research program "Kunskapsbildning och organisering" and from VINNOVA's Actor Department. We gratefully thank VINNOVA for its support. We also thank Jens Laage-Hellman and Annika Rickne for their willingness to discuss their experiences in studying the Biomedicine West initiative, the authors of this book for their insightful comments during workshops, the interviewees at Microwave Road and Biomedicine West, and Marcia Halvorsen for editing the text.
2. For example, in the early 1990s, the OECD issued policy recommendations on managing National Innovation Systems. Finland was the first country to incorporate these recommendations into national economic and innovation policies. At the EU level, the 2001 Lisbon Strategy, as set out by the European Council, and the recently proposed Cluster Memorandum 2008 reflect a similar direction in innovation policy.
3. In total, there were 66 encounters with various actors linked to the initiative. Of these, 56 were observations (including participation in meetings, workshops, seminars, etc.), 7 in-depth interviews and 3 shadowing occasions where the process leader was followed and observed in his daily work.
4. The scientific leader for IBCT, the process leader for CVM in BMV and a representative for Astra Zeneca.
5. Jens Laage Hellman and Annika Rickne.
6. IVF is a research institute in the SWEREA Group that is jointly owned by the Swedish Government and several industry actors. IVF's business model is to initiate, perform and transform R & D projects in manufacturing. Such projects include product development, production systems and process optimization, among others.
7. The concept "editing space" may be related to the "macro policy" and "micro practice" distinction in implementation studies – a distinction that has been used to emphasize the necessity of creating the capacity to adapt policies to local contexts.

References

Akrich, M. (1992) 'The De-Scription of Technical Objects', in *Shaping Technology/ Building Society: Studies in Sociotechnical Change,* edited by W. E. Bijker and J. Law, 205–224. Cambridge, MA: MIT Press.

Akrich, M. and Latour, B. (1992) 'A Summary of a Convenient Vocabulary for the Semiotics of Human and Nonhuman Assemblies', in *Shaping Technology/Building Society: Studies in Sociotechnical Change,* edited by W. E. Bijker and J. Law, 259–264. Cambridge, MA: MIT Press.

Czarniawska, B. and Joerges, B. (1996) 'Travels of Ideas', in *Translating Organizational Change*, edited by B. Czarniawska and G. Sevón. Berlin: de Gruyter.

Czarniawska, B. and Sevón, G. (eds) (1996) *Translating Organizational Change.* Berlin: de Gruyter.

Etzkowitz, H. (2002) 'Networks of Innovation: Science, Technology and Development in the Triple Helix Era'. *International Journal of Technology Management and Sustainable Development* 1 (1): 7–20.

Fischer, F. (2003) *Reframing Public Policy: Discursive Polities and Deliberative Practices*. Oxford: Oxford University Press.

Freeman, C. (1987) *Technology Policy and Economic Performance: Lessons from Japan*. London: Pinter.

Goffman, E. (1959) *The Presentation of Self in Everyday Life*. New York: Doubleday.

Joerges, B. and Czarniawska, B. (1998) 'The Question of Technology, or How Organizations Inscribe the World', *Organization Studies* 19 (3): 363–385.

Laage-Hellman, J. and Rickne, A. (2007) 'Hur blir man en vinnare? Fallet biomedicinsk utveckling i Västsverige', in *Regional Växtkraft i en Global Ekonomi*, edited by S. Laestadius, C. Nuur and H. Ylinenpää. Sweden: Santérus Academic Press.

Latour, B. (1986) 'The Powers of Association', in *Power, Action and Belief. A New Sociology of Knowledge?* edited by J. Law, 261–277. London: Routledge.

Latour, B. (1992) 'Where are the Missing Masses? The Sociology of a Few Mundane Artifacts', in *Shaping Technology/Building Society: Studies in Sociotechnical Change,* edited by W. E. Bijker and J. Law, 225–257. Cambridge, MA: MIT Press.

Lavén, F. (2008). *Organizing Innovation: How Policies are Translated into Practice*. Göteborg: BAS Publishing.

Nelson, R. R. (ed.) (1993), *National Innovation Systems: A Comparative Analysis*. New York: Oxford University Press.

Orton, D. J. and Weick K. E. (1990) 'Loosely Coupled Systems: A Reconceptualization', *Academy of Management Review* 15: 203–223.

Porter, M. E. (1998) 'Clusters and the New Economics of Competition', *Harvard Business Review* 76 (6): 77–90.

Sahlin-Andersson, K. (1996) 'Imitating by Editing Success: The Construction of Organizational Fields', in *Translating Organizational Change*, edited by B. Czarniawska and G. Sevón (eds), 69–92. Berlin: de Gruyter.

Saxenian, A. (1990) 'Regional Networks and the Resurgence of Silicon Valley', *California Management Review* 33 (1): 89–112.

Simmie, J. (1997) *Innovation, Networks, and Learning Regions?* London; Bristol: Jessica Kingsley Publishers; Regional Studies Association.

Stone, D. A. (1988) *Policy Paradox and Political Reason,* Glenview, IL: Scott, Foresman and Co.

Suchman, L. (1987/2007). *Human-Machine Reconfigurations: Plans and Situated Actions*, 2nd edn, Cambridge: Cambridge University Press.

Van de Ven, A. H. (1986) 'Central Problems in the Management of Innovation', *Management Science* 32 (5): 590–607.

VINNVÄXT program description (2002) *Regional Growth Through Dynamic Innovation Systems: Program Description as a Basis for Internal Support and Decision-Making*.

Weick, K. E. (1979) *The Social Psychology of Organizing,* 2nd edn, New York: McGraw-Hill.

Wenger, E. (1998) *Communities of Practice: Learning, Meaning, and Identity*. Cambridge, New York: Cambridge University Press.

Wigren, C. and Melin, L. (2007) 'Vinnväxt-tävlingen: Spelregler för vinnare, och förlorare'. In *Regional Växtkraft i en Global Ekonomi*, edited by S. Laestadius, C. Nuur and H. Ylinenpää. Sweden: Santérus Academic Press.

Woolgar, S. (1991) 'Configurating the User: The Case of Usability in Trials'. In *A Sociology of Monsters: Essays on Power, Technology and Domination*, edited by J. Law, 58–99. London: Routledge.

10 Can regional innovation systems be "constructed"?

Johan Johansson and
Håkan Ylinenpää

Introduction

Innovations and technical progress are becoming topics of growing interest for policy-makers wishing to promote economic growth.[1] Cooke and Memedovic explain:

> that the economic growth and competitiveness of their regions depends largely on the capacity of indigenous firms to innovate, offering the appropriate support to these indigenous firms to be more competitive through innovation is a rising star on the regional policy agenda.
>
> (Cooke and Memedovic, 2003: 8)

As often is stated, however, firms today seldom innovate in splendid isolation, and therefore the concept of regional innovation systems (RIS) has evolved into a widely used analytic framework and a foundation for innovation policy-making since it offers a method to understand innovation processes within a socially embedded and regional framework (Cooke and Memedovic, 2003; Nilsson and Uhlin, 2002).

Innovation systems may be defined and analyzed in several ways (Carlsson *et al.*, 2002). In this chapter, a starting point is that place-specific and regional resources stimulate the innovation capability and competitiveness of firms and regions (Asheim *et al.*, 2003), and an RIS emerges from the acknowledgement that innovation is a geographically bounded phenomenon. The importance of proximity, usually expressed in terms of geographical proximity, arises from the complexity of the processes of knowledge where social interaction and personal contacts are important in the learning processes and therefore geographical proximity within a region provides an optimal mechanism for the transfer and exchange of knowledge (Asheim and Coenen, 2005; Cumbers and MacKinnon, 2004; Simmie, 2003). Following these thoughts, an RIS consists of:

> a set of interacting private firms and public authorities, research organizations, and other bodies that function according to organizational and

> institutional arrangements and relationships conducive to the generation, use and dissemination of knowledge. In this conception, the environment of an innovating firm consists of a multitude of actors involved in the innovation process.
>
> (Hommen and Doloreux, 2003: 2)

From this, it can be stated that innovation, to a large extent, seems to be a localized and interactive process between sets of actors. As such, it is assumed that competitiveness occurs in places where localized capabilities such as institutional endowment, built structures, knowledge and skills exist. To stimulate and to promote innovativeness among companies, policy initiatives connecting them with the regional innovation support infrastructure may be regarded as a deliberate attempt to create a regional innovation system according to Asheim *et al.* (2003), since it involves a systemic perspective on innovation that seeks to enhance collaboration between innovating partners. Given these characteristics and the notion that innovation systems are constituted by their actors and the context in which they operate, it is important to understand how regional innovation systems are shaped by industry-specific characteristics and region-specific institutions.

The aim of this chapter is to describe and analyze a single RIS (ProcessIT Innovations) and its origins. The field ProcessIT includes automated treatment of process, production- and product information. ProcessIT Innovations is an RIS focusing on research and development of new products and services based on needs in the region's process industries and the potential to meet these needs through the development of new products and services based on information technology. At the time of this study, the innovation system consisted of 61 IT companies, larger industries, universities and public actors, primarily from the two northernmost counties in Sweden (Norrbotten and Västerbotten) and followed from the early initiatives to build an RIS in the late 1990s. By using ProcessIT Innovations as a longitudinal and illustrative example, we answer the question as to whether and/or to what degree a regional innovation system may be planned from the top or if such successful collaborative structures instead emerge from bottom-up initiatives and circumstances.

The chapter is structured as follows. First, we outline the theoretical base that later will be used for "making sense" of our empirical data, followed by a section in which we describe our methodological approach. Then, we present our empirical case ProcessIT Innovations and its prehistory as a basis for the succeeding section where we, building on the theories and theoretical concepts presented earlier, arrive at a better understanding of factors involved in the development of a regional innovation system. This chapter then concludes with a final section in which more general lessons to be learned are suggested.

Theoretical points of departure

As theoretical points of departure and building blocks for forthcoming empirical analyses, we will here address (1) innovation systems as embedded and complex systems; (2) innovation systems as learning communities; and (3) innovation systems as path-dependent constructions.

Innovation systems as embedded and complex systems

Adopting Laestadius and Rickne's definition of innovation systems (see Chapter 2 in this book), an innovation system can be defined as an interrelated structure of institutional (including cultural) and actor-based condensations in an economic space that has specific industrial and innovative properties related to its environment with which it interacts. Following this view implies an ambition to see, for example, an RIS through the lenses of the actors involved in developing a specific initiative (a point to which we will return in our method section). It also builds on the basic understanding that the development of an RIS depends not only on the actors involved but also on the situational and historical context in which it is embedded. Here, the successful development of an innovation system may be as regarded as much a product of its social context as are unsuccessful ones, requiring not only specific factors or parameters within the system itself, but also fit to, acceptance of, and support from the specific context in which it is embedded (Bijker *et al.*, 1987; Pinch and Bijker, 1984). Nilsson and Uhlin (2002) characterize innovation processes and the systems in which they are embedded as complex systems, where, for example, interactions among actors involved in an innovation system are influenced by laws, regulations, treaties, or by social, political and economic forces.

A complex system is composed of diversity of actors that interact with and affect each other, and in doing so generate behavior in the system as a whole (Lewin and Regine, 1998). The patterns of behavior are not constant because when a system's environment changes, so does the behavior of the system. The system is thus constantly adapting to the conditions around it. The major building blocks of an innovative system include actors such as firms, universities, and other public research organizations involved in research, education, and training. Market demand is also an important building block since the level and quality of customer demand serve as pull-factors for firms to innovate. In order to improve the innovative performance, activities to strengthen only a single building block normally are not enough. Instead the entire system should be stimulated, since each building block interacts with the others (Groenewegen and van der Steen, 2006; Nilsson and Uhlin, 2002).

An innovation system may further be regarded as a self-organized complex system that consists of actor networks and clusters of firms active in specific lines of industry and where knowledge flows in various directions. The outstanding features of such a system are, according to Nilsson and Uhlin

(2002), incomplete information, the ability to self-organize activities, and non-reducibility. Information is incomplete in the sense that no individual actor has full information, and it is assumed that actors that possess more information in areas vital for the innovation system to prosper can receive advantages. Activities to enhance the information flow within a system can speed up the development pace of the whole system. However, a complex system is open in the sense that information flows across its boundaries, which in turn are difficult to identify with any clarity. This causes difficulties in distributing information effectively. As any system, an RIS strives for outputs larger than the inputs that involved actors invest. In a complex system, however, such linearity, according to Loye and Eisler (1995), is not always easy to detect or demonstrate, because minor changes and variations can lead to highly unexpected and unpredicted effects that exponentially grow in magnitude over time. These dynamic effects are caused by the interaction among the system's individual elements, but also between the individual and contextual elements that comprise a system. Underlying this dynamic behavior is a process of self-organization that makes a complex system to evolve in an unpredictable way. Self-organization is therefore not subject to rules imposed from above, but from the rules emerging in the process of interaction.

Because all of the actors are independent and given the incomplete nature of information, no single actor can be expected to govern the system. It is also debatable if a common steering-organization staffed with representatives from the different actors represented in the system has the ability to govern such a system. Nilsson and Uhlin (2002) therefore underline the importance of self-organization; that complex systems have the ability to self-organize activities and where a precondition is that collaboration is based on a voluntary basis. Moreover, complex systems cannot, as simple systems, be reduced because they are open, dynamic and non-linear. This means that they cannot be dismantled into pieces and then be put together again (Kiel and Elliot, 1997). As Nilsson and Uhlin (2002) and Stacey (2001) point out, social systems are complex but we often view them as simple, which often results in inadequate analyses and misleading implications.

A first theoretical building block we will adopt and use in our empirical analyses is a view of an RIS characterized as an embedded complex system.

Innovation systems as learning communities

Learning is closely connected to innovations since innovation processes can be understood as interactive learning, which implies close co-operation among multiple actors, and where the result depends upon the actors' skills to adopt and utilize other actors' experiences and competencies (Lundvall and Johnson, 1994). One of the main concepts used when studying learning in innovation systems or regions is Camagni's (1991) concept of innovative milieu. An innovative milieu is characterized by geographical proximity, informal relationships between firms and collective learning processes. Geographical proximity facilitates information and knowledge exchange, as well as creating a social structure on which to build. A weakness with the innovative milieu is that it on a system level focuses on the system components necessary for support or for allowing learning to take place; it does not focus on the actual individual and organizational processes that lead to learning. Nonaka and Takeuchi (1995) highlight this weakness when describing knowledge creation as interplay between tacit and explicit knowledge on different organizational levels, forming a spiral of interplay between different kinds of knowledge (epistemological dimension) and knowledge levels (ontological dimension). Nonaka's and Takeuchi's (1995) model of knowledge creation can be viewed as a link between the individuals' learning and the organizational learning, and towards an inter-organizational learning where the creation of knowledge is a continuous reciprocal activity between tacit and explicit knowledge where organizational or inter-organizational knowledge creation, as distinct from individual knowledge creation, takes place when different phases of knowledge creation are put into action and form a continuous cycle, a spiral (Nonaka and Takeuchi, 1995).

To distinguish in what kind of areas the actors in an innovation system generate learning, Sanchez's (1996) description of three knowledge modes (know-how, know-why and know-what) that describes levels of understanding is applicable. *Know-how* builds on Simon (1962) and can be described as the "practical understanding" about the current state of a system. According to Johansson (2005), and following Grant's (1991) ideas, the current production systems and its routines may be seen as expressions of organizational know-how, and therefore organizational know-how can be understood as an organization's process technology. *Know-why* is the theoretical understanding of the principles governing the function of a system. In order to change the state of a system, know-why knowledge is critical. Know-why consists of both explicit scientific knowledge and of individual or organizational experience and intuition. Dutton and Thomas (1985) called the process of acquiring this kind of knowledge learning-by-studying. Sanchez (1996) proposes that product development is the critical function that an organization needs to develop and manage in order to capitalize on this specific knowledge mode. This implies that an organization's know-why forms the base for the organization's product technology, since it is in this area that existing products will be refined and new products will be developed. *Know-what*, eventually, refers to what the market demands or will demand. Know-what is generated through interactions between actors (Sanchez, 1996) and is based on experience and intuition that enables managers to predict what kind of products are suitable to develop and what the outcomes are likely to be. Know-what is hence the strategic understanding of the purposes to which specific forms of know-why (product technology) and know-how (process technology) knowledge might be applied in e.g. a firm's value creation process to meet the customers' needs and wants.

In an innovations system, it can be difficult for the individual actors to identify where and how the different knowledge modes are distributed. Therefore, as Lundvall and Johnson (1994) put it, it also becomes important to *know-who*: to know which actor who knows how and/or why and especially knows to do what.

As a second theoretical building block, we will adopt the view of RIS as learning communities combining individual and organizational learning in exchange and development processes involving different kinds of knowledge modes.

Innovation systems as path-dependent organizations

Evolutionary theory based on Nelson and Winter's *An Evolutionary Theory of Economic Changes* (1982) is often employed in innovation system studies, addressing the path-dependent nature of learning and development. Carlsson and Jacobsson (1997) hence point out that the learning taking place in innovation processes is rooted in a factual economic structure and in a local context that include strong elements of path dependency.

Innovation systems are not developed in isolation, but are dependent on their situational, geographic and historical context. Being socially constructed complex systems embedded in a factual and historic context and involved in continuous development/learning activities, this should imply that such systems to a significant degree are affected by their past and present track record. As Arthur (1989) puts it, "where industries" (or innovation systems) "can and do go is partly determined by where they are and therefore where they have been" (Guerrieri and Tylecote, 1997: 125). This implies that the structural capital of an innovation system in the form of accumulated expertise and experiences possessed by individual actors in the system are important, as well as the collective experiences and the level of trust between actors that emanate from its previous history. This also involves the risk of specific regional lock-in effects when studying regional economic evolution (Martin and Sunley, 2006) or, as in this case, the development of a specific RIS.

Utilizing Mahoney's (2000) path-dependence approach implies tracing a given outcome back to a specific set of historical events, and showing how these events are themselves contingent occurrences that may be understood in the light of relevant past and present contingencies. In a study of more than 400 larger corporations and their firm-specific technological competences, Patel and Pavitt (1997) concluded that the degree of complexity and path dependency are important factors for understanding the development of technological competence over time.

Recognizing the relevance of evolutionary theory and path dependency, in this chapter we will use as a third theoretical point of departure an evolutionary perspective when trying to make sense of our empirical case and the conclusions that may be drawn from this.

Method

As already noted, viewing a regional innovation system as a phenomenon constructed by its actors implies a desire to understand a regional innovation system through the lenses of the actors involved in developing the specific initiative. It also builds on the basic understanding that the development of a regional innovation system depends not only on the actors involved, but also on its situational and historical context. In order to understand such a phenomenon empirically, a case study approach was considered as the most appropriate to gain a both detailed and holistic understanding of the phenomenon.

The research that this chapter builds on has adopted an inductive approach, starting with studies of the empirical reality, followed by a phase trying to make sense of this reality by the use (or creation) of theory. Such an inductive approach is often seen as beneficial in qualitative and case-oriented research since it allows us to reach a deeper level of understanding (Lincoln and Guba, 1985). It also implies that the research setting is approached without any preconceived notions, since such preconceptions might hinder the researcher to see beyond the limits of his or her own theoretical framework. Presenting the results from such an inductive (and, in practice, normally abductive) research process is however not unproblematic. The design of this chapter therefore follows a more traditional outline and in order to simplify it for the reader follows a conventional deductive logic.

The empirical data to this chapter consist of primary and secondary sources such as interviews, written applications for public funding, reports and evaluations. Some of these reports and evaluations have been written by the authors. The written reports and evaluations that have been produced during the evolvement of ProcessIT Innovations and its pre-history are used as an empirical basis in an attempt to reconstruct the evolutionary trajectory of this innovation system. Personal interviews with key actors active during different phases of the development of the regional innovation system under study constituted another main source of data. Overall, the main strategy during data collection has been to understand why the innovative collaboration was started, who the actors involved were, what purposes and aims guided its actions and activities during the development process, and how the system with a base in these activities and in interaction with its context emerged as a regional innovation system.

When the "reconstruction" of the evolution of ProcessIT Innovations as an innovation system was accomplished and transferred into text describing "the story," a second round of interviews with key actors in each course of events was conducted to validate and/or question the reconstructed image of how ProcessIT Innovations evolved into an innovation system. As a result of these interviews, some minor changes were made to provide an accurate image of the evolutionary process.

In the next section, we will describe how ProcessIT Innovations has developed in four phases as an innovation system from its early start in 1999.

The evolution of the innovation system ProcessIT Innovations

Phase 1: InternetBay inventory (1999–2000)

Northern Sweden and Northern Finland have reached a leading position in IT, especially within wireless technology and its applications but also in general usage of IT-based services. InternetBay was established in 1999 as an industrial network with the aim to support its members' growth strategies in distance spanning and mobile applications. Today the cluster has developed into more IT-related industries and consists of nearly 100 members. The InternetBay region stretches around the Bay of Bothnia, from Vaasa in Finland to Umeå in Sweden.

In this InternetBay region, several large and competitive process industries are located. Maintaining market positions, however, requires continuous improvement and further development. In order to delineate how the IT industry and the universities in the region could support such a development, a needs inventory was initiated by Luleå University of Technology and InternetBay, and was conducted by InternetBay during the winter 2001/2002 at more than 20 of the region's process industries. Over 100 ideas for development were identified and then prioritized as short- or long-term needs for several of the companies (Johansson and Örtqvist, 2006). This need-driven inventory, in combination with the knowledge that the region's universities and IT companies possessed, constituted the most significant basis for what we today know as ProcessIT Innovations. As we will see, however, the endeavor to establish a need-driven innovation system was hardly a straightforward journey from A to B.

Phase 2: The SISU Pilot Project (2001–3)

The Swedish governmental agency for innovation systems (VINNOVA) was founded in 2001 with objective to promote innovation, growth and prosperity throughout Sweden. During VINNOVA's first year, the SISU-Pilot was launched as a learning project with a focus on collaboration among universities, industries and society in the Bay of Bothnia region (the northern part of Finland and Sweden). This project was one of five pilots that were supported by VINNOVA as a start-up for the coming VINNVÄXT program, whose objective was to provide learning for both VINNOVA and the participants while the basic conditions for a future innovation system were being mapped out.

The SISU-Pilot was a knowledge-driven development initiative targeting the region's IT industry. It consisted of four projects. The first project was a joint PhD program in Telecom and Internet research where Luleå University of Technology (Sweden), University of Lapland (Finland), University of Oulu, (Finland) and Meri-Lappi Institute (Finland) were the participating actors. Supervision, PhD courses and joint projects was the core of this project as was the possibility to stay as a visiting researcher or PhD student at a partnering university. The second project created a joint application for the sixth European framework program for research on functional product development. The third project established an annual Internet and cyber community conference that would be hosted biannually in Finland and Sweden. The fourth project was an interactive research project of the other three projects with the aim of generating more knowledge of the problems and opportunities for this innovation system.

Since the focus was on activities intended to strengthen the region's IT industry, InternetBay and its members were contributing to and benefiting from the SISU-Pilot. The lessons learned from the SISU-Pilot and the other pilot projects were important input into the forthcoming VINNVÄXT program. This program required the participation of allied actors from industry, government and university in joint efforts to create strong regional innovation systems by developing internationally competitive research and innovation environments in fields with high growth potential. This was to be achieved by funding need-driven research and development in order to strengthen cutting-edge competence and applications related to different growth fields and by strategic efforts to develop an effective RIS. From the outset, VINNVÄXT took the form of a competition in which "the best" submissions won, and where the winning projects/regions received up to 10 Million SEK per annum for 10 years. The SISU pilot project was one of several competing RIS initiatives to win the VINNOVA competition and thereby obtain access to valuable and critical funding for its development.

The co-operation among actors in the SISU-Pilot was, however, not without friction. Ylinenpää (2002) noted the challenges that the different actors had to handle, and observed that most pertained to cultural and institutional differences between how Swedes and Finns saw some issues: the meeting and the management culture differed between the countries and there was a language barrier. The SISU-pilot phase ended in 2003 and the vision of creating an innovation system bridging both sides of the Gulf of Botnia faded away due to the difficulties mentioned above but also because a two-national innovation system had to consider and adapt to both VINNOVA's and Tekes' (the Finnish governmental funding agency for technology and innovation) demands which required too much complexity and too heavy a bureaucratic burden for a start-up phase. Instead the universities, with the larger process industries in the northern part of Sweden, decided to create a regional innovation system on a purely national basis. This ambition took the form of a new regional innovation system: the Northern Light project.

Phase 3: Northern Light (2002–3)

As a result of InternetBay's inventory and the SISU-pilot, Luleå University of Technology and Umeå University as the leading actors, together with the county administrative boards, the county councils and the larger cities and some of the large industrial companies in Northern Sweden in 2002 started to develop a new joint, multifaceted innovation system under the brand name of "Northern Light." For this phase, four existing or potential target areas/ technology bases were identified as a base for further development: ProcessIT, field test of vehicles and components, e-health, and media technology and learning; all areas related to information technology:

- ProcessIT included automated treatment of process, production and product information. The idea was to develop new products and services based on the needs previously identified in the regional process industry. The focus was on measurement technology, mobile workplaces, automated process optimization, and maintenance systems.
- Field test of vehicles and components was by 2002 an established area in the inland of the region exposing an annual growth rate between 15 and 20 percent, historically based on the cold snowy climate and the sparsely populated area that provided some secrecy for the manufacturers in or related to vehicle production.
- e-Health was a third area where the region had a good position. Almost 50 percent of joint projects between the IT industry and the county councils in Sweden had been conducted in the region. e-Health was a rapidly growing industry, and the focus of Northern Light was to develop new e-health applications.
- Media technology explained how to design, implement and evaluate prototype multimedia systems that included storage, distribution, and presentation of interactive media such as audio, video, music and animations, as well as non real-time media such as text and graphics that could be used for learning.

The main actors behind Northern Light were the universities. The innovation system was introduced through a classic top-down approach, manifested by pamphlets presenting Northern Light as "a winning innovations system" that companies were urged to join. A major reason why the universities tried to squeeze as much as possible into Northern Light was their adaptability to the regional growth program that the government had introduced at this time, and where transparency with this policy program on national and regional levels was regarded as a gateway for future EU structural funds and complementary regional funding.

The development of the Northern Light initiative, however, turned out to be a very complicated. The four clusters had few similarities, implicating that economy of scale or scope was difficult to obtain in the project's core activities. The ambition was to establish think tanks in each cluster, dedicated to innovative business models building on distance spanning technologies and where new business logic would be introduced and applied. These business models could then be transferred to other areas as learning cases, where the administrative board of Northern Light viewed business intelligence, coordination and planning, and knowledge transfer between and within the Northern Light initiative as its main task.

The application from Northern Light to VINNOVA was rejected on the grounds that the innovation system design was too broad and lacked clear synergies among the areas. The Northern Light therefore ceased to exist as an RIS initiative, but the areas included in the Northern Light concept are still developing on their own merits.

Phase 4: ProcessIT Innovations (2003–)

Of the four target areas in the Northern Light project, a majority of the regional actors involved regarded ProcessIT as the most promising soil for further development because of InternetBay and Luleå University of Technology's needs inventory in the process industries and the work that had already been done by the SISU-pilot and Northern Light. Moreover, the ideas behind ProcessIT was also consistent with VINNOVA's ideas that establishing a dynamic innovation system focusing on regional growth required need-driven research that would result in new products, services, and new companies.

ProcessIT is based on automated treatment of process-, production- and product information and includes modeling, measurement technology, automated processes, signal processing and information transfer technology. The prerequisites for an innovations system in the region were already there since the large process industries in Northern Sweden were, as already described, strong and well established on in their respective world markets. In addition, the IT-sector was in a good position, especially within wireless technology and applications but also in offering general IT-based services. Moreover, an insight that IT had strengthened the large process industries' positions on their markets was at this time established, and IT was expected to be even more important for future competitiveness. With this background, the region had excellent and unique conditions for an effective innovation system initiative based on the knowledge possessed by the IT companies and the universities, combined with the demand deriving from the large process industries. Their collective effort in this area was expected to produce economic growth, based on solutions such as new technical products and services leading to increased competitiveness for the process industry, and to new product applications within the regional IT industry. As a result of this reasoning, in 2003 ProcessIT Innovations started to support research projects for the process industry with the objective of contributing to the creation and growth of companies by developing new products and services based on the process industry's needs.

With the knowledge gained from both the SISU-pilot and the innovation system, Northern Light, ProcessIT Innovations received support from the VINNVÄXT programme in 2004. As winner of the VINNVÄXT competition, they were guaranteed 60 million SEK in financial support over ten years. With matching funds from the county administrative boards of Norrbotten and Västerbotten, the municipalities of Luleå, Piteå, Skellefteå and Umeå, the universities of Luleå and Umeå, and the process industries in the region, ProcessIT Innovations had attracted 120 million SEK in financial capital to meet the process industry's needs for new products and services based on information and communication technology. The support from VINNOVA was, moreover, not only financial. VINNOVA's aim was to establish innovation systems that should be recognized as world-class in their fields during VINNVÄXT's 12-year program period. In order to facilitate the development of such excellent innovative milieus, VINNOVA initiated the utilization of a five phase model initially developed and introduced by Cluster Navigator Ltd when training management teams from the winning innovation system initiatives. The model consisted of 12 steps to develop the cluster or innovation system and can be classified as a linear process model. According to this normative model, cluster and innovation system development is a planned and deliberate process conducted by a neutral and independent process leader that is supported by a financier such as local government (Ffowcs-Williams, 2004). The management team of ProcessIT Innovations has since then received training and support from VINNOVA and participates regularly in activities in which learning from other VINNVÄXT winners is an important element.

ProcessIT Innovations is governed by its board of directors, which is well-established in the region's industry, universities and local communities. The board consists of representatives from the process industries, international suppliers to the process industry, ICT companies, the universities of Luleå and Umeå, and the county administrative boards. The executive management team of ProcessIT Innovations includes the process leader and representatives from the universities of Luleå and Umeå, and from the region's major process industries. Their main task is to execute the board of directors' strategic decisions. A research management team, supported by an industry advisory board, also plays an important role in the way activities develop. Their task is to assess and take responsibility for long-term plans and the scientific quality of the research done in R&D projects and in research programs. Several pre-studies in different technology areas have today been conducted, where many of these feasibility studies has resulted in new research projects that have become mature enough to spin out new products and new companies.

The recent development of this RIS links the present ProcessIT Innovations to its pre-history. Since both the large process industries in the region and the majority of the IT companies are located along the coast of the Bay of Bothnia, the old vision to build an innovation system consisting of actors from both sides of the bay that was a basic idea during phase 2 (the SISU pilot project) is now being implemented again by InternetBay and by actors related to ProcessIT Innovations. A pilot project, Profi, was therefore launched in an attempt to expand the regional innovation system to both sides of the bay (Johansson, 2006).

Analysis

Figure 10.1 depicts how ProcessIT Innovations have developed as a regional innovation system and how its scope has changed during this process.

In this section, we will analyze the evolution of ProcessIT Innovations with its pre-history by using the building blocks introduced in the second section above, and describe our case as an embedded and complex system, as a learning community, and as a path-dependent construction.

ProcessIT Innovations as an embedded and complex system

The innovation system that has emerged as ProcessIT Innovations has, as we have depicted above, developed through a number of initiatives or phases. The leadership has shifted in the course of this journey. Initially the organization InternetBay was in the driver's seat and launched the idea of an innovation system as an outcome of its needs inventory. Thereafter, in the SISU-pilot project, the universities took the wheel. After the SISU-pilot phase the universities kept this position and with the country administrative boards they attempted but failed to launch a more diverse and multifaceted innovation system initiative (Northern Light). Building on these experiences, the universities, in close collaboration with the process industry in the region and with support from regional government, joined forces when establishing the innovation system ProcessIT Innovations to which a neutral and independent process leader was recruited. The recruited process leader had a background in the IT industry, was highly respected by the region's process industry, and today has his office at one of the universities.

In ProcessIT Innovations, all actors have different objectives and different roles in the innovation system (Johansson *et al.*, 2007). The objective of the process industry is to strengthen its position by increasing its effectiveness by enhancing the use of information technology in all production processes. Their role is to identify needs and to participate in the entire product development process with a strong commitment to pilot projects and pilot installations. The main objective of the IT companies is to develop commercial products for the international market with support from the process industry, and their role in

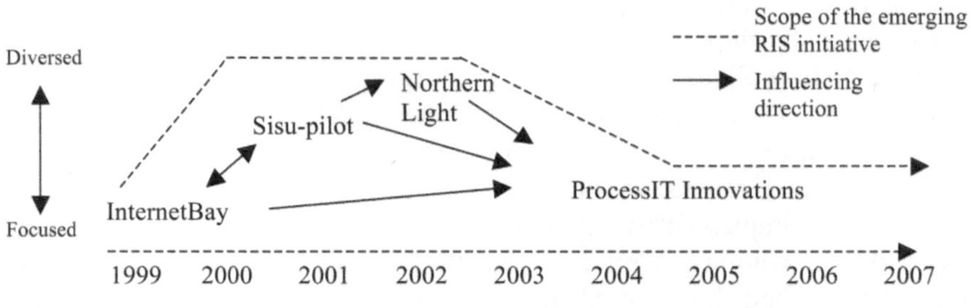

Figure 10.1 Scope and infuencing directions in the evolution of ProcessIT as an RIS.

the innovation system is to be responsible for developing, packaging and commercializing these products and services. The main objective for the universities is to establish themselves as a leading research environment in Europe in areas covered by ProcessIT Innovations and to gain a leading position among European universities with respect to commercializing research results. The universities have an important role to add new knowledge that facilitates innovative solutions that fulfills the needs derived from the process industry. In so doing, the universities can earn recognition as leading research institutes, and the industry can take advantage of engineers and PhDs. The public actors are to promote growth in the region, to create the right conditions for collaboration among the actors in the innovation system, and to maintain a long-term focus when making decisions. This is expected to lead to new innovative products, services and processes that make the process industry more efficient and that enables new products to be developed and sold on new markets by the IT companies. Figure 10.2 illustrates how the work in ProcessIT Innovations reaches these objectives.

As noted earlier, the starting point was an up-to-date list with identified needs of the industry, originating from an inventory by InternetBay that identified both short- and long-term needs for several companies. The main idea is that ProcessIT Innovations shall work with needs that are of great importance for the process industries and initiate research and development in areas where potential services and products based on IT solutions have been identified. Around each identified need, interested parties form research and development project teams that perform research and development, market research, packaging, pilot testing and evaluation. The outcome of the projects in the form of products and services shall after a project is completed be transferred to a product providing company. This way of collaboration questions the traditional triple-helix metaphor introduced by Etzkowitz and Leydesdorff (1997) since it widens the concept to involve a critical collaboration among small firms, large firms, and universities. Moreover, the

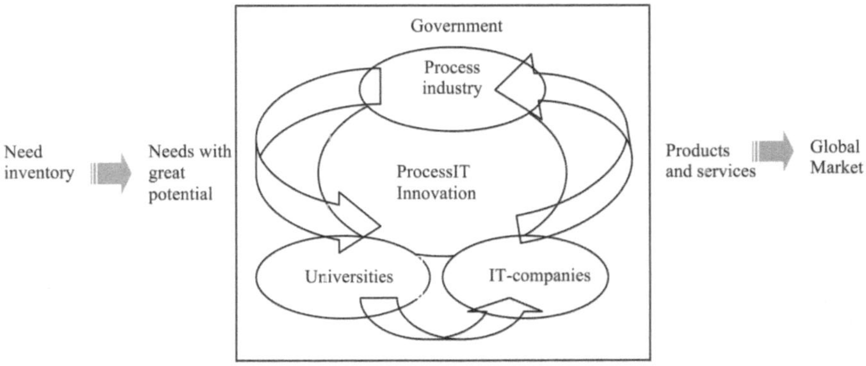

Figure 10.2 ProcessIT Innovations' work method.

mode for collaboration in ProcessIT Innovations demonstrates that government's role is limited to setting the institutional rules and to providing initial funding (see also Johansson *et al.*, 2007 for a more detailed discussion).

Both the SISU-Pilot and the initiative Northern Light are examples of development patterns that are not constant because when the system's environment changes, so does the behavior of the system as a whole since the system is constantly adapting to new conditions. It is therefore doubtful that a common steering-organization has the ability to, in a classic sense, *govern* an RIS initiative. Their role is important but restricted to preparing for unexpected events, selecting projects, facilitating meetings, and acting as a lubricant in the system when creating awareness of the potential benefits of a successful innovation system for the actors. In fact, according to Sarasvathy (2001), a mode of decision-making employed during different phases of the ProcessIT Innovations initiative could better be understood as "effectuation" (which is action-oriented and departs from available/accessible resources) rather than as "causation" (which is a long-term and planning-oriented mode of making decisions focused on risk reduction and control). As the evolution of this RIS from its start until today demonstrates, functional innovative collaboration engaging different actors and sectors of society normally develop the ability to self-organize. Initiatives emerging in the process of interaction therefore are less problematic and more rooted among the actors as compared to initiatives imposed from above or by a single category of actors (here illustrated by the experiences from the Northern Light project phase).

ProcessIT Innovations as a learning community

ProcessIT Innovations' work method offers excellent opportunities for learning on different levels in the Nonaka spiral of knowledge creation. Many of the organizations and individuals within each organization have been active in the innovation system from the outset. This fosters know-who knowledge since knowing who is learned from social interaction via activities in the innovation system, day-to-day dealings with customers, suppliers and university researchers. All these interactions increase the knowledge about partners' strength and weaknesses and foster trust among the actors.

Organizational know-how, such as a firm's process technology, relies on practical experience through tacit learning, but also increasingly on network relationships with different partners in the innovation system. Collaboration in an innovation system can also encourage actors to be even more specialized since they can take advantage of other actors' know-how knowledge and thereby lower their internal transaction cost, while the other actors benefit from economies of scale or scope. The focus on know-how knowledge was clarified when ProcessIT Innovations was established in 2003; before that, most of the focus had been on generating know-why knowledge as an outcome from the universities and their research.

As noted already, know-why is at the heart of an organization that aims to develop innovations since know-why is closely connected to product technology. In the early phases of the evolution of the innovation system initiative, much focus was put on know-why knowledge when effort was put on developing ProcessIT as a field of research. In addition, in today's ProcessIT Innovations, much effort is on developing a better theoretical understanding of the field of technology that ProcessIT represents; knowledge that later can be applied in new products and services. As Johansson *et al.* (2007) report, the university researchers' focus in ProcessIT Innovations was and is on new knowledge generation where commercialization of this new knowledge is of minor interest for these actors. Commercialization, however, is the driver for market-driven companies to engage in cross-sector collaboration.

In ProcessIT Innovations the process leader and the research management team are central actors when it comes to deciding which projects are going to be prioritized and which new know-why knowledge that is going to be produced. The process leader also has to deal with the question of how the system that governs the innovation initiative should be designed. Governing a complex system is, as already noted, not trivial, and may at best materialize when the management team influences development processes in a specific direction (Groenewegen and van der Steen, 2006; Nilsson and Uhlin, 2002). This has also motivated VINNOVA to support process leaders by providing different fora where experiences across project boarders can be shared and obstacles can be discussed. Moreover, VINNOVA finances interactive research on all VINNVÄXT winners where the research aim is to contribute new knowledge to innovation systems and to advise the process leaders on the basis of theoretical knowledge on innovation systems and empirical observations made through interactive innovation system research (Laestadius *et al.*, 2007).

When it concerns know-what knowledge, or what the market really demands or will demand, ProcessIT Innovations is well situated because its work methods seem to provide this information in an efficient way. This work method has so far been very successful since it clearly communicates the objective of the innovation system initiative and how this objective will be achieved. This is probably the main reason for the functionality of the emerging innovation system since all actors have some strategic understanding of the purposes to which specific forms of know-why (product technology) and know-how (process technology) knowledge might be applied in the innovation system's value creation process to meet the process industry's needs and wants.

ProcessIT Innovations as a path-dependent construction

As the history and pre-history of ProcessIT Innovations demonstrates, the evolution of a regional innovation system initiative is a process that is far from linear but is rather a result of a dynamic learning and search process that involves interplay between several actors and factors that no single actor alone can control. Unpredictable occurrences such as high levels of uncertainty, unexpected events, and inter-firm rivalry allow for the recombination of knowledge and the stimulation of creativity. Uncertainty and complexity should from this perspective be viewed as positive phenomena in entrepreneurial organizations and innovation systems (Ylinenpää, 2007). The roles of the process leader and the formal organization are then to make sense of a complicated context and to create a sense of belonging and understanding (Weick, 1995), to facilitate meetings and to act as a lubricant in the system. Such a role is facilitated by following the classical management cookbook, but also involves considerable amounts of learning, parrying and effectuation (Sarasvathy, 2001).

Continuous learning through accumulation of experiences and knowledge during the evolution of an innovation system such as ProcessIT Innovations invariably implies a form of path dependency that will affect the way in which a specific innovation system develops and the activities of its actors. In our empirical case, several examples of path dependencies may be observed. Referring again to Arthur's (1989) statement that the construction of industries and their development trajectories is affected by "where they are and where they have been," it is obvious that established regional strongholds materialized as important industries and advanced and preferably applied research expertise have had a major impact on the construction of the ProcessIT Innovations initiative. Besides noting that the history of a specific region matters, our case illustrates the path-dependent impact from previous experiences in the history of this RIS initiative. This is marked by how experiences from the diversified and top-down-oriented Northern Light project was transferred into a more focused innovation system concept anchored in real needs articulated by regional industry. Another example is how this innovation system employing elements of opportunistic behavior incrementally learns what it takes to be successful in competing for VINNOVA funding, and (in phase four) manages to "break through" the firewalls of the government agency (cf. Wigren and Melin, 2007, for a somewhat ironic illustration of "the rules of the game" in competing for funding from the VINNVÄXT program). A third example is how ProcessIT Innovations today has returned to and to some degree "reuses" the concept of collaborating with Finland underlying the SISU pilot project, but today materialized as two specific but inter-related projects (ProcessIT Innovations and the Profi project).

Conclusions and suggested implications

The analysis of our empirical case ProcessIT Innovations forms a basis for several conclusions and implications referring to the development of regional innovation systems. First, development of innovation systems by using a top-down approach seems to be a somewhat dangerous route to success, since the development of an innovation system (as demonstrated by the empirical example in this chapter) is seldom a linear process. As suggested by Westerberg *et al.* (2007), innovation system development may start when the public actors, in consultation with industry representatives and university representatives, set regional priorities, allocate funding, and influence the actors to be involved in the dialogue concerning the innovation system's initiative and also which actors that are to be represented in steering boards and committees. Once regional priorities have been set and the steering boards and committees are staffed, the public actors' intervention declines and the other actors are freer to choose ways to reach the objective with the innovation system. However, innovation systems do not follow a linear development and a bottom-up approach seems more suitable since the system is better described as complex systems that, over time, develop on the basis of interaction, learning and experiences gained by the actors that constitute the system. The development of the innovation system is actor-dependent in the sense that the result is driven by characteristics of the actors and the actors' ability to discover and use contingencies inherent in and external to the innovation system. An important driver of the progress of the innovation system is the actors' learning from previous experiences in the innovation system.

Instead of a structured and well-ordered process, the evolution of regional innovation system initiatives such as ProcessIT Innovations may better be understood as interaction among actors engaged in a learning community. Occasional meetings among actors in seminars or other events in the innovation system are important since these activities facilitate social contacts among the actors. However, such activities are not enough since learning and trust-building among the actors involved normally emerge from running joint projects over a longer time period. Co-operation in an innovation system hence builds on mutual trust and this takes a long time to build; a conclusion well in line with earlier research on inter-organizational collaboration (Vangen and Huxham, 2003). Mutual dependence also has an important impact on the innovation system's outcome. Ahlström-Söderling (2003) states that mutual dependence seems to be weaker when competencies from different types of actors are not needed in the projects. The design of ProcessIT Innovations' work method takes this issue into consideration since the work methods are very stringent and where all actors have well defined roles based on their specific competences, which fosters a mutual dependence among the main actors in the innovations system. During operative phases of the evolution of an RIS, the role of public authorities (as illustrated by our empirical case) is limited to being supporters and facilitators for overcoming institutional barriers to the evolutionary process, and as a gatekeeper for communication with national government bodies. Adopting such an insight would focus on market-induced initiatives where the public actors more act as facilitators and moderators focusing on removing market imperfections and to serve as an interface for communication with national bodies such as VINNOVA.

Our analysis also indicates that know-who is important in an innovation system, where know-who refers very much to knowing individuals since individuals constitute organizations. An advantage in the studied case was that several of the key actors were active in the innovation system from its early beginning. A focus on need-driven research combined know-how and know-why and here offered an arena for IT companies, expected to commercialize the system's new products and services, to contribute with their expertise. As the empirical example ProcessIT Innovations also indicates, know-what knowledge is closely connected with need-driven research since it incorporates both know-why and know-how. By combining the different motives of actors, various motives for their participation in the system – motives that in turn are related to different modes of knowledge – necessary commitment from different stakeholders may be achieved. A dominant focus on only one of these knowledge modes would probably have made the support and engagement from other actors more difficult to obtain.

The endeavor of developing a regional innovation system is also strongly characterized by path dependency. As most similar initiatives, our case ProcessIT Innovations tries to exploit and develop existing regional strengths manifested by the region's industrial structure and specific research expertise in the region's universities. As noted by Martin and Sunley (2006), such a strategy however also runs the risk of regional lock-in effects which may counteract a functional restructuring of an outdated regional industry base. Building on our empirical case, the story of ProcessIT Innovations seems to support the understanding that it is favorable for a regional innovation system to build on what a region "already has" rather than trying to develop something new in areas exposing a low degree of contingencies with the established industrial and knowledge-based structure. A need-based innovation system, as compared to a science-driven innovation system, is also easier to manage when the idea behind the innovation system is easy to communicate, when the actors can easily see their role in the system, and when the perceived outcome and usefulness of activities in the innovation system is known in advance. As a potential path-dependent effect, however, the scope of the innovation activities for which the system strives tend to be limited, implying that the innovation system might only give rise to incremental innovations.

Our observation from our illustrative case is that system actors such as universities and regional and local public bodies use the rhetoric of the innovation system to gain legitimacy and to secure resources that otherwise would have been difficult to obtain also underpins a conclusion related to path dependency, since these two categories of actors seems to adapt more quickly to the requirements that public financing agencies like VINNOVA state in their calls for project funding. A benefit of being so adaptive is that the chance for funding increases. A drawback can be that the innovation system becomes short-term and that the evolution of the innovation system will be based on the opportunistic possibility to receive funding and not on more sustainable underlying conditions such as end customer needs and demands. An implication, then, is that it is important to involve stakeholders close to market early in the evolution of regional innovation system initiatives.

This chapter has addressed the question of whether and/or to what degree a successful RIS initiative may be planned from the top or if such initiatives emerge from more bottom-up oriented initiatives and circumstances. By using a combination of three theoretical perspectives or lenses and analyzing our RIS initiative as a complex, learning and path-dependent system, the answer seems obvious: the evolution of RIS initiative is a complex process embedded in a factual and historic context where path-dependence has a major impact and where continuous learning characterizes the evolutionary process. One argument for stimulating planned innovation systems is that it allows policy-makers to pinpoint and prioritize selected parts of a region's economy in order to spend public investments more efficiently. Castells and Hall (1994), however, compared several initiatives around the world and concluded that this strategy was problematic. Porter (1998) convincingly showed that the future growth of a region is rooted in present strengths and determinants of competitive advantage, and assigns a very limited (and sometimes even dysfunctional) role to government and policy-makers.

Yet, one conclusion drawn from previous research and the empirical case presented in this chapter is that innovation systems may be regarded as complex systems that, over time, develop based on the learning and experiences gained by the actors that constitute the system. Such systems may only to a very limited degree be planned or directed from a top-down perspective, but evolve and develop as a result of path-dependences and interactive learning processes among a variety of different actors. Building on such an insight, policy actors such as VINNOVA are advised not only to rely on management practices inspired by causation and manifested as formal planning and linear thinking in their efforts to stimulate innovation system development, but also develop support structures that recognizes more entrepreneurially and effectuation-oriented systems (Cook and Leydesdorff, 2004; Ylinenpää, 2008).

Future research in the field would benefit from more longitudinal studies of regional innovation system evolution while also addressing empirical contexts that we have not been able to study here. An interesting question is if and how path dependencies in the form of a specific regional (and according to some authors, "colonial") history affects the way in which a regional innovation system is constructed. Another interesting area is to investigate whether the development of such systems differs when comparing non-metropolitan areas (which is the topic of our study) and metropolitan areas. A third important avenue for future research pinpointed by Laestadius *et al.* (2007) and by Teräs and Ylinenpää in another chapter of this book is to address how regional innovation systems interact and benefit from collaboration with actors and competences outside the region – a theme we have only very briefly touched upon in this study.

Note

1. The authors gratefully acknowledge the fruitful collaboration we have had with actors related to the empirical case used in this chapter, as well as the insightful comments and suggestions we have received from the editors of this book.

References

Ahlström-Söderling, R. (2003) "SME strategic business networks seen as learning organizations." *Journal of Small Business and Enterprise Development*, 10 (4): 444–54.

Arthur, B. (1989) "Competing technologies, increasing returns, and lock-in by historical events." *Economic Journal,* 99: 116–31.

Asheim, B. T. and Coenen, L. (2005) "Knowledge bases and regional innovation systems: Comparing Nordic clusters." *Research Policy* 34, 1173–90.

Asheim, B. T., Coenen, L. and Svensson Henning, M. (eds) (2003) *Nordic SMEs and regional innovation system*, Oslo: Nordic Industrial Fund.

Bijker, W.E., Hughes, T.P. and Pinch, T.J. (eds) (1987) *The social construction of technological systems: New directions in the sociology and history of technology*, Cambridge, MA: MIT Press.

Camagni, R. (1991) *Innovations, Networks,* London: Belhaven London Press.

Carlsson, B. and Jacobsson, S. (1997) "In search of a useful technology policy: General lessons and key issues for policy makers," in Carlsson, B. (ed.), *Technological systems and industrial dynamics,* Boston, Dordrecht, London: Kluwer Academic Publishers.

Carlsson, B., Jacobsson, S., Holmén, M. and Rickne, A. (2002) "Innovation systems: Analytical and methodological issues," *Research Policy*, 31: 233–45.

Castells, M. and Hall, P. (1994) *Technopoles of the world: The making of 21st century industrial complexes*, London: Routledge.

Cooke, P. and Leydesdorff, L. (2004) "Regional development in the knowledge-based economy: The construction of advantage," *Journal of Technology Transfer*, 31: 5–15.

Cooke, P. and Memedovic, O. (2003) *Strategies for regional innovation systems: Learning transfers and applications*, Vienna: Policy papers, UNIDO.

Cumbers, A. and MacKinnon, D. (2004) "Introduction: Clusters in urban and regional development," *Urban Studies*, 41 (5/6): 959–69.

Dutton, J.M. and Thomas, A. (1985) "Relating technological change and learning by doing," in R.S. Rosenbloom (ed.), *Research on technological innovation, management and policy*, Greenwich, CT: JAI Press, 187–224.

Etzkowitz, H. and Leydesdorff, L. (1997) *Universities and the global knowledge economy, a triple helix of university–industry–government*, London: Pinter.

Ffowcs-Williams, I. (2004) *Klusterutveckling: fem faser, tolv steg,* in Christensen, L. and Kempinsky (eds) *Att mobilisera för regional tillväxt.* Lund: Studentlitteratur.

Grant, R. M. (1991) "The resource-based theory of competitive advantage," *California Management Review*, 33: 114–35.

Groenewegen, J. and van der Steen, M. (2006) "The evolution of national innovation systems," *Journal of Economic Issues*, 40 (2): 277–85.

Guerrieri, P. and Tylecote, A. (1997) "Interindustry differences in technical change and national patterns of technological accumulation," in Edquist, C. (ed.), *Systems of innovation, technologies, institutions and organizations*, London, Washington: Pinter.

Hommen, L. and Doloreux, D. (2003) *Is the regional innovation system at the end of its life cycle? Innovation in Europe: Dynamics, institutions and values*, Roskilde University, Denmark.

Johansson, J. (2005) *Mindre verkstadsföretags jakt på konkurrensfördelar genom strategiska allianser*, Licentiate Thesis, Luleå University of Technology.

Johansson, J. (2006) *Innovationssamverkan inom ProcessIT Innovations*. Rapport, Avdelningen för företagsekonomi och verksamhetsutveckling (Entreprenörskap), Luleå tekniska universitet.

Johansson, J. and Örtqvist, D. (2006) *An evaluation of InternetBay*, Research Report, Division of Entrepreneurship, Luleå University of Technology.

Johansson, J., Wincent, J. and Ylinenpää, H. (2007) "Triple-Helix som koncept för att förstå sektorsöverskridande samverkan: funktionell modell eller retorisk återvändsgränd? En analys av rollförväntningar i ProcessIT Innovations," in Laestadius, S., Nuur, C. and Ylinenpää, H. (eds) *Regional tillväxtkraft i en global ekonomi*, Stockholm: Santérus Academic Press.

Kiel, L.D. and Elliott, E. (1997) *Chaos theory in the social sciences*, Ann Arbor, MI: University of Michigan Press.

Laestadius, S., Nuur, C. and Ylinenpää, H. (eds) (2007) *Regional tillväxtkraft i en global ekonomi*, Stockholm: Santérus Academic Press.

Lewin, R. and Regine, B. (1998) *The soul at work: Unleashing the power of complexity for business success*, London: Orion Books.

Lincoln, Y.S. and Guba, E.G. (1985) *Naturalistic inquiry*, Newbury Park, CA: Sage Publications.

Loye, D. and Eisler, R. (1995) "Chaos and transformation: Implications of non-equilibrium theory for social science and society," *Behavioral Science*, 32: 53–65.

Lundvall, B.-Å., and Johnson, B. (1994) "The learning economy," *Journal of Industry Studies*, 1 (2): 23–42.

Mahoney, J. (2000) "Path dependence in historical sociology," *Theory and Society*, 29: 507–48.

Martin, R. and Sunley, P. (2006) "Path dependence and regional economic evolution," *Journal of Economic Geography*, 6 (4): 395–437.

Nelson, R. and Winter, S. (1982) *An evolutionary theory of economic change*, Cambridge, MA: Harvard University Press.

Nilsson, J-E. and Uhlin, Å. (2002) *Regionala Innovationssytem: En fördjupad kunskapsöversikt*, VINNOVA Report VR 2002:03.

Nonaka, I. and Takeuchi, H. (1995) *The knowledge-creating company*, Oxford: Oxford University Press.

Patel, P. and Pavitt, K. (1997) "The technological competencies of the world's largest firms: Complex and path-dependent, but not much variety," *Research Policy*, 26: 141–56.

Pinch, T.J. and Bijker, W.E. (1984) "The social construction of facts and artefacts: Or how the sociology of science and the sociology of technology might benefit each other," *Social Studies of Science*, 14: 399–441.

Porter, M. (1998) "Clusters and the new economics of competition," *Harvard Business Review*, November–December: 77–90.

Sanchez, R. (1996) "Strategic product creation: Managing new interactions of technology, markets and organizations," *European Management Journal*, 14 (2): 121–38.

Sarasvathy, S.D. (2001) "Causation and effectuation: Towards a theoretical shift from economical inevitability to entrepreneurial contingency," *Academy of Management Review*, 26: 243–64.

Simmie, J. (2003) "Innovation and urban regions as national and international nodes for transfer and sharing of knowledge," *Regional Studies*, 37 (6/7): 607–20.

Simon, H. (1962) "The architecture of complexity," *Proceedings of the American Philosophical Society*, 106: 467–82.

Stacey, R.D. (2001) *Complex responsive processes in organizations*, London: Routledge.

Vangen, S. and Huxham. C. (2003) "Nurturing collaborative relations: Building trust in inter-organizational collaboration," *Journal of Applied Behavioral Science*, 39: 5.

Weick, K.E. (1995) *Sensemaking in organizations*, Thousand Oaks, CA: Sage Publications.

Westerberg, M., Wincent, J. and Ylinenpää, H. (2007). "Triple-Helix och utvecklingen av innovationssystem," in Laestadius, S., Nuur, C. and Ylinenpää, H. (eds) *Regional tillväxtkraft i en global ekonomi*, Stockholm: Santérus Academic Press.

Wigren, C. and Melin, L. (2007) "Vinnväxttävlingen – spelregler för vinnare, och förlorare," in Laestadius, S., Nuur, C. and Ylinenpää, H. (eds) *Regional tillväxtkraft i en global ekonomi*, Stockholm: Santérus Academic Press.

Ylinenpää, H. (2002) *Gränsöverskridande triple helix samverkan mellan Sverige och Finland*, working paper 2002:51, Luleå University of Technology.

Ylinenpää, H. (2007) "Innovationssystem ur ett aktörsperspektiv. Nätverk och entreprenörer," in Laestadius, S., Nuur, C. and Ylinenpää, H. (eds) *Regional tillväxtkraft i en global ekonomi*, Stockholm: Santérus Academic Press.

Ylinenpää, H. (2008) "Entrepreneurship and innovation systems: Towards a development of the ERIS/IRIS concept," in Johannisson, B. and Lindholm Dahlstrand, Å., *Bridging the functional and territorial views on regional entrepreneurship and development*, Stockholm: Swedish Foundation for Small Business Research.

11 Gender in governance of regional innovation

Why gender matters and is mainstreamed in the Swedish case

Malin Lindberg

Introduction

Viewing policy as "management of change," this chapter sets out to explore how the governance of regional innovation induces change of gendered practices in regional innovation systems (RISs). One of the main topics of this book is to discuss how regions can introduce and realize innovative strategies and the role of innovation governance in such processes – setting national innovation policy in relation to specific initiatives to spur regional innovation hubs. This forms the basis for this chapter's analysis of how efforts to mainstream gender in a selection of RISs in Sweden have been affected by a national program managed by the Swedish Governmental Agency for Innovation Systems (VINNOVA). The analysis – based on common gender-scientific theories – casts a light on what motivates RISs to mainstream gender and how successful different modes of implementation are. It will be illustrated how gender is interesting, specifically in relation to multi-actor governance of regional innovation, by introducing ways to expand the range of actors, areas and innovations involved in such processes.

The gender-scientific theories employed in this chapter are "doing gender" and "gender mainstreaming." In the theory of "doing gender," gender is seen as continuously constructed in organizations and society (West and Zimmermann, 1987; Fenstermaker and West, 2002). The theory of "gender mainstreaming" provides classifications of how a gender perspective can be systematically integrated into all policies, processes and organizations (Squires, 2005; Walby, 2005). The main research question to be answered in this chapter is to what extent the main elements of the theories of "doing gender" and "gender mainstreaming" are present in the motivation and implementation of RIS efforts to mainstream gender.

The chapter is initiated by an account of the research design of the performed study, scrutinizing the data selection and the prospects of theoretical generalization. A short introduction to the theories of "doing gender" and "gender mainstreaming" is then provided, suggesting that gender can be regarded as continuously constructed in interaction between people in organizations and exposing how gender can be mainstreamed in various manners in organizations. This is followed by an overview of theoretically identified incentives for gender mainstreaming in RISs and regional innovation policies. The gender-scientific theories are thereafter employed for analyzing the data, exposing why and how gender has been mainstreamed by RISs supported by VINNOVA. Finally, conclusions are drawn concerning why gender matters and how it is mainstreamed in the governance of regional innovation.

Research design

The study presented in this chapter uses Sweden as an empirical case. The selection of Sweden underpins the investigation with relevant and substantial data since Sweden can be regarded as a forerunner in its efforts to implement a gender perspective in RIS practices, innovation research and innovation policies (Danilda and Granat Thorslund, 2011). Three RISs are focused on in the study: Triple Steelix, Fiber Optic Valley and Innovation at Interfaces. These have been chosen since they all have proved to be ambitious in their efforts to mainstream gender in their organizations and, as VINNVÄXT winners, provide fruitful illustrations of the interplay between national innovation policies and RIS practices. The national policy program BEGINN, managed by VINNOVA, is employed as a policy context to the RIS efforts to mainstream gender.

The data informing the study consists of text documents describing: (1) the RIS efforts to mainstream gender in their organizations; and (2) the contextual frames for their efforts set by the national policy program financing their efforts. Specifically, text documents from the official websites of the three RISs have been used, as well as from the official websites of VINNOVA's programs BEGINN and VINNVÄXT. These documents describe the efforts made by the RISs to mainstream gender and mediate the assumptions and prerequisites underpinning these efforts.

The data covers the period of 2003–9 for Innovation at Interfaces, 2004–9 for Fiber Optic Valley and 2004–9 for Triple Steelix. Fiber Optic Valley is located in the mid region of Sweden, working to make Sweden into the world leader in the development of products and services based on fiber optics. This goal will be reached by "cutting-edge" research projects and by assisting the growth of new and existing businesses. The vision of Innovation at Interfaces – located in the region of Skåne in Southern Sweden – is to create good and nutritious food for more people through co-operation and innovation. They will achieve this by arranging meetings at the interfaces between different areas of knowledge and by enhancing the development of innovative products, services and projects. Triple Steelix – situated in the region of Bergslagen in the mid-west of Sweden – strives to improve the exchange of information between the private, public and academic sector concerning the existing base of knowledge around steel. New networks, new meeting points and stimulated entrepreneurship are three desired results.

The data from VINNOVA's policy program BEGINN covers the period of 2004–8 and the data from VINNVÄXT covers the period of 2003–8. BEGINN was initiated in 2004 in order to further needs-driven gender research for innovation, thus aligning the areas of gender research and gender equality work. Two calls have been made within BEGINN. The first call was "Gender perspective on innovation systems and gender equality" (2004–8), encouraging the development of methods for gender equality work in organizations, as well as critical examinations of the concepts of "innovation" and "innovation system" from a gender perspective. The second call was TIGER (2007–8), furthering applied gender research in those strong research and innovation milieus that had already been granted long-term financial support from VINNOVA – e.g. in the program VINNVÄXT – or any other financier. The program VINNVÄXT was initiated by VINNOVA in 2002, aiming to promote RISs with a maximum of 10 million SEK (approximately €940,000) per year up to 10 years. Funding was exclusively granted to triple helix constellations gathering actors from industry, research and the state. Three calls have been made within VINNVÄXT singling out winners in 2003, 2004 and 2008 – including the three RISs focused on in this paper.

Doing gender in the governance of regional innovation

In this section, the theory of "doing gender" will be discussed, since it provides a framework for understanding how gender is operationalized in efforts to integrate a gender perspective in policy and organizations.[1] This theoretical stream has come to constitute the main approach employed in contemporary gender research (Gunnarsson *et al.*, 2003). In the theory of "doing gender," the gender dimension is understood as a constitutive part of organizational processes and organizations (West and Zimmermann, 1987; Acker, 1999; Fenstermaker and West, 2002). This view is advocated by the Swedish Research Council's Committee on Gender Research, claiming that:

> Gender is manifested at many levels: in the case of individuals, as an aspect of their own personalities; at the cultural level, in figures of speech, metaphors, categories; at the social level, as a principle for organisation of work and decision-making processes.
>
> (2005: 3)

According to the theory of "doing gender," gender can be regarded as an ongoing activity and interaction performed among and between women and men in organizations and in the society as a whole (West and Zimmermann, 1987). Everyday practices and activities are thereby related to an institutional and structural level (Acker, 1999). One of the classic contributions of gender research relies on the principles of "doing gender" as it exposes how gender often is done in ways that creates dichotomies, e.g. between "men" and "women" or between "femininity" and "masculinity" (Butler, 1990). This leads not only to segregation – e.g. on the labor market – but also to hierarchies where areas associated to "men" and "masculinity" are often ascribed higher value – e.g. by higher wages, faster careers and political prioritization (Gunnarsson *et al.*, 2003). This implies an uneven distribution of power and resources between women and men (Acker, 1999).

One of the main critiques of the theory of "doing gender" is that it fails to acknowledge the impact of other power dimensions, such as ethnicity and class (Fenstermaker and West, 2002). However, since contextual variations are emphasized in studies of "doing gender," it is quite possible to include all kinds of dimensions that appear as important in a specific case (Gunnarsson, 2007). In order to distinguish contextual variations, four different analytical aspects have been suggested where gendering in organizations can occur (Gunnarsson *et al.*, 2003):

- gender-differentiated division of work (structures);
- images, language and ideology (symbols);
- relations and patterns of interaction (relations); and
- individual identity (individuals).

These four levels of the ongoing construction of gender can also be detected within RISs and innovation policies. In the promotion and practices of RISs, gender is done on the level of division of work when different actors and sectors are distinguished in a manner that is congruent with the sex-segregated labor market, thus reinforcing segregating gender categories. On the level of images and language, gender is done when differing value is ascribed to the distinguished actors and sectors in relation to innovation and economic growth, reinforcing hierarchical gender categories (Pettersson, 2007; Lindberg, 2009). Internally in RISs, the doing of gender can be perceived on the levels of interaction and identity when women and men are treated differently and when their contributions to the systems are esteemed in differing manners (Pettersson, 2007; Andersson *et al.*, 2009). These findings are similar to the ones made in other organizational contexts as well (Acker, 1999; Gunnarsson *et al.*, 2003).

Another aspect of "doing gender" that has been problematized is the subject of change. This aspect is principal in the theory of "doing gender," as the focus on everyday practices underlines the possibility of doing things differently (Gunnarsson *et al.*, 2003). According to "doing gender," it is not compulsory that the doing of gender ends up in segregating and hierarchical patterns. Quite the opposite, since it – at the individual level – is possible to act in ways that break this trend, opening up – at the structural level – for more dynamic and nuanced perceptions of the world. Concerning promotion and development of RISs, such a change in the doing of gender could imply that sites of innovation are decentered "from singular persons, places and things to multiple acts of everyday activity" (Suchman, 2007: 1), since this would open up for both men and women as inventors and promoters of innovation.

However, it is not always easy to bring about change, since the dominating structures can be tenacious, reacting with resistance when they are challenged. The resistance reveals that there is power relations involved, requiring negotiations about "what men and women are allowed to do, how they are allowed to behave and how men and women are to be ranked and valued" (Kvande, 2003: 37–38).

Since the theory of "doing gender" provides a framework for understanding how gender is operationalized in efforts to mainstream gender in the governance of regional innovation, this theory will be utilized to answer the main research question to what extent the main elements of the theories of "doing gender" and "gender mainstreaming" are present in the motivation and implementation of RIS efforts to mainstream gender. This is analyzed by evidence of that the "doing gender" aspects of gender and change have been addressed by the RISs funded by VINNOVA. Two sub-questions will be of guidance in that analysis:

- Is the "doing gender" view of gender as continuously constructed advocated by the RISs?
- Is the contraposition between change and resistance as addressed in the theory on "doing gender" acknowledged by the RISs?

Gender mainstreaming in the governance of regional innovation

In this section, the theory of "gender mainstreaming" will be examined as a way to understand and classify how gender has been mainstreamed in the governance of regional innovation. As Walby (2005: 321) observes, there are many different definitions of "gender mainstreaming" and many different ways of implementing it in practice. One definition is provided by Rees (2005: 560), stating that "gender mainstreaming" is "the promotion of gender equality through its systematic integration into all systems and structures, into all policies, processes and procedures, into the organization and its culture, into ways of seeing and doing." Another definition is maintained by True and Mintrom (2001: 28), proposing that "gender mainstreaming" is to be understood as "efforts to scrutinize and reinvent processes of policy formation and implementation across all issue areas to address and rectify persistent and emerging disparities between men and women." According to the Swedish government, "gender mainstreaming" means that "decisions in all policy areas are to have a clear gender equality perspective."

Walby (2005: 323–324) highlights two general ways in which gender is mainstreamed. First, there is a strategy of "integrationism" that "introduces a gender perspective without challenging the existing policy paradigm, instead 'selling' gender mainstreaming as a way of more effectively achieving policy goals." Second, there is strategy of "agenda setting" that implies "the transformation and re-orientation of existing policy paradigms, changing decision-making processes, prioritizing gender equality objectives, and rethinking policy ends." In this context, the term "transformation" is to be understood as "neither the assimilation of women into men's ways, nor the maintenance of a dualism between women and men, but rather something new, a positive form of melding." As Walby notes, while the strategy of "agenda setting" means that "gender mainstreaming" is less likely to be rejected, it is more likely to have only a limited impact. In relation to the theory of "doing gender," the strategy of "agenda setting" seems to share more features than the strategy of "integrationism." This is since the first-mentioned strategy tends to bridge dualistic gender constructions and challenge their static appearance in a way that accords with the understanding of gender as continuously constructed as advocated in "doing gender."

Squires (2005: 368–371) presents a different way of categorizing efforts to mainstream gender. She discerns three types of strategies: "inclusion," "reversal" and "displacement." She connects the strategy of "inclusion" to a liberal type of feminism, seeking gender-neutrality and conceiving people as autonomous. What is mainstreamed in this category is primarily formal equality of opportunities, realized by bureaucratic policy processes. The strategy of "reversal" is associated to the stream of radical feminism, seeking recognition for a specifically female gendered identity and explicitly talking of "women" as a given category. Mainstreaming is here seen as a way to include women's voices. Squires links the strategy of "displacement" to a post-modern type of feminism, treating "gender" as a verb rather than a noun and seeking to deconstruct discursive regimes that engender people. "Gender mainstreaming" is, then, equivalent to norm diffusion. This grammatical and deconstructive stance is similar to the understanding of gender as continuously constructed and thus possible to change, as proclaimed in the theory of "doing gender." The strategy of "displacement" is thus the strategy that is most congruent with "doing gender." Squires underlines that the three strategies are to be regarded as complementary, rather than competing or incompatible. Ultimately, she concludes that "gender mainstreaming" is most likely to be truly transformative when all three strategies are implemented side by side.

Since the theory of "gender mainstreaming" provides a framework for understanding how gender is mainstreamed in governance of regional innovation, this theory will in this chapter be utilized to analyze to what extent the efforts of the RISs funded by VINNOVA can be classified in the light of this theory. This analysis helps answering the main research question to what extent the main elements of the theories of "doing gender" and "gender mainstreaming" are present in the motivation and implementation of RIS efforts to mainstream gender, since it reveals how different strategies reflect different understandings of gender. Two sub-questions will be of guidance in this analysis:

- To what extent can the RIS efforts to mainstream gender be classified in the light of Walby's two types of "gender mainstreaming" strategies: "integrationism" and "agenda setting"?

- To what extent can the RIS efforts to mainstream gender be classified in the light of Squires' three types of "gender mainstreaming" strategies: "inclusion," "reversal" and "displacement"?

Why gender matters in governance of regional innovation

In this section, theoretical incentives for applying a gender perspective on RISs and innovation policy will be scrutinized as a framework for understanding why gender matters in the governance of regional innovation. Surveying existing research, several incentives for employing a gender perspective in RISs and innovation policies are discerned. Gender has been advocated as an interesting aspect in relation to innovation, entrepreneurship and growth within a number of research fields, e.g. Feminist Science and Technology Studies, Political Science, Business Economics and Human Geography (Hacker, 1989; Bacchi, 1999; Ahl, 2002; Blake and Hanson, 2005; Pettersson, 2007). There it has been highlighted how the dominating image of innovation and innovators is based on stereotypical notions of gender, promoting men and masculinity as the norm. This gender imbalance makes the national innovation system – as well as its regional equivalents – inefficient and leaves innovative opportunities unexploited, thus hampering political prospects of continuous economic growth (Pettersson, 2007; Nyberg, 2009). In order to make better use of these innovative opportunities, Ghaye and Gunnarsson (2009) suggest the creation of "cultures of appreciation," enhancing innovation in organizations. Such cultures effectively address gender issues since the positive and creative sides of the organization and the employees are accentuated, reaching beyond hampering gender stereotypes. According to Blake and Hanson (2005: 681), gender stereotypes in innovation policies and innovation systems can be challenged by acknowledging that innovative activities occur also "in economic sectors and by actors that are typically ignored or undervalued by current research and by policy."

Ever since the initiation of financial support to RISs within the Swedish regional growth agreements in 2000 and the establishment of VINNOVA in 2001, the gendered aspects of these endeavors have been addressed in research, policy and practice. In research, the ability to make use of the innovative potential among both men and women in RISs and policy programs has been scrutinized (Lindberg, 2007; Pettersson, 2007). On a practical level, structured gender equality work has been initiated within a number of RISs, challenging their internal organization, as well as their innovative processes (Andersson *et al.*, 2009). In policy, several initiatives have been taken by VINNOVA to encourage RISs to mainstream gender, e.g. in the policy program BEGINN furthering needs-driven gender research for innovation. Internationally, the topic of gender in RISs has seldom been explicitly addressed, neither in research nor in practice or policy. It has merely been examined indirectly in studies of gender and innovation in general (Blake and Hanson, 2005; Aufhauser, 2007).[2]

Blake and Hanson (2005: 686) are concerned about a distorted understanding of innovation processes if gender is not accounted for. They point out how a gender perspective can illuminate "how and why certain geographic contexts encourage some kinds of innovations to emerge and develop while discouraging or preventing others." Their concern is motivated by existing studies revealing that gender is rarely addressed by RISs and innovation policies, thus implicitly promoting men and male as the norm when highlighting best practices, prioritizing among RISs and deciding upon funding (Pettersson, 2007; Lindberg, 2010; Danilda and Granat Thorslund, 2011). The consequences are described by Pettersson:

> there is a risk that only certain people, e.g. highly educated men in technical sciences, or men entrepreneurs, are supported by innovation policies. This, in turn, can lead to other people not being seen as important for the development of innovations, the economy and society – nor as important to support.
>
> (Pettersson, 2007: 9)

Blake and Hanson (2005: 697) discern a potential for an increased understanding of how innovation can contribute to the welfare of places, by broadening "the scope of contexts, economic sectors and actors that they consider as potential sites and creators of innovation." The broadening of scope is also a concern for Cameron and Gibson-Graham (2003: 146), who want their research to prevent "the exclusion of feminized activities from the economy by challenging and shifting the boundary between what is considered economic and non-economic." On the topic of economy, Blake and Hanson (2005) add that the general economic development, with declining manufacturing and rising services, calls for an adaptation of contemporary innovation policies. Cameron and Gibson-Graham (2003: 152) launch the idea of "feminising the economy," in order to replace the view that "the economy is a whole comprised of a pre-established number of parts or sectors," with a view where the economy is "an open-ended discursive construct made up of multiple constituents."

The theoretical incentives presented above can mainly be claimed to adhere to one single category, namely the one of "innovation incentives." These incentives rely on the conviction that RISs and innovation policies run the risk of neglecting important factors for innovation and growth if these are too narrowly defined. However, it is thereto possible to discern a second category, namely the one of "democratic incentives." In Pettersson's (2007) account, it is perceived as "a risk" that only certain people – highly educated men in technical sciences and male entrepreneurs – are supported by innovation policies. Here, the problem is depicted as concerning an unequal distribution of resources between women and men. This can be interpreted as a democratic view just as much as a matter of enhancing innovation. Cameron and Gibson-Graham's (2003) ambitions to "feminize" the economy can be regarded as a democratic incentive as well, in that they want to include presently excluded feminized activities in the economy. In addition to these two types of incentives in research, a third one is mentioned, namely the one of "political incentives." Pettersson (2007: 12–13) dedicates a whole chapter in her report about the innovation policies of the Nordic countries to the issue of "gender mainstreaming" as a political strategy concerning all policy areas – innovation policy included – thus motivating her examination of how gender is mainstreamed in these countries.

In relation to the theory of "doing gender," the understanding of gender as continuously constructed is evident in the categories of "innovation incentives" and "democratic incentives." This is since they both highlight how the promotion of men and male as the norm contributes to a segregated and static understanding of gender, underlining the importance of "men" and downplaying the importance of "women" in a way that hampers innovation and citizens' participation. The contraposition between change and resistance being addressed in the theory on "doing gender" is primarily acknowledged in the categories of "innovation incentives" and "political incentives." Both these types of incentives observes the tenacious structures of the prevalent policy paradigm defending the exclusion of feminized activities from the economy and the view that the economy relies on a pre-established number of parts or sectors, thus contributing to a distorted understanding of innovation processes and avoiding the implementation of "gender mainstreaming" in all policy areas. At the same time, these incentives discern a potential for change by accounting for social relationships and involved societal processes for in an appropriate way, by broadening the scope of contexts, economic sectors and actors being considered as potential sites and creators of innovation and by implementing a view where the economy is regarded as an open-ended discursive construct made up of multiple constituents.

Since the theoretical incentives presented above provide a framework for understanding why gender matters in governance of regional innovation, these incentives will in this chapter serve to analyze to what extent they correspond to the incentives advocated by the RISs supported by VINNOVA. This analysis helps answering the main research question to what extent the main elements of the theories of "doing gender" and "gender mainstreaming" are present in the motivation and implementation of RISs efforts to mainstream gender, since it reveals how different incentives reflect different understandings of gender. Two sub-questions will be of guidance in this analysis:

- To what extent can the theoretical incentives of innovation, democracy and policy be discerned in the RIS efforts to mainstream gender?
- To what extent can any additional incentives be discerned in their efforts?

Why gender matters and how it is mainstreamed in the Swedish case

In this section, the Swedish case will be analyzed as an example of gender mainstreaming in governance of regional innovation. With guidance from the research questions posed in the introduction, it will be examined why and how RISs have applied a gender perspective in their areas of activity, guided by the incentives from VINNOVA's program. First, it will be scrutinized to what extent the incentives for mainstreaming gender in RISs correspond to the incentives distinguished in theory. Second, it will be analyzed to what extent the RIS efforts can be classified in the light of the theory of "gender mainstreaming." Third, the two aspects discussed will serve as evidence of that the "doing gender" aspects of gender and change have been addressed by the RISs.

Incentives for integrating gender in RISs

The growing interest for gender in RISs and innovation policies reflects two processes: one initiated bottom-up and another top-down, evoking interplay between practice and policy. The bottom-up process is due to the perceived lack of women in many of those industries embraced by the VINNOVA-funded RISs, mainly representing industries based on natural resources, manufacturing and hi-tech (Lindberg, 2009; 2010; 2011). This gender imbalance has created a shortage of manpower as these industries keep evolving, requiring strategies for how to expand the recruitment range to include both women and men. Two of the three RISs in this study motivate their gender equality efforts by the shortage of labor. Innovation at Interfaces wants the gender perspective to strengthen the attractiveness of the food industry – which is depicted as suffering from gender stereotype values – among competent manpower. Triple Steelix motivates their gender equality work with the impediment on growth caused by increasing difficulties to recruit persons with suitable competences as a consequence of the sex-segregated labor market and the demographic development in the region.

The top-down process is due to national and international policy strategies prescribing the mainstreaming of gender in all policy areas and policy efforts. During the last decades, the Swedish government has advocated a gender equality policy aiming to: (1) combat and change systems that preserve the gender-based distribution of power and resources at societal level; and (2) create the conditions for women and men to enjoy the same power and opportunities to influence their own lives. "Gender mainstreaming" is the government's primary tool for achieving this goal, implying that the Swedish gender equality policy is to be implemented in all policy areas (Sterner and Biller, 2006: 32). This provides a background to the government stating – in its instructions from 2007 – that VINNOVA "shall promote gender equality between women and men within their area of activity" (SFS, 2007:1114: 1).

This is reflected in VINNOVA's description of their gender equality work, emphasizing the goals and guidelines provided by the state. As a consequence, the RISs funded by VINNOVA are encouraged to mainstream gender in their organizations. Primarily, this is managed via the program BEGINN, providing additional funding to the RISs gender equality efforts. The main incentive for these efforts, as stated by VINNOVA, is the correlation between gender equality and sustainable growth, based on the conception that the ability to innovate is hampered by normative understandings of gender. RISs are thereby encouraged to use a gender perspective in order to "strengthen" their areas of activity. A similar incentive is advocated by Fiber Optic Valley, arguing a gender perspective is indispensable for the innovation and open-mindedness required to reach their ambitions of growth.

Addressing the research questions if the theoretical incentives of innovation, democracy and policy can be discerned in the RISs efforts to mainstream gender and if there are any additional incentives to be discerned, the recitation above exposes a slight divergence to the incentives identified in theory. The incentives for integrating a gender perspective can be claimed to adhere to two of the categories advocated in theory, namely the ones of "innovation incentives" and "policy incentives." The first category of "innovation incentives" seeks to maximize the opportunities for innovation among Swedish citizens, opening up for a broad scope of actors, areas and innovations as contributors. Such incentives are maintained in VINNOVA's understanding, linking innovative ability to defiance of normative understandings of gender. These incentives are also present in the example of Fiber Optic Valley, depicting the gender perspective as essential for innovation and open-mindedness. The formulation of "innovation incentives" differs somewhat in practice and policy compared to the research studies. As practice and policy claim that innovation requires norm-breaking action, the research argument is that innovation is affected by social relations determining how different actors and sectors are esteemed by RIS managements and policy-makers/civil servants.

Apart from the differing formulations accompanying the "innovation incentives," all parties accord with the theory of "doing gender" in the understanding of gender as continuously constructed. This is because these types of incentives highlight how the promotion of men and male as the norm in innovation policy contributes to a segregated and static understanding of gender. Also, the contraposition between change and resistance being addressed in "doing gender" is acknowledged in the "innovation incentives," exposing both the tenacious structures of the prevalent innovation paradigm and the potential for change by increasing the ability to innovate in a broadened scope of contexts, economic sectors and actors.

The second category of "policy incentives" relies on the explicit demand from the Swedish government to public authorities to mainstream gender into their areas of activity, which is reflected in their instructions to VINNOVA and subsequently in VINNOVA's encouragement of RISs addressing gender in their organizations. The "policy incentives" are presented in a similar manner in practice, policy and theory, underlining the role of the state to impose a gender perspective in innovation policy. In relation to the theory of "doing gender," this type of incentives accords with the aspect of the contraposition between change and resistance, since it exposes whether "gender mainstreaming" truly has been implemented in all policy areas or not. Regarding the category of "democratic incentives" identified in theory, the recites from the RISs and VINNOVA reveals that this type of incentives is absent in practice and policy. This absence means that the accordance between "democratic incentives" and "doing gender" regarding the understanding of gender as continuously constructed is left unexploited.

An additional category of incentives can be perceived among the RISs, namely the one of "recruitment incentives." These types of incentives target the problem of attracting sufficient manpower to secure the planned business ventures, as maintained in the examples of Triple Steelix and Innovation at Interfaces. The theory of "doing gender" is appropriate in this context in the sense that it underlines the possibility of change, blurring the normative boundaries for which occupations women and men are expected to choose. However, the understanding of gender as continuously constructed is not evident in the case of "recruitment incentives," since these cling to the segregating categories of "women" and "men." Theory thus differs somewhat from practice and policy regarding the third category of incentives; while "democratic incentives" are advocated in theory, "recruitment incentives" are conveyed by some of the funded RISs. Maybe the reason for this is to be found in the differing prerequisites characterizing practice, policy and theory, inclining a focus on the economic outcomes in RISs, on overarching policy goals in policies and on analyses of inter-human relations in theory.

Ways of mainstreaming gender in RISs

In this section, it will be explored how successful different modes of implementation are for attaining gender equality in RISs, by classifying the made efforts in the light of the theory of "gender mainstreaming." This theory is constituted of Walby's two types of "gender mainstreaming" strategies: "integrationism" and "agenda setting" and of Squires' three types of "gender mainstreaming" strategies: "inclusion," "reversal" and "displacement."

The recite of the gender equality efforts performed by the three RISs focused in this study reveals that the gender equality work in Innovation at Interfaces has implies measures to increase gender equality among women and men in the organization as well as to strengthen the preconditions for the development of innovations and new markets. They also increase the gender awareness in procedures of selection and in allocation of project funding. As a part of this, processes that exclude women within the food industry and as researchers or entrepreneurs are scrutinized. Triple Steelix has approached the issue of gender equality by efforts to increase the number of women in the steel industry. This is to be realized by augmenting the attractiveness of the workplaces, including the work environment, the external image of the companies and women as role models. In Fiber Optic Valley, the gender equality work has focused how middle managers contribute to the doing of gender in their organizations and how they can act in order to change this. Their process of change has been described in the following manner:

> Fiber Optic Valley's member organisations have recognised the potential in working on gender equality and gender issues on a daily basis. At overall level, greater knowledge and awareness means that concepts such as gender and gender equality are now used in internal discussions within the organisations and in their planning. Active gender equality work has opened the doors to a whole new way of thinking, partly by questioning prevailing norms. A more open working climate and a broader approach to developing products and services boost the competitiveness of the individual organisations and the regional innovative environment.
>
> (Andersson *et al.*, 2009: 4)

As stated earlier, the RIS gender equality efforts have been affected by the framework set by VINNOVA's program BEGINN, financing these efforts. VINNOVA's way of handling the program forms associations with Rees' (2005: 560) definition of "gender mainstreaming," aspiring to systematically integrate gender aspects in all their areas of activity. But, as Walby (2005: 321) observes, there are several different ways of implementing "gender mainstreaming" in practice. According to the presented survey, two different approaches to this matter are expressed in BEGINN; one encouraging RISs to initiate gender equality work in their organizations and the other furthering critical examinations of the concepts of "innovation" and "innovation system." The first-mentioned approach is manifested in the TIGER call, where RISs and other strong research and innovation milieus got the chance to profit from applied gender research. VINNOVA's encouragement of gender equality work in existing RISs can be classified as a strategy of "integrationism." "Gender mainstreaming" is then introduced as a way of more effectively achieving policy goals, e.g. increased growth and innovation, which is discernable in VINNOVA's ambition to "strengthen" the RISs within their areas of activity by means of gender equality work. As recited earlier, the gender equality work of Fiber Optic Valley is claimed to "boost the competitiveness of the individual organizations and the regional innovative environment" and Innovation at Interfaces expects their gender equality work to strengthen the development of innovations and new markets.

Even if it might be tactical to implement "gender mainstreaming" in RISs as a tool for increased innovativeness, there is a risk that this focus fails to challenge existing innovation paradigms, where segregating and hierarchical gender constructions determine how different persons, sectors and innovations are ranked and valued. This risk is highlighted in the theory of "doing gender."

In relation to this theory, the strategy of "agenda setting" seems to accord to a greater extent than the strategy of "integrationism." This is since the strategy of "agenda setting" bridges dualistic gender constructions and challenges their static appearance in a way that accords with the understanding of gender as continuously constructed as advocated in "doing gender." The strategy of "agenda setting" is most obvious in the second approach to mainstreaming gender discerned in the data – furthering critical examinations of the concepts of "innovation" and "innovation system" – characterizing the call "gender perspective on innovation systems and gender equality" in its support to projects aiming to scrutinize the logic and effect of the dominant discourses in innovation policies and innovation system practices. This effort can be categorized as a strategy of "agenda setting" since these examinations have a potential to transform and re-orient existing policy paradigms by challenging segregating and hierarchical constructions of gender influencing how men and women are ranked and valued in innovation policy and innovation systems. This implies that VINNOVA's support to critical examinations of the concepts of "innovation" and "innovation system" is more in accordance to "doing gender" than their encouragement of gender equality work in existing RISs.

However, employing Squires' three types of "gender mainstreaming" strategies – "inclusion," "reversal" and "displacement" – the analysis becomes more nuanced. The two strategies applied by the RISs and in VINNOVA's two calls can then be interpreted as embracing all three of Squires' strategies. The strategy of "inclusion" is to be seen in the efforts to initiate deliberate processes of change where the gender awareness is raised in the organizations, thus striving to create equal opportunities for women and men. This strategy is also discernable in the efforts made by Triple Steelix to increase the number of women in the steel industry by ameliorating the work environment and the external image of the companies. The strategy of "reversal" – highlighting "women" as a category and striving to include women's voices – is absent in VINNOVA's calls. In the funded RISs this strategy is perceivable, however. Innovation at Interfaces claims to scrutinize processes that exclude women within the food industry and as researchers/entrepreneurs. Triple Steelix wants to highlight women as role models in order to attract other women to work in the steel industry. Neither the strategy of "inclusion" nor the strategy of "reversal" are congruent with the theory of "doing gender," in that they both mediate an understanding of gender as static rather than continuously constructed. The gender stereotype categories of "men" and "women" remain unchallenged when the last-mentioned group is to be included or reversed. However, "doing gender" might be applicable in the sense that the inclusion and reversal of women acknowledge the possibility of change, blurring the normative boundaries for which life paths women and men are expected to choose.

Finally, the strategy of "displacement" – treating "gender" as a verb rather than a noun and seeking to challenge prevalent norms – is evident in some of the RISs, e.g. in the focus on the doing of gender by middle managers in Fiber

Optic Valley and in the aspiration to increase the gender awareness in procedures of selection and in allocation of project funding in Innovation at Interfaces. Concerning the challenge of prevalent norms, which is another element in the strategy of "displacement," Fiber Optic Valley has contributed to this by questioning prevailing norms in their gender equality work, which they claim to have "opened the doors to a whole new way of thinking." The ambition within Innovation at Interfaces to scrutinize processes that exclude women within the food industry and as researchers/entrepreneurs might also be a way to challenge prevalent norms, if it is implemented in a manner that highlights those power relations that influence how different actors and innovations are valued in the RIS. The strategy of "displacement" is congruent with the theory of "doing gender," in that it highlights gender as continuously constructed.

Ways of "doing gender" in RISs

In this section, the two aspects of incentives and modes of gender integration discussed above will serve as evidence of that the "doing gender" aspects of gender and change have been addressed by the RIS gender equality efforts, thus linking back to the research questions of this chapter. The questions posed are if the "doing gender" view of gender as continuously constructed is advocated in the RIS gender equality efforts and the policy framework of VINNOVA and if the contraposition between change and resistance, as addressed in the theory on "doing gender," is acknowledged. The results from the analysis of incentives and modes indicate that some of the incentive categories and "gender mainstreaming" strategies used by the RISs and VINNOVA correspond to "doing gender" and that others do not. Concerning the understanding of gender as continuously constructed, the data reveals that this aspect of "doing gender" is solely prevalent in one of the categories of incentives advocated, namely the one of "innovation incentives." There, it has been highlighted how the promotion of men and male as the norm in innovation policy contributes to a segregated and static understanding of gender. In the different strategies of "gender mainstreaming," this aspect of "doing gender" is evident in the maintenance of the strategy of "agenda setting." There, dualistic gender constructions have been bridged and their static appearance has been challenged in the support to critical examinations of the concepts of "innovation" and "innovation system."

Concerning the contraposition between change and resistance being addressed in "doing gender," this aspect is discernable in the "innovation incentives" conveyed in the RISs and VINNOVA's acknowledgement of the resistance of the prevalent innovation policy paradigm and the change offered by a broadened scope of contexts, economic sectors and actors. This aspect is also detectable in VINNOVA's advocacy of "policy incentives," where they recognize the importance of "gender mainstreaming" truly being implemented in all policy areas. The possibility of change is thereto partly prevalent in the "recruitment incentives" being maintained by some RISs, blurring the normative boundaries for which occupations women and men are expected to choose. In the strategies of "gender mainstreaming," this aspect of "doing gender" is mainly discernable in the strategy of "displacement," where the ambition to challenge prevalent norms of innovation policy is expressed. This aspect is thereto present in the references to the strategies of "inclusion" and "reversal," challenging the normative prescriptions for men and women's life paths.

Returning to the research question if the "doing gender" aspects of gender and change have been addressed by the RISs and VINNOVA, the answer is mainly yes and partly no. They have addressed the understanding of gender as continuously constructed in their advocacy of "innovation incentives" and the strategies of "agenda setting" and "displacement," but not in their maintenance of "policy incentives" and the strategies of "inclusion" and "reversal." Moreover, the contraposition between change and resistance has been addressed in their references to their "innovation incentives" and "policy incentives," but only partly in relation to their "recruitment incentives." This is true also for their advocacy of the strategy of "displacement," but merely to some extent for their maintenance of the strategies of "inclusion" and "reversal." The overall pattern of incentives and modes is summarized in the upcoming section.

Conclusions

In this last section of the chapter, conclusions are drawn concerning how the RISs array of incentives and manners for integrating gender in their organizations, supported by VINNOVA, could inspire the design of measures for gender mainstreaming in governance of regional innovation. As the above analysis show, the different incentives and manners vary in their coherence with the theories of "doing gender" and "gender mainstreaming." Crucial features of "doing gender" are prevalent – though inconsistently – in the RISs gender equality efforts and in the policy frames set by VINNOVAs policy program. Summarizing the overall pattern it is evident that the incentives and modes most likely to change gendered patterns in RISs and innovation policies are the "innovation incentives" and "policy incentives" – making gender relevant in all areas of activity and reducing the risk of neglecting important factors for innovation and growth by widening the definition of these concepts – and the strategies of "agenda setting" and "displacement" – challenging prevalent norms of innovation and critically examining the concepts of "innovation" and "innovation system."

This link between theoretical findings and practically employed incentives and manners could inspire the design of measures for gender mainstreaming in governance of regional innovation on three areas: the impact of combined strategies, the influence of government prescriptions and the importance of critical examinations.

First, the impact of combined strategies is crucial when mainstreaming gender in governance of regional innovation. Such governance could be designed by embracing a wide array of incentives and measures, including "innovation incentives," "recruitment incentives" and "policy incentives" as well as the strategies of "integrationism," "agenda setting," "inclusion," "reversal" and "displacement." This implies motivating gender equality efforts in RISs by: (1) the innovative potential among a broad range of actors, areas and innovations; (2) the increased range of manpower available when considering both men and women; and (3) top-down policy goals and strategies prescribing the integration of a gender perspective in all areas of activities. It further implies performing gender equality efforts in RISs by: (1) promoting gender equality as a way to achieve other goals, e.g. growth; (2) re-orienting existing innovation paradigms to prioritize a gender perspective; (3) ensuring formal equality of opportunities for men and women; (4) including women's voices and experiences; and (5) deconstructing gendered discourses of innovation prescribing men and male as the norm. The key factor is to involve all these measures, not leaving anyone out, in order to truly highlight and change gendered patterns of definitions, prioritization and measures. This is confirmed by Squires' (2005) conclusion that "gender mainstreaming" is most likely to be transformative when all of the strategies are implemented side by side. This transformative effect is illustrated by Fiber Optic Valley's claim that their gender equality work has "opened the doors to a whole new way of thinking."

Second, the importance of critical examinations is essential when mainstreaming gender in governance of regional innovation. In order not to leave certain parts of gendered aspects of innovative processes and institutionalization out when designing and implementing gender equality measures, both the present and the potential state of the organizations and policies must be addressed. The present state concerns gendered patterns of structures, symbols, relations and individuals (Gunnarsson *et al.*, 2003). These four aspects of gendered organizations and policies could be mapped, discussed and altered. This has been the main procedure of the RISs studied in this chapter and also of the policy program supporting them. In contrast, the potential state concerns the possible range of actors, industries and innovations that *could* be included in the core activities of the organizations and policies if the gendered patterns would have looked different. If those industries employing many women – such as services and creative industries – would have been as highly esteemed in innovation promotion as industries employing mostly men – such as manufacturing and hi-tech industries, the gendered patterns of the RISs and policies would be easier to change. The same goes for if a wider range of innovations – such as social, organizational and service innovations – would have been as highly esteemed as product and hi-tech innovations (Lindberg, 2009; 2010; 2011). As long as the general pattern of prioritization and estimation of actors, areas and innovations are permeated by a male norm, it is hard to change the gendered relations within present organizations and policies. Mainstreaming of gender in governance of regional innovation thus requires efforts addressing both the present and the potential state of organizations and policies.

Third, the influence of government prescriptions is an effective ingredient when mainstreaming gender in governance of regional innovation. It can be concluded that the Swedish government has had a certain influence on the initiation of gender equality measures in the studies of RISs by its formulation of policy goals and its instructions to VINNOVA as a public authority, resulting in a program enhancing the RIS gender equality efforts. The apprehension that innovation can benefit from a gender perspective – inducing a wider view on how and where innovations may arise – permeates the governmental prescriptions, the policy program and the studies of RISs. The understanding of gendered norms in innovation as possible to change is also visible throughout the line. There is one aspect that is downplayed by all actors, though. That is the restrictions imposed by tenacious power relations for the prospect of succeeding with gender equality efforts in RISs. There is no evidence of anyone acknowledging the contraposition between different interests regarding the issue of how different actors, areas and innovations are to be esteemed. Rather, the efforts depict a consensus around the target of gender equality, even if altered gender relations would imply a redistribution of resources from certain organizations, networks and sectors to others.

Viewing policy as "management of change," this chapter has explored how the governance of regional innovation induces change of gendered practices in RISs. By discussing how regions can introduce and realize innovative strategies and the role of innovation governance in such processes this chapter has cast a light on the efforts to mainstream gender in some Swedish RISs supported by a national policy program. Based on common gender-scientific theories, the analysis have discerned different incentives for RISs to mainstream gender and esteemed the successfulness of different modes of implementation. It has thereby been illustrated how gender is interesting specifically in relation to multi-actor governance of regional innovation, by introducing views and ways to expand the range of actors, areas and innovations involved in such processes.

Notes

1. The origin and developments of "doing gender" is found in works by West and Zimmermann (1987) and Fenstermaker and West (2002). A Nordic overview of research on "doing gender" in organizations is provided in Gunnarsson *et al.* (2003).
2. The topic of gender and innovation has been addressed at several of the triple helix conferences, e.g. in Glasgow 2009, Madrid 2010 and Stanford 2011.

References

Acker, J. (1999). "Gender and organisation." In Saltzman Chafetz J (ed.) *Handbook of the sociology of gender*. New York: Plenum Publishers.

Ahl, H. (2002). *The making of the female entrepreneur – a discourse analysis of research texts on women's entrepreneurship*. Jönköping International Business School.

Andersson, S., Amundsotter, E., Svensson, M. and Däldehög, A.-S. (2009). *Middle managers as change agents – action oriented gender research*. Hudiksvall: Fiber Optic Valley.

Aufhauser, E. (2007). *Gender mainstreaming regional policy in Europe – the Austrian case*. Paper presented at a conference in Daegu. May 30.

Bacchi, C.L. (1999). *Women, policy and politics – the construction of policy problems*. London: Sage Publications.

Blake, M.K. and Hanson, S. (2005). "Rethinking innovation – context and gender." *Environment and Planning A*. Vol 37: 681–701.

Butler, J. (1990). *Gender trouble – feminism and the subversion of identity*. London and New York: Routledge.

Cameron, J. and Gibson-Graham, J.K. (2003). "Feminising the economy – metaphors, strategies, politics." *Gender, place and culture*. Vol 10. No 2: 145–157.

Danilda, I. and Granat Thorslund, J. (2011). *Innovation and gender*. VI 2011:03. Stockholm: VINNOVA.

Fenstermaker, S. and West, C. (eds) (2002). *Doing gender, doing difference – inequality, power and institutional change*. London: Routledge.

Ghaye, T. and Gunnarsson, E. (2009). "Creating cultures of appreciation – organisational innovation through employee well-being." In Döös, M. and Wilhelmson, L. (eds). *Organising work for innovation and growth*. VINNOVA.

Gonäs, L. and Karlsson J.C. (eds) (2006). *Gender segregation – divisions of work in post-industrial welfare states*. Farnham: Ashgate.

Gunnarsson, E. (2007). "The other sides of the coin – a feminist perspective on robustness in science and knowledge production." *International journal of action research*. Vol 3. No 3: 349–363.

Gunnarsson, E., Andersson, S., Vänja Rosell, A., Lehto, A. and Salminen-Karlsson, M. (eds) (2003). *Where have all the structures gone? Doing gender in organisations, examples from Finland, Norway and Sweden*. Stockholm University.

Hacker, S. (1989). *Pleasure, power and technology*. London: Unwin Hyman.

Kvande, E. (2003). "Doing gender in organizations – theoretical possibilities and limitations." In Gunnarsson, E., Andersson, S., Vänja Rosell, A., Lehto, A. and Salminen-Karlsson, M. (eds). *Where have all the structures gone? Doing gender in organisations, examples from Finland, Norway and Sweden*. Stockholm: Stockholm University.

Lindberg, M. (2011). "Genusgränser i Sveriges innovationspolitik – en fråga om prioritering." In Freidenvall, L. and Rönnbäck, J. (eds). *Bortom Rösträtten*. Södertörn: Södertörns högskola.

Lindberg, M. (2010). *Samverkansnätverk för innovation – en interaktiv och genusvetenskaplig utmaning av innovationspolitik och innovationsforskning*. Doctoral thesis. Luleå: Luleå University of Technology.

Lindberg, M. (2009). "Män och manlighet i Sveriges innovationspolitik." *Tidskrift för genusvetenskap*. No 2–3: 29–43.

Lindberg, M. (2007). *Deconstructing gender and innovation – four regional networks challenging Sweden's innovation policy*. Paper presented at the conference Past, Present, Future in Umeå, Sweden, June 14–17.

Nyberg, A.-C. (2009). *Making ideas matter – gender technology and women's invention*. Luleå University of Technology.

Pettersson, K. (2007). *Men and male as the norm? A gender perspective on innovation policies in Denmark, Finland and Sweden*. Stockholm: Nordregio.

Rees, T. (2005). "Reflections on the uneven development of gender mainstreaming in Europe." *International Feminist Journal of Politics*. Vol 7. No 4: 555–574.

SFS 2007:1114. *Förordning med instruktion för Verket för innovationssystem*. Svensk författningssamling.

Squires, J. (2005). "Is mainstreaming transformative? Theorizing mainstream in the context of diversity and deliberation." *Social Politics*. Vol 12. No 3: 366–388.

Sterner, G. and Biller, H. (2006). *Gender mainstreaming in the EU member states – progress, obstacles and experiences at governmental level*. Ministry of Integration and Gender Equality.

Suchman, L. (2007). *Agencies in technology design – feminist reconfigurations*. Lancaster, UK: Lancaster University.

Swedish Research Council's Committee on Gender Research (2005). *Gender studies – a summary*. Stockholm: Vetenskapsrådet.

True, J. and Mintrom, M. (2001). "Transnational networks and policy diffusion – the case of gender mainstreaming." *International Studies Quarterly*. Vol 45. No 1: 27–57.

Walby, S. (2005). "Gender mainstreaming – productive tensions in theory and practice." *Social Politics*. Vol 12. No 3: 321–343.

West, C. and Zimmermann, D.H. (1987). "Doing gender." *Gender and Society*. Vol 1: 125–151.

12 Entrepeneurship and public policy in emerging clusters

Bo Carlsson

Introduction

The purpose of this paper is to examine the role of entrepreneurship and public policy in the formation of technology-based industrial clusters. The emphasis is on the early high-growth stage rather than the mature or declining stages in the life cycles of clusters. The idea is to identify common patterns or regularities in the role played by human actors – entrepreneurs and policy-makers – in cluster formation. Having surveyed the burgeoning literature on how clusters emerge, I concentrate on eight cases in different areas of the world and different time periods that allow sufficient depth to discuss the role of individual actors as well as the role of public policy. I draw on previous literature, as well as the case studies in Braunerhjelm and Feldman (2007). The cases studied are the emergence of the automobile industry in Detroit in the early part of the twentieth century; the motion picture industry in Hollywood in the period 1915–30; Route 128 in Boston in the 1940s to 1980s; Silicon Valley in the 1950s and 1960s; the biotech industry in the US Capitol region in the 1970s and 1980s; the venture capital industry in Israel in the 1980s and 1990s; the ICT clusters in Ireland during the same time period; and the regional biotechnology clusters along the east coast of China, also in the 1980s and 1990s. In each of these cases, I focus on the historical background and on how pre-existing agglomerations were transformed into rapidly growing clusters.

Only successful clusters have been selected; success is understood as rapid growth sustained over at least a couple of decades. Hence, a certain historical distance is necessary. This precludes very recent clusters. Other cases (such as Research Triangle Park in North Carolina and the concentration of pharmaceutical companies in the mid-Atlantic region of the US) have been excluded because they do not meet the criteria for industry clusters as defined below. Five of the clusters on which this paper is based are in the US, reflecting the fact that there have been a relatively large number of studies of US clusters.

The paper is organized as follows. In the next section, I discuss the theoretical background and define the terms used in the paper. Following that, I present a brief synopsis of the emergence of five of the eight clusters.[1] From these case histories, I then extract some common elements that form the core of a theoretical framework. This allows me, in a subsequent section, to discuss the role of entrepreneurs, as well as public policy in each case. That discussion draws on a theoretical framework for policy analysis, developed jointly with a group of Swedish researchers (Bergek *et al.*, 2008). The paper concludes with a discussion of the functions of public policy and entrepreneurship with respect to cluster formation.

Theoretical background

The term "industry cluster" became prevalent in the economic literature around 1990 (Krugman, 1991; Porter, 1990; 1998), but it has been used rather loosely. Porter (1990) defined an industry cluster as a geographically proximate group of firms and associated institutions in related industries, linked by economic and social interdependencies. This is the definition most commonly used in the literature. Gordon and McCann (2000) distinguish between three different interpretations of industry cluster: the classic model of "pure agglomeration" based on the (neo-)classical tradition in economics, the industrial complex model of tight integration and stable relationships among firms, and the social network model built on interpersonal trust and relationships transcending firm boundaries. There are many different types of clusters, depending on the type of economic activity involved, as well as stage of development. Much of the literature refers to Alfred Marshall's *Principles of Economics* (1920).

As Marshall pointed out, many industrial activities tend to cluster in certain geographic regions. For example, he noted that the English iron industries located at first near oak forests in order to get access to charcoal. Later, as charcoal became more expensive and as new processes made it possible to use coal instead, they sought to be near coalmines. Eventually, as new technologies reduced the coal consumption in iron production, the industry moved to the coast to get access to imported iron ore. In contrast, industries that are not dependent on heavy materials may need special labor skills that are not easily acquired. Marshall noted that Sheffield in England and Solingen in Germany "have acquired industrial 'atmospheres' of their own . . . which yield gratis to the manufacturers of cutlery great advantages, that are not easily to be had elsewhere; and an atmosphere cannot be moved" (Marshall, 1923: 283–4).

Marshall distinguished between regional agglomerations and "industrial districts." He referred to the former as "elementary localization of industry," which "gradually prepared the way for many of the modern developments of division of labor in the mechanical arts and in the task of business management" that characterize industrial districts (Marshall, 1920: 268). According to Marshall, there are three primary causes of localization of industries: non-tradable inputs (physical conditions such as climate, soil, and access to raw materials), "patronage of a court" (demand for goods of high quality), and "the presence of a town" (urbanization economies, i.e. a sufficient number of customers) (ibid.: 268–9). Once an agglomeration has emerged, it may be transformed, over time, into an industrial district if certain advantages are acquired: a local market for special skills that can be passed on to the next generation (mysteries of the trade are no longer mysteries but are "in the air"); growth of subsidiary trades; and use of highly specialized machinery (Belussi and Caldari, 2009: 337). The resulting industrial district is the locus of economic activity that makes up a large fraction of an industrial economy; it represents the ordinary growth process – what Schumpeter would refer to as "economic growth" in the stationary state.

To get to a more dynamic stage ("economic development" in Schumpeter's terminology), two additional factors are needed – emphasized by Marshall in both *Principles* and *Industry and Trade*, although not specifically in connection with his discussion of industrial districts: what he calls "industrial leadership" (i.e. entrepreneurship) and "introduction of novelties" (i.e. innovation). These additional elements make it possible to break out of mere "organic" growth into a more dynamic phase, transforming an industrial district into what we may call a rapidly growing technology-based industrial cluster. These are the types of clusters with which this paper is concerned.

According to this interpretation of Marshall, there are two key elements to look for in the formation of clusters: a pre-existing local or regional agglomeration ("industrial district") of economic activity and a scaling-up of that activity through entrepreneurship and innovation. In addition, in this paper we are also interested in what role, if any, public policy has played.

Case studies of emerging clusters

The emergence of Detroit as the automobile capital of the US

Several authors (e.g. Carroll *et al.*, 1996; Szudarek, 1996; Klepper, 2002; 2003; 2004) have studied the emergence of Detroit as the leading automobile-producing region in the US. The question in this research is: why did Detroit emerge as the leader, even though the area was a latecomer in the sense that none of the early entrants located there? Initially, entry into the automobile industry was quite dispersed geographically, reflecting the fact that many firms were the result of diversification from related industries that were widely dispersed. Eleven states in the Northeast and Midwest accounted for 90 percent of all the entrants. The first Detroit-area firm to produce more than one car was Olds Motor Works, which was not established until 1901 (Klepper, 2004: 5).

Klepper's analysis of the automobile industry in Detroit is basically a story of competence (knowledge spillovers) and path dependence. The key is the experience of the people starting new firms. Klepper studied the backgrounds of all the 725 firms that entered the automobile industry in the US between 1896 and 1966. He found that firms founded by people with experience either in related industries (carriages and wagons, bicycles, and engines) or in incumbent firms in the auto industry performed better than firms founded by inexperienced entrepreneurs.[2] The latter ("inexperienced firms") had lower survival rates everywhere than spin-offs, and they actually performed worse in Detroit than elsewhere. Detroit spin-offs, however, outperformed spin-offs elsewhere. Location per se in Detroit does not seem to have mattered; the important thing was the experience of the founders. "The superior performance of Detroit firms was largely due to the disproportionate number of spin-offs in the industry that located around Detroit, whose superior performance in turn was related to their better pedigree" (Klepper 2004: 2). In Detroit, new firms founded by employees (experienced entrepreneurs) of the leading incumbents (foremost of these being Olds Motor Works) were competitive with firms (diversifiers) coming out of the most closely related industries.

Detroit offered no particular geographic advantage. Several cities were as well prepared as Detroit. The reason the industry came to concentrate in Detroit was essentially the presence there of the Olds Motor Works, an engine manufacturer organized by Random Olds. Olds was the first to realize the opportunity of integrating the whole automobile manufacturing process, by designing his own components for manufacture by various suppliers (rather than simply buying parts from bicycle, engine, and wagon manufacturers) and then doing the final assembly in his plant in Detroit. As a result, Olds became the first mass producer of automobiles, producing 5,000 cars as early as 1904. Among his suppliers were the Dodge brothers, who supplied engines (and later started their own automobile company) and Henry M. Leland (who became the co-founder, with Henry Ford, of Cadillac). When Ford was ousted from Cadillac, he formed his own company (first the Henry Ford Company, which failed, and then the Ford Motor Company). Dodge became the engine supplier to Ford, before being spun off as a separate company. Thus, as demonstrated by Klepper (2004), while there were hundreds of automobile companies started in the US in the period 1896–1966, spin-offs dominated among the survivors in the eventual shakeout. Spin-offs accounted for 48 percent of the entrants in Detroit versus only 15 percent of the entrants elsewhere, and nearly all the spin-offs in the Detroit area had parents located there. The most prolific parents were the industry leaders, Olds, Cadillac, Ford, and Buick/General Motors. All the companies that survived the subsequent shakeout had close links to Oldsmobile. Thus, Olds played much the same role in the formation of the automobile industry in Detroit as Fairchild Semiconductor did later in the formation of the semiconductor industry in Silicon Valley: it was a repository of industrial competence and source of several spin-offs that formed the core of the new industry (Moore and Davis, 2001). In sum, the entrepreneurship of Olds and Ford, in combination with innovation in the form of new organization of production (mass production), transformed an existing agglomeration into a rapidly growing industry cluster. No mention is made of public policy in the studies cited here.

The origins and growth of the motion picture industry in Hollywood

The emergence of Hollywood as the dominant center of the motion picture industry is described by Musser (1990) and Scott (2007). At the beginning of the twentieth century, the production of motion pictures was concentrated in the New York-New Jersey metropolitan area, with subsidiary centers in Chicago and Philadelphia. In the early days, most shooting of films was carried out outdoors. This led producers to look for more suitable locations than the northeastern US during the winter months. Southern California was one such location, with its sunny winters, warm climate, and diversity of landscapes. Several other places in Florida and southern California offered similar advantages. Scott argues that a mature motion-picture industrial complex might have sprung up virtually anywhere in the US in the early years of the twentieth century. He provides two reasons why the Los Angeles area emerged as a leading movie production center, even though the main creative and commercial activities of the motion picture industry remained in New York. One was the growing accumulation of movie production facilities in Hollywood, coupled with increasing concentration as new facilities were built and others disappeared through bankruptcy and merger. This created a sufficient agglomeration of activity ("critical mass") for takeoff. The other reason was the arrival (in 1911) in Southern California of Thomas Ince, who established a new studio and re-organized the production process.

> Ince was the first producer to attempt to industrialize the whole film-making process and to push it beyond the rather simple set of craft practices that had largely constituted it up to that point . . . Above all, he radically separated conception from production, and broke the shooting process down into disconnected segments that were then reassembled into the final film at the post-production stage. The continuity script could thus be used much like an industrial blueprint. On these foundations, an advanced division of labor started to make its appearance in the motion-picture industry.
>
> (Scott, 2007: 26)

Thus, the emergence of Southern California as the dominant location of the movie industry was linked to both geographic advantages and a re-organization of the industry. As the motion picture industry re-located to Southern California, it was re-organized and vertically integrated to resemble that of large-scale manufacturing industries with routinized production processes (Storper and Christopherson, 1987: 106). By 1919, 80 percent of the world's motion pictures were made in California, and by 1921 there were 5,329 workers in 68 establishments, compared with 3,922 workers in 20 establishments in New York (Scott, 2007: 28). Thus, similarly to the Detroit automobile cluster, the key elements in the formation of the Hollywood motion picture industry were organizational innovation and entrepreneurship.

The emergence of Route 128

The foundations of Route 128 were laid before World War II. Several large pre-war electronic firms, such as the Raytheon Corporation, as well as branch plants of General Electric and GTE Sylvania, were located in Massachusetts. Many of these companies later specialized in military and space contracts (Dorfman, 1983: 302).

MIT played a crucial role in the formation of Route 128. For a long time, MIT stood virtually alone as a university that embraced rather than shunned industry.

> From its start MIT developed close ties with technology-based industrialists, like Edison and Alexander Graham Bell, then later with its illustrious alumnus Alfred P. Sloan, during his pioneering years at General Motors, also with close ties to the growing petroleum industry. In the 1930s, MIT generated The Technology Plan, to link industry with MIT in what became the first and is still the largest university–industry collaborative, the MIT Industrial Liaison Program.
>
> (Roberts, 1991: 33)

MIT's involvement with industry was not confined to established companies, however:

> The traditions at MIT of involvement with industry had long since legitimatized active consulting by faculty about one day per week, and more impressive for its time had approved faculty part-time efforts in forming and building their own companies, a practice still questioned at many universities. Faculty entrepreneurship, carried out over the years with continuing and occasionally heightened reservations about potential conflict of interest, was generally extended to the research staff as well, who were thereby enabled to "moonlight" while being "full-time" employees of MIT labs and departments. The result is that approximately half of all MIT spin-off enterprises, including essentially all faculty-initiated companies and many staff-founded firms, are started on a part-time basis, smoothing the way for many entrepreneurs to 'test the waters' of high-technology entrepreneurship before making a full plunge.
>
> (Roberts 1991: 34)

Vannevar Bush, an electrical engineering professor at MIT, helped start the company in 1925 that later became Raytheon. In 1932, he was appointed vice president and dean at MIT. As World War II began, he was recruited by President Roosevelt to head up the newly formed Office of Scientific Research and Development (OSRD), the first federal agency dedicated to science and research.

> In this role, Bush revolutionized the relationship between science and government by funding universities rather than government labs to pursue basic military research. He also cemented the ties between MIT and Washington by using his friends in the local industrial and research communities to ensure that MIT graduates dominated the OSRD's committees.
>
> (Saxenian, 1994: 13–14)

MIT's president during World War II was Karl Taylor Compton, who brought MIT into close contact with the war effort as he himself headed up all national R&D coordination in Washington. It is not surprising, therefore, that MIT laboratories received fully one-third of the $330 million in contracts awarded by the OSRD during the 1940s and 1950s. Other universities in the Boston area, including Harvard and Tufts, also received millions of dollars for research in emerging fields such as radar, missile guidance and navigational systems, and submarine warfare. This massive government funding fueled the industrial revitalization of the New England economy (Saxenian, 1994: 14).

In the immediate post-war years, Compton pioneered efforts to commercialize military products, among other things by helping to create the first institutionalized venture capital fund, American Research and Development (ARD), set up in 1946. It was largely the brainchild of Compton, who became a board member, along with three MIT department heads. ARD's first several investments were in MIT developments, and some of the emerging companies were initially housed at MIT (Roberts, 1991: 33–4).

Thus, by the early 1950s, the foundation had been laid for a regional agglomeration of (mostly defense-related) hi-tech businesses in the Boston area.

> Massachusetts inherited a technological infrastructure of special importance to new and growing electronics firms. The state has historically supported a large machinery industry, begun in the nineteenth century to serve textiles and other nondurable mills. During and after World War II, its emphasis shifted to supplying defense and space contractors, increasingly linked to the electronics industry. As the electronics industry grew, the infrastructure expanded to meet its needs.
>
> (Dorfman, 1983: 306)

Part of this infrastructure consisted of relatively easy access to venture capital. According to Dorfman (1983: 307), among the top 100 venture capital investors in the US in 1981, the value of total investment by Boston firms was fourth in the nation, after New York, San Francisco, and Chicago. Other resources in relative abundance were a highly educated labor force and agglomeration externalities, especially in the form of specialized services. What put Route 128 (along with Silicon Valley) in a unique position to exploit new opportunities in electronics was the existence of entrepreneurial firms.

> The nationwide high tech boom was . . . driven mainly by technological opportunities opened up by advances in semiconductor and computer technologies that led to smaller and cheaper computers with higher performance and to an explosion of possibilities for new applications. With minor exceptions, advances in these fields were led by firms that had been founded after the invention of the transistor at Bell laboratories and the development of the computer in the 1940s. New firms such as Fairchild and Intel in California were at the forefront of advances in semiconductor technology . . . In minicomputers, DEC, Data General, Wang, Prime and Computervision in Massachusetts, and Hewlett Packard in Palo Alto, are all children of the post-war electronics revolution. Almost 80 percent of all minicomputer sales in 1980 and 1981 were accounted for by post-war firms.
>
> (Dorfman, 1983: 309)

> Each of these faculty-based enterprises exemplifies the importance of direct transfer of advanced technology to the commercial marketplace. But although these professors and their early new enterprises became visible very quickly, most firms in the Greater Boston area that have been formed and grown in areas of high-rate-of-change technology are not founded by academics. Rather, they are created by engineers who had worked for a major MIT or industrial lab. [For example], Digital Equipment Corporation (DEC), was founded by two former employees of MIT's Lincoln Laboratory, a major government-sponsored research and development organization established by MIT in the late 1940s to focus on the problems of air defense of the United States. Like the faculty-based companies DEC reiterates the importance of transfer of advanced technology to the initial product lines of an entrepreneurial firm. But DEC also demonstrates the critical contribution toward corporate "take-off" of a continuing flow of highly skilled professionals from a closely connected university into a rapidly growing and exciting organization.
>
> (Roberts, 1991: 5)

The emergence of Silicon Valley

The Silicon Valley case is similar in many ways to that of Route 128. One similarity is that a key leader in a great university located in the region played a major role in creating favorable conditions for a regional agglomeration that proved invaluable as a foundation of an electronics-based industry cluster. That key leader was Frederick Terman, dean of the engineering school at Stanford University. Terman was actually a disciple of Vannevar Bush at MIT and had gained first-hand exposure to the close ties between MIT and industry. He joined the electrical engineering department at Stanford and eventually became dean of the engineering school. He enlarged and strengthened Stanford's engineering program and persuaded the university to establish Stanford

Research Institute and to institute the Honors Co-operative Program (encouraging engineers at electronics companies to enroll in graduate courses). He also promoted the development of Stanford Industrial Park. The attitudes he had developed at MIT led him to encourage and guide his former students, such as William Hewlett and David Packard and the Varian brothers, to start their high-technology firms and eventually to locate them next to the university in Stanford Research Park (Roberts, 1991: 5).

Similarly to Route 128, the Federal Government played a supportive role via military research expenditures and other research grants. However, thanks in large measure to Terman's efforts to use government grants and contracts to finance what he called "steeples of excellence," he managed to build both outstanding research capability in microelectronics at Stanford while also persuading aerospace and electronics firms to locate near the university. As a result, federal government grants and contracts became central, not just supportive, to the development of Silicon Valley (Leslie, 2000; Lenoir *et al.*, 2003). These firms then received federal (largely military) research funding, boosting the local research base. But while these efforts were instrumental in founding what later became known as "Silicon Valley," they did not follow the dominant Greater Boston pattern of direct fostering of new firms from MIT labs and departments. Instead, the proliferation of firms in Silicon Valley came from multiple spin-offs of other companies, such as Fairchild.

> Only eight out of 243 new technical firms studied in the Palo Alto area have their origins in Stanford University . . . probably due in part to Stanford's lack of major government-sponsored laboratories. Indeed, despite the distance from their alma mater, MIT alumni are surprisingly the founders of over 175 companies in northern California, accounting for 21 percent of the manufacturing employment in Silicon Valley.
>
> (Roberts, 1991: 35)

All this created an environment in which new industries (including semiconductors) could prosper. But the environment that existed in the late 1950s in what later became known as Silicon Valley was not unique; similar conditions existed in Boston and New York, for example. And, indeed, none of the scientists and engineers that led the cluster formation in Silicon Valley came from the area; they were all "imported" from the East Coast of the US. The reason they located in the Bay Area was, to some extent, incidental: William Shockley, one of the co-inventors of the transistor at Bell Labs in New Jersey, wanted to be near his mother. Having identified new market opportunities in transistors, he therefore set up his company, Shockley Semiconductor, in Silicon Valley and attracted a set of extraordinarily talented group of scientists and engineers to his new company (Moore and Davis, 2001). But clearly Terman's institution building in the form of research capability and infrastructure at Stanford made the area attractive for companies, both existing and new, to locate there. It was also a part of Terman's strategy to design an environment in which university researchers could interact with and learn from industrial scientists (Lenoir *et al.*, 2003; Lécuyer, 2005).

In his account of the emergence of Silicon Valley, Sturgeon (2000) traces its roots to the regional agglomeration of radio and telegraph firms that formed prior to World War I. Other authors pick up the story with the formation of the cluster of firms in the semiconductor industry after the establishment of Shockley Semiconductor in 1956, thus implicitly distinguishing between the industry cluster and the regional agglomeration. Moore and Davis (2001) emphasize the importance of Fairchild. The scientists and engineers who started Fairchild were all originally employees of Shockley Semiconductor who soon found it impossible to work for Shockley. They then became the "traitorous eight," who jointly left Shockley to found Fairchild and then eventually left Fairchild to start their own individual companies (e.g. Intel and Advanced Micro Devices, Inc. (AMD)). They were not really entrepreneurs initially, but they learned while at Fairchild the importance of managing a business and converting science into commercializable products. They learned to organize themselves and their businesses differently from existing ones, thereby transforming science into business, to take advantage of a significant technological opportunity (semiconductors).

Moore and Davis downplay the role of both military spending and Stanford University. They claim that neither Stanford nor military spending were necessary for the formation of Silicon Valley. They contend that the success and structure of modern Silicon Valley (the industry cluster, not the regional agglomeration) stem more from an incremental learning process than from any one person, company, or organization.

Concerning Stanford, Moore and Davis take the view (confirmed by Bresnahan *et al.*, 2001) that Stanford University was helpful, but neither necessary nor sufficient, for the formation of Silicon Valley.

> Like most American universities, Stanford was transformed by the Cold War, becoming in many ways an institution in service to society rather than apart from it. Fred Terman's role in Stanford's particular transformation and rise to prominence was considerable. The years following World War II saw an explosion of federal funds for research in defense, and Terman was a successful advocate for the exploitation of postwar federal patronage. It is true that some electronics firms did grow out of this university research, and the San Francisco Bay Area did find itself home to a growing electronics industry. But the clustering of high-technology firms into "science regions" around universities was notable throughout the country, and yet no other regions experienced Silicon Valley-style transformation in the decades hence. Likewise, Stanford was not alone in setting up an Industrial Park and attempting to directly bring firms to the university community. Finally, the personal efforts of Fred Terman in funding local ventures, were personal (not institutional) efforts not unlike those taking place in countless other places around the country.
>
> (Moore and Davis, 2001: 10)

Concerning the role of the military:

> In no way did the government represent an "angel" investor in the semiconductor marketplace, simply funding companies through targeted demand until they get off their feet, or even guiding the direction of research or product development. These expenditures purchased semiconductor products, but did not guide their development, at least not at Fairchild. Military product requirements, it turned out, placed greater emphasis on reliability, measurability, and proven products than did the marketplace. As a result the products the military purchased were rarely at the leading edge of product development.
>
> The role of the government was nonetheless significant, in at least two ways [namely, by creating spillovers and spreading the benefits]. Very early on, it was the government (as part of an anti-trust settlement) that forced AT&T to share their findings on semiconductors freely. All four firms discussed above [Sprague Electric, Motorola, Shockley, and Fairchild], save Fairchild (whose founders were yet too young) were formed under key personnel who attended a government funded Bell Labs symposium in 1951, where Bell released all their collected knowledge of transistors to the scientific community. This was the true birth of the industry.
>
> (Moore and Davis, 2001: 17)

The other contribution of the government, according to Moore and Davis, was that by broadly scattering expenditures on finished technology products, it created large potential benefits.

Kenney and Florida (2000) studied the role of venture capital in Silicon Valley and found that entrepreneurial activity and venture capital co-evolved to shape a unique environment. Moore and Davis also point out that Fairchild preceded the VC industry, noting that venture capital partnerships really started in Silicon Valley about three years after Fairchild was born (Moore and Davis, 2001: 28).

Thus, although there remain different views of the emergence of Silicon Valley, it seems fair to conclude that this is a case of co-evolution and interdependence among scientific research, federal funding, company formation and re-location, and the emergence of new funding mechanisms such as venture capital.

The creation of a venture capital industry in Israel

The emergence of a venture capital (VC) industry in Israel is closely linked to the rise of hi-tech industry (particularly information technology) and is rooted in policy changes and other developments starting in the late 1960s. In particular, innovation policy in the form of government support for industrial R&D in Israel has played an important role. The policy changes took place in several stages. The first phase started in 1968, when the Office of the Chief Scientist (OCS) was established at the Ministry of Industry and Commerce. The mandate of the OCS was to subsidize commercial R&D projects undertaken by private firms (Honig, 2001; Breznitz, 2007). The next key development was the enactment of the "Law for the Encouragement of Industrial R&D" in 1985. The goal of the legislation was to develop science-based, export-oriented industries in order to promote employment and improve the balance of payments (Trajtenberg, 2000: 5). In this phase, a technological incubator program was initiated to give fledgling entrepreneurs an opportunity to develop their innovative technological ideas and set up new businesses in order to commercialize them.

The third phase started in 1993 with the establishment of the Yozma Venture Capital program, a government program aimed at promoting venture investments in Israel. Its first fund, Yozma I, founded 10 funds, each capitalized with more than $20 million (but only $8 million from Yozma, the rest from private investors). This marked the beginning of a professionally managed venture capital market in Israel. 58 private VC and investment companies existed before 1993. They were invited to participate in the Yozma ventures from the beginning, as were US venture capitalists. European investors were also attracted. Additional Yozma funds were established over the next few years; the whole program was privatized in 1997. From 1999 on, the foreign share of VC invested has been about 60 percent. By 2003, there were 558 private VC, private equity, and investment companies active in Israel that had raised about $8 billion and helped create about 2,500 new start-up companies. As a result, Israel has more hi-tech companies and start-ups than any other country relative to its size, is second only to the US on US stock exchanges (including Nasdaq), and has more venture capital than any region outside California and Massachusetts (Israel Ministry of Industry and Trade website).[3]

In their studies of the emergence of a venture capital (VC) industry in Israel, Avnimelech and Teubal (2004, 2006, 2007) show that the Israeli VC cluster emerged in conjunction with a transformation of a pre-existing electronics industry from a defense-dominated industry in the 1970s and 1980s into a "new" industry dominated by a large number of start-up companies. It is important to note that the conditions that eventually formed a solid base for the emergence of a VC cluster started to evolve soon after the Six-Day War in 1967 – i.e. 25 years before the establishment of the Yozma program. Foremost among the favorable background conditions were the prior existence of a large pool of qualified scientists and engineers, as well as universities that could add to this pool, a civilian hi-tech industry with experience in R&D and innovation, strong prior presence of multinational firms such as Motorola, IBM, DEC, and Intel, as well as a number of local communications equipment companies that generated spin-offs during the 1990s (Avnimelech and Teubal, 2004: 43). Another favorable factor was the porous and flexible structure of the Israeli bureaucracy and its close interaction with industry and private entrepreneurs (Breznitz, 2007). In addition, the culture of teamwork characterizing the military establishment and the initial reliance of Israeli R&D efforts on military technologies fostered a climate of public-private co-operation (Honig, 2001). As a result, the Yozma program was launched after a couple of decades of government policy reform and in favorable conditions for hi-tech entrepreneurial ventures.

Thus, the emergence over several decades of both a hi-tech cluster and a venture capital industry in Israel was the result of a policy-led process in the sense that it was triggered by a deliberate and targeted policy directed to this objective. But the targeting was limited in the sense that the aim of the technology policies was to create a venture capital industry in order to stimulate innovative activities and capabilities throughout industry without targeting any particular industries or technologies (Breznitz, 2007).

Theoretical observations

It is clear from all the case studies that cluster formation is not an event; it is a process spread out over several decades. The process involves an early stage in which a regional agglomeration of activity in a certain sector is formed, usually quite gradually – and, in successful cases, a second stage in which a viable industry cluster emerges and grows rapidly.

The first phase can happen either spontaneously, even randomly or by chance (as in Detroit and Hollywood) or as a result of public policy intervention – whether intentional and targeted on a specific sector as in China, intentional and non-specific as in Ireland, or inadvertent (not directly intended but co-evolving with other actors and institutions) as in Route 128, Silicon Valley, and the US Capitol region. The result of this early phase is a gradually emerging regional agglomeration whose characteristics can vary a great deal even within the same industry, as shown by Prevezer and Tang (2007) for the Chinese coastal regions, by Romanelli and Feldman (2007) for biotherapeutics in US regions, and by Owen-Smith and Powell (2007) for biotechnology in Boston and the San Francisco Bay area. The development in the early phase then conditions the development in the later phase. This is the sense in which the process is path dependent. As Wolfe and Gertler (2007) state, the lesson here is that the path dependencies for cluster creation are highly variable, that the random events which provide the trigger for cluster formation can come from many sources, and that there is a strong element of serendipity in virtually all cases (Wolfe and Gertler, 2007: 262).

In order to enter the second phase, i.e. in order for a regional agglomeration to be converted into an industry cluster, some triggering event coupled with an entrepreneurial spark is needed. Clusters emerge as an entrepreneurial/organizational response to technological opportunity. This response often involves some kind of knowledge transfer or spillover from incumbent firms involved in similar activities. The literature (as well as the case studies presented here) suggests that rapid, sustainable growth is obtained only when second-generation start-up occurs, i.e. when incumbents start to spin off new enterprises or when technology or experience is brought in from outside the region.[4] This suggests further that knowledge spillovers or transfers from incumbent firms or from outside the region are crucial. It is only when the lessons from early attempts (and perhaps failures) are learned and applied in new ventures that rapid growth begins. As pointed out by Romanelli and Feldman (2007), it is these second generation start-ups that differentiate the successful clusters that grow rapidly from the less successful ones that stagnate or fail to develop. This further suggests the importance of an entrepreneurial culture.[5] What seems to be required is not so much a certain "critical mass" or density of activity – although that certainly helps – as an entrepreneurial spark in combination with competence in the form of entrepreneurial, organizational, and management know-how to convert technological opportunities arising from technical and scientific discoveries into successful business ventures.

Path dependence and knowledge spillovers

Path dependence can be both a positive and a negative factor. Pre-existing competence in the same or related industries can help, especially in cases of incremental innovation, but can also block radical innovation (technological discontinuities). Incumbent firms typically carry out and absorb incremental and cumulative technological changes while they resist technological discontinuities. New entrants, on the other hand, are often carriers of radical innovations (Baptista and Swann, 1999: 395). The most important aspect of path dependence may be the existing entrepreneurial climate resulting from pre-existing conditions.

The emergence of clusters has more to do with competence than with geography/location. Location plays a role in that local conditions favor certain industries and that entrepreneurship is a local activity. As Storper and Walker argue, geography is not a set of static, pre-existing conditions but rather an integral and co-evolving part of the industrialization process:

> Industrial location patterns are created through the process of growth rather than through a process of efficient allocation of plants across a static economic landscape. That is, industries produce economic space rather than being hostage to the pre-existing spatial distribution of supplies and buyers.
>
> (Storper and Walker, 1989: 70)

Thus, the industrialization process, with its increasing returns over time, drives location, freeing industry from the prison of the past and the given distribution of factors of production. (Storper and Walker, 1989: 97)

In each of the eight cases studied here, path dependence in the form of a pre-existing regional agglomeration has played a critical role. In Detroit (as well as in many other cities), the existence of related industries (bicycles, engines, and wagons) provided the technical competence needed for the auto industry. What set Detroit apart from other regions was that one particularly competent firm (Olds), and several closely connected spin-offs, happened to locate there. The story is quite similar in the motion picture industry. Hollywood was one of several possible locations, but once a New York-based company came to Los Angeles for a winter haven and then established a permanent studio there, other companies followed. Within a few years a self-sustaining industry cluster had been created.

In Route 128, the policies of MIT to promote entrepreneurship and to build research labs to conduct federally sponsored research, as well as the coupling of these via huge federal defense spending, laid the foundation for the electronics industry. In Silicon Valley, the efforts of Frederick Terman and Stanford University had created sufficiently favorable conditions for the semiconductor industry to grow out of the regional agglomeration, but the technical and scientific ideas came from the East Coast. They could have been exploited elsewhere, but once the scientists and engineers happened to locate in Silicon Valley, the technological opportunities and the spillovers they generated were sufficiently large to create enormous success.

In the US Capitol region, the biotechnology competence in federal agencies and the number of entrepreneurial experiments were sufficiently large to overcome the absence of venture capital and the lack of an entrepreneurial culture (Feldman, 2001; Feldman and Francis, 2003). In Israel, the need to scale down the defense research establishment while absorbing a large number of Russian immigrants, many of whom were scientists and engineers, led to a re-orientation of R&D towards industrial and commercial development. This paved the way for the subsequent implementation of the Yozma venture fund program. In Ireland, a whole range of policy changes in conjunction with the Irish entry into the EU, especially the efforts to attract foreign direct investment, set the stage for the emergence of ICT clusters (Barry, 2007; Breznitz, 2007). And, in China, a series of institutional and policy reforms created the conditions needed for transition to rapid development of biotechnology clusters (Wang and Bradbury, 1986; Prevezer and Tang, 2007).

Triggering event and entrepreneurial/organizational response

While it is clear that path dependence plays an important role in cluster formation, there is also a role for path creation (West and Bamford, 2005: 434). There is a need for a triggering event and an entrepreneurial spark to create something new, to launch a new trajectory.

New technological opportunities create possibilities of spillovers. In the Detroit auto industry it was the integration of automobile component suppliers with the assembly process and the resulting mass production of automobiles by Olds that led to a re-organization of the industry. Once the possibilities of mass production and specialization had been identified, the newly gained knowledge spilled over via the creation of new companies such as the Ford

Motor Company. Similarly, in the Hollywood motion picture industry it was Ince's attempt to industrialize and re-organize the whole film-making process that permitted increased division of labor and gave rise to new opportunities and led to rapid growth.

In Route 128, MIT and federal defense spending during and after World War II had laid the foundation for the electronics industry. The triggering event was the build-up of defense spending during and after World War II, and the entrepreneurial sparks were provided by Vannevar Bush, Karl Taylor Compton, and other MIT faculty who started new companies to commercialize the new military technology. In Silicon Valley, the triggering event was the diffusion of transistor technology from Bell Labs to the scientific community via the establishment of Shockley Semiconductors. The formation of Fairchild by the 'traitorous eight' and subsequent spin-offs created the spillovers on which the Silicon Valley electronics industry is based. In the biotechnology industry in the US Capitol region, the triggering event was the layoffs from federal labs and agencies that forced spillovers into the commercial sector via the creation of new firms. Many of the entrepreneurs came from government agencies and others from private laboratories and companies.

In Israel, the triggering event was the creation of the Yozma funds, which provided hundreds of potential private entrepreneurs new opportunities. In Ireland, it seems to have been the Single European Market that gave the impetus to a rapid rise in foreign direct investment that also spawned indigenous entrepreneurship. In the Chinese coastal regions, the triggering event was the change in government policies to assist start-ups, attract potential entrepreneurs back to China from abroad, and develop regional clusters around science parks (Prevezer and Tang, 2007). Most of the new companies were formed by the government and by scientists returning from abroad, effecting knowledge spillovers from both government labs and foreign universities and companies.

A striking common feature in several of these cases is that once a triggering event had occurred, finance followed. In the pre-World War II cases, wealthy individuals were attracted by the new opportunities being formed. After World War II, the supply of private investments became more formalized, organized first in the Route 128 case as American Research and Development (ARD), the first venture capital fund. In the formation of Silicon Valley and later clusters, venture capitalists as we now know them have played a crucial role.

The role of public policy

What role has public policy played in the development of each cluster?

In the Detroit automobile industry, there seems to have been no public policy intervention. Certainly Klepper makes no mention of any government policies, nor do other authors. The same is true for the motion picture cluster in Hollywood.

In Route 128, government policies were hugely important, even though they were not designed primarily to build a regional agglomeration or industry cluster. Federal defense funding was crucial for the formation of Route 128, but without MIT's policies of promoting entrepreneurship and of organizing military research in separate laboratories, it is likely that the resulting economic activity would have taken a very different shape.

In Silicon Valley, federal defense funding played a less direct role than in Route 128, but the co-evolution of research capability and infrastructure at Stanford University and interaction between the university and firms in the area, as well as emergence of venture capital, nevertheless gave public policy an important indirect role.

In the Capitol region, the federal government role included setting a national agenda for research in biotechnology, with broad funding priorities through its laboratories and grants, large funding for the industry as a whole (not only in the Washington area), as well as establishing regulations and standards for the industry as a whole. State and local governments have played a limited role by issuing building permits for special laboratory facilities and by establishing policies (tax incentives, grants, incubators, and other business services) to attract firms to the region (Feldman and Francis, 2003: 785).

In the Israeli case, the whole development of the VC industry has been policy-led, starting in the early phase with the build-up of government policies in support of industrial R&D and continuing in the later phase with the establishment of the government-initiated Yozma venture capital fund. The Israeli case appears unique in two respects: it illustrates that government policy can indeed be successful in creating an industry cluster; it also shows that it is possible to design a policy that engages and leverages private resources in such a way that government support is not permanently needed.

The Irish government has played a crucial role in creating the "Celtic Tiger," primarily via institutional and policy reforms targeted at attracting direct foreign investment. These reforms have been successful in creating industry clusters in several sectors, including ICT.

In China, finally, the governments in the three coastal regions have played a leading role, first in carrying out institutional and policy reforms in the early phase and then in promoting as well as engaging in start-ups of biotechnology companies. It is too early yet to tell how successful these companies will be (Prevezer and Tang, 2007).

Thus, public policy has played an important role in six of the eight cases studied here. The two exceptions are Detroit and Hollywood. These are also the cases that most closely resemble the process of agglomerations transformed into Marshallian industrial districts and eventually into knowledge-based industry clusters. However, the role of public policy was inadvertent (in the sense that the formation of a regional agglomeration or industry cluster was not the primary policy target) in three cases: defense funding in Route 128 and Silicon Valley, and funding for biotechnology research in the US Capitol region. The policy was intentional but not sector-specific in Ireland and both intentional and sector-specific in Israel and China.

It is important to note, however, that many of the sources used for this study are not focused specifically on public policy. Also, policy intervention is often viewed as an event rather than a process co-evolving with the actors and institutions in the emerging cluster over extended periods of time. In particular, pre-existing policies and conditions may not come into view. As a consequence, a deeper examination of the role of public policy may yield different results. What is at issue is a systems cum evolutionary approach to policy-making. The formation of an industry cluster that can sustain rapid growth for at least a couple of decades involves the creation of an innovation system, which means avoiding systemic obstacles and overcoming system failures. As shown by Avnimelech and Teubal (2007), there are many causes of system failures.

In joint work with Swedish co-authors, I have explored these issues (Bergek *et al.*, 2008; Carlsson *et al.*, 2002). It is based on 15 years' joint work on innovation systems (summarized in Carlsson, 1995; 1997; 2000). In this work, we have identified six functional requirements that must be fulfilled if an innovation system or industry cluster is to evolve and perform well. How these functions are fulfilled depends on structural features in the innovation system or cluster, i.e. the characteristics of actors, networks, and institutions. Sometimes, the functions are fulfilled by private actors, sometimes through fortuitous government policies, and sometimes through proactive and intentional public policies. Let us examine each of the industry clusters with respect to each function.

Functions of public policy

(1) Knowledge development and diffusion

In each case, what was the knowledge base required to build the cluster? This refers to all the scientific, technical, and practical knowledge related to all activities within the cluster.

In the Detroit auto industry, knowledge was transferred from pre-existing related industries and in the Hollywood motion picture industry from company headquarters elsewhere. In Route 128, the results of military research spilled over into the commercial arena. In Silicon Valley, the transistor technology was diffused by forcing Bell Labs to disseminate the technology (developed in New Jersey) to the national scientific community. In the US Capitol region, the know-how existed in federal agencies and labs and was commercialized via new start-up companies. In Israel, much of the technical knowledge came from the military and from Russian immigrants and industry knowledge from foreign venture capitalists. In Ireland, most of the knowledge came via direct foreign investment and, in China, from both government labs and scientists returning from abroad.

It is interesting to note the different mechanisms by which the biotechnology clusters in Boston and the San Francisco Bay area acquired knowledge (see

Owen-Smith and Powell, 2007). In Boston, the biotechnology companies were rooted in public research organizations such as Harvard and MIT, whereas the Bay Area was dominated by venture capitalists. Both clusters have been successful, but their networks and trajectories over time are quite different.

(2) Creating incentives and providing legitimacy

Incentives may be used to reinforce positive forces or to overcome negative forces. This may take the form of reducing uncertainty or providing legitimacy. In the case of Route 128, large amounts of federal defense spending created powerful positive incentives for defense contractors in the Boston area. The biotechnology cluster in the US Capitol region shows that negative incentives (such as layoffs by federal agencies) can spur the formation of new entities in related businesses. In Israel, the setting up of venture capital funds with partial but substantial government support created vastly improved funding possibilities for potential entrepreneurs.

(3) Promoting entrepreneurial experiments

In some cases, the technical opportunities (when identified) provide enough incentives, and the entrepreneurial climate is sufficiently favorable, to stimulate existing firms to innovate and new firms to emerge. When such incentives are lacking, public policies may help to promote entrepreneurial experiments. It is interesting to note that the promotion of entrepreneurship has been an important component of public policy in the three non-US regions studied (Ireland, Israel, and China), but not generally in the US.

(4) Creating markets or appropriate market conditions

In the Detroit auto industry and Silicon Valley, new start-ups were able to create new markets on their own. In Route 128, government policies in the form of defense contracts and trade barriers, respectively, created new markets for domestic firms. In Israel, one of the main drivers was the desire to convert military R&D capability into commercial products in both foreign and domestic markets. In Ireland and China, institutional and policy reforms to improve the market conditions have been extremely important.

(5) Creating resources (financial as well as human capital)

Lack of human, technical, and financial resources may impede the formation of new industry clusters. Sometimes such resources co-evolve with the business opportunities in the cluster, as happened, for example, with respect to venture capital in Silicon Valley and the US Capitol region. But sometimes the injection of new resources can help new clusters to form, as happened in Route 128 as a result of federal defense spending during World War II and in the early post-war period. As noted by Kenney and Patton (2007), the needs for supporting services such as legal and financial services vary among industries. Kenney and Patton suggest that mobilizing local law firms to support the entrepreneurial process is particularly important, since these are the single most localized support network actors.

(6) Promoting positive externalities

The formation of a successful industry cluster involves positive externalities within the system, such as economies in the form of common labor markets, an elaborated division of labor, and knowledge spillovers. The Detroit auto cluster is an example of this happening spontaneously, without any public policy intervention. On the other hand, Moore and Davis (2001) argue that it was the government forcing AT&T to share their findings on semiconductors freely (as part of an anti-trust settlement) that represented the true birth of the semiconductor industry in Silicon Valley. Similarly, federal military research expenditures channeled via research labs set up by MIT, in conjunction with MIT's policies of encouraging entrepreneurial activity, had positive spillover effects in Route 128.

Each of the six functions outlined above relates to one or more resource requirements that must be satisfied for a system or cluster to develop successfully. An analysis of the resource requirements could make use of a checklist of the following sort. With respect to the existing knowledge base, some examples are: What type of knowledge exists in the system currently (scientific knowledge and know-how relating to systems, materials, components, production, design, etc.)? Are there important gaps? Who are the carriers of knowledge, how many, how diverse, and how well connected? Is the whole value chain represented? How does this system compare with others elsewhere? Is the knowledge base expected to change in the next decade? With respect to other resources, how much financial capital is available (seed, angel, and venture capital, as well as other risk capital)? How many and how diverse are the sources of finance? Similar questions can be raised with respect to each functional requirement. The policy-maker will then have to decide where the greatest needs exist and develop policies to address those needs. But the existence of a particular need does not necessarily require public policy involvement. Various types of organizations can fulfill each of the six functions.

While all of these functions are necessary for the formation of a successful industry cluster or innovation system, they may not be sufficient. Public policy can support and sometimes even initiate a cluster in its early phase – although spontaneous development and serendipity seem to be more prevalent mechanisms. A more frequent role of public policy is to provide support and reinforcement in the later phase of cluster formation. As shown in the preceding analysis, the policy requirements in each phase vary a great deal both over time and between clusters. No single policy is universally applicable, and even sustained and substantial policy efforts may fail to yield the desired results. In the end, success or failure depends on the creativity and persistence of the entrepreneur – with an element of luck, as well.

Notes

1. For considerations of space, the cases on the biotech industry in the US Capitol region, the ICT cluster in Ireland, and the biotech clusters in China are not presented here.
2. Carroll *et al.* (1996: 133) show, somewhat surprisingly, that firms coming from bicycle and carriage manufacturers had higher survival rates than those coming from engine manufacturers. They also suggest that early in an organization's life the skills, knowledge, and other assets accumulated earlier in related industries give the organization an advantage over other organizations without such assets. Later in the organization's life these factors diminish in importance and may even constitute a drag.
3. An important driving force behind the strong representation of Israeli companies on US stock exchanges and the close links with foreign (particularly American) finance is the recognition that the domestic market is too small to support extensive R&D investments for the domestic market alone. As a result, Israeli companies tend to export components for manufacturing and distribution elsewhere by foreign companies (Honig, 2001).
4. Bathelt *et al.* (2004) distinguish between "local buzz" and "global pipelines" as mechanisms to generate rapid cluster development.
5. This is in line with Minniti's (2005) argument that entrepreneurship tends to concentrate geographically in part because of the social environment. When making decisions, individuals follow social cues and are influenced by what others have chosen, especially when facing ambiguous situations. Minniti refers to this as a non-pecuniary network externality.

References

Avnimelech, G. and Teubal, M. (2004) "Venture Capital Start-Up Co-Evolution and the Emergence and Development of Israel's New High Tech Cluster," *Economics of Innovation and New Technology,* 13(1): 33–60.

—— (2006) "Evolutionary Innovation and Technology Policy: A Four-Phase High Tech Policy Model," paper presented at the DRUID summer conference, Copenhagen, June 18–20.

—— (2007) "Israel's Innovation and Technology Policy (ITP) Cycle: How It Led to the Emergence of Venture Capital (VC) and to the Re-configuration of Its High Tech Cluster," in Braunerhjelm, P. and Feldman, M. (eds) *Cluster Genesis: Technology-Based Industrial Development,* Oxford: Oxford University Press.

Baptista, R. and Swann, P. (1999) "A Comparison of Clustering Dynamics in the US and UK Computer Industries," *Journal of Evolutionary Economics* 9(3): 373–99.

Barry, F. (2007) "FDI and the Emergence of Ireland's ICT Clusters," in Braunerhjelm, P. and Feldman, M. (eds) *Cluster Genesis: Technology-Based Industrial Development,* Oxford: Oxford University Press.

Bathelt, H.A., Malmberg, A. and Maskell, P. (2004) "Clusters and Knowledge: Local Buzz, Global Pipelines and the Process of Knowledge Creation," *Progress in Human Geography* 28(1): 31–56.

Belussi, F. and Caldari, K. (2009) "At the Origin of the Industrial District: Alfred Marshall and the Cambridge School," *Cambridge Journal of Economics* 33: 335–55.

Bergek, A., Jacobsson, S., Carlsson, B., Lindmark, S., and Rickne, A. (2008) "Analyzing the Functional Dynamics of Technological Innovation Systems: A Scheme of Analysis," *Research Policy* 37(4): 407–29.

Braunerhjelm, P. and Feldman, M. (eds) (2007) *Cluster Genesis: Technology-Based Industrial Development*, Oxford: Oxford University Press.

Bresnahan, T. F., Gambardella, A. and Saxenian, A. (2001) "'Old Economy' Inputs for 'New Economy' Outcomes: Cluster Formation in the New Silicon Valleys," *Industrial and Corporate Change* 10(4): 835–60.

Breznitz, D. (2007) *Innovation and the State: Political Choice and Strategies for Growth in Israel, Taiwan, and Ireland*, New Haven, CT: Yale University Press.

Carlsson, B. (1995) *Technological Systems and Economic Performance: The Case of Factory Automation*, Boston, MA: Kluwer Academic Publishers.

—— (1997) *Technological Systems and Industrial Dynamics*, Boston, MA: Kluwer Academic Publishers.

—— (2000) *Technological Systems in the Bio Industries: An International Study*, Boston, MA: Kluwer Academic Publishers.

Carlsson, B., Jacobsson, S., Holmén, M., and Rickne, A. (2002) "Innovation Systems: Analytical and Methodological Issues," *Research Policy* 31(2): 233–45.

Carroll, G.R., Bigelow, L.S., Seidel, M-D.L., and Tsai, L.B. (1996) "The Fates of De Novo and De Alio Producers in the American Automobile Industry 1885-1981," *Strategic Management Journal*, 17 (Special Issue): 117–37.

Dorfman, N. (1983) "Route 128: The Development of a Regional High Technology Economy," *Research Policy* 13: 299–316.

Feldman, M.P. (2001) "The Entrepreneurial Event Revisited: Firm Formation in a Regional Context," *Industrial and Corporate Change* 10(4): 861–91.

Feldman, M.P. and Francis, J. (2003) "Fortune Favours the Prepared Region: The Case of Entrepreneurship and the Capitol Region Biotechnology Cluster," *European Planning Studies* 11(7): 765–88.

Gordon, I.R. and McCann, P. (2000) "Industrial Clusters: Complexes, Agglomeration, and/or Social Networks?" *Urban Studies* 37(3): 513–32.

Honig, B. (2001) "Israel and the ICT: The Case of Industrial Transformation," in International Labour Organization, CD: World Employment Report, 2001, Life at Work in the Information Economy, Country Studies, background paper. Geneva: ILO.

Kenney, M. and Florida, R. (2000) "Venture Capital in Silicon Valley: Fueling New Firm Formation," in Kenney, M. (ed.) *Understanding Silicon Valley: The Anatomy of an Entrepreneurial Region,* Stanford, CA: Stanford University Press.

Kenney, M. and Patton, D. (2007) "Supporting the High-Technology Entrepreneur: Support Network Geographies for Semiconductor, Telecommunications Equipment, and Biotechnology Start-ups," in Braunerhjelm, P. and Feldman, M. (eds) *Cluster Genesis: Technology-Based Industrial Development,* Oxford: Oxford University Press.

Klepper, S. (2002) "The Capabilities of New Firms and the Evolution of the US Automobile Industry," *Industrial and Corporate Change* 11(4): 645–66.

—— (2003) "The Geography of Organizational Knowledge." Working paper, Carnegie Mellon University.

—— (2004) "Agglomeration Through Spinoffs: How Detroit Became the Capital of the US Automobile Industry." Working paper, Carnegie Mellon University.

Krugman, P. (1991) *Geography and Trade*, Leuven: Leuven University Press; Cambridge, MA and London: MIT Press.

Lécuyer, C. (2005) "What Do Universities Really Owe Industry? The Case of Solid State Electronics at Stanford," *Minerva,* 43: 51–71.

Lenoir, T., Rosenberg, N., Rowen, H., Lécuyer, C., Colyvas. J. and Goldfarb, B. (2003) "Inventing the Entrepreneurial University: Stanford and the Co-Evolution of Silicon Valley,"www.siepr.stanford.edu/programs/SST_seminar/Lenoir.pdf.

Leslie, S.W. (2000) "The Biggest 'Angel' of Them All: The Military and the Making of Silicon Valley," in Kenney, M. (ed.) *Understanding Silicon Valley: The Anatomy of an Entrepreneurial Region,* Stanford, CA: Stanford University Press.

Marshall, A. (1920) *Principles of Economics,* 8th edn, vol. 1, London: Macmillan and Co.

Marshall, A. (1923) *Industry and Trade,* London: Macmillan and Co.

Minniti, M. (2005) "Entrepreneurship and Network Externalities," *Journal of Economic Behavior and Organization*, 57(1): 1–27.

Moore, G. and Davis, K. (2001) "Learning the Silicon Valley Way," Stanford Institute for Economic Policy Research discussion paper 00-45.

Musser, C. (1990) *The Emergence of Cinema: The American Screen to 1907,* Berkeley, CA: University of California Press.

Owen-Smith, J. and Powell, W. (2007) "Accounting for Emergence and Novelty in Boston and Bay Area Biotechnology," in Braunerhjelm, P. and Feldman, M. (eds) *Cluster Genesis: Technology-Based Industrial Development*, Oxford: Oxford University Press.

Porter, M.E. (1990) *The Competitive Advantage of Nations*, New York: Free Press.

—— (1998) "Clusters and the New Economics of Competition," *Harvard Business Review* 76 (November–December): 77–90.

Prevezer, M. and Tang, H. (2007) "Policy-Induced Clustering: The Genesis of Biotechnology Clustering on the East Coast of China," in Braunerhjelm, P. and Feldman, M. (eds) *Cluster Genesis: Technology-Based Industrial Development*, Oxford: Oxford University Press.

Roberts, E. (1991) *Entrepreneurs in High Technology: Lessons from MIT and Beyond*, Oxford: Oxford University Press.

Romanelli, E. and Feldman, M. (2007) "Anatomy of Cluster Development: Emergence and Convergence in the US Human Bio-Therapeutics, 1976–2003," in Braunerhjelm, P. and Feldman, M. (eds) *Cluster Genesis: Technology-Based Industrial Development*, Oxford: Oxford University Press.

Saxenian, A. (1994) *Regional Advantage: Culture and Competition in Silicon Valley and Route 128*. Cambridge, MA: Harvard University Press.

Scott, A. (2007) "Origins and Growth of the Hollywood Motion-Picture Industry: The First Three Decades," in Braunerhjelm, P. and Feldman, M. (eds) *Cluster Genesis: Technology-Based Industrial Development*, Oxford: Oxford University Press.

Storper, M. and Christopherson, S. (1987) "Flexible Specialization and Regional Industrial Agglomerations: The Case of the US Motion Picture Industry," *Annals of the Association of American Geographers*, 77 (1): 104–17.

Storper, M. and Walker, R. (1989) *The Capitalist Imperative: Territory, Technology, and Industrial Growth*, Oxford: Blackwell.

Sturgeon, T.J. (2000) "How Silicon Valley Came to Be," in Kenney, M. (ed.) *Understanding Silicon Valley: The Anatomy of an Entrepreneurial Region,* Stanford, CA: Stanford University Press.

Szudarek, R.G. (1996) *How Detroit Became the Automotive Capital,* Detroit, MI: Typocraft Co.

Trajtenberg, M. (2000) "R&D Policy in Israel: An Overview and Reassessment," National Bureau of Economic Research (NBER) Working Paper 7930, Cambridge, MA.

Wang, J. and Bradbury, J.H. (1986) "The Changing Industrial Geography of the Chinese Special Economic Zones," *Economic Geography*, 62(4), 307–20.

West III, G.P. and Bamford, C.E. (2005) "Creating a Technology-Based Entrepreneurial Economy: A Resource Based Theory Perspective," *Journal of Technology Transfer*, 30: 433–51.

Wolfe, D.A. and Gertler, M.S. (2007) "Path Dependency and Discontinuities in Cluster Formation," in Braunerhjelm, P. and Feldman, M. (eds) *Cluster Genesis: Technology-Based Industrial Development*, Oxford: Oxford University Press.

13 The "start-up factor"

Regional innovation policy convergence between the US and Sweden

Henry Etzkowitz

Introduction: The start-up factor

There is a global movement towards encouraging the formation of knowledge-based firms, irrespective of regional industrial or research intensity.[1] Whether starting from a forward linear advanced research base or a reverse linear perspective of meeting unmet needs, policy-makers, from Leuven to Lagos, seek economic development and employment growth by capitalizing old or new knowledge resources. Policy-makers still offer tax breaks and other incentives to retain or attract branches of large firms, exemplified by the recent struggle between New York and New Jersey states to relocate elements of the UK Pearson Educational firm from one state to another that resulted in a reshuffling of assets and a loss of tax revenues to both states but no overall increase in jobs in the New York metropolitan region. Governments are still at the mercy of savvy companies able and willing to play them off against each other for their own benefit. Indeed, much governmental resources are still committed to large employers in the search for quick, highly visible, results.

Nevertheless, in recent years there has been a significant shift towards focusing on the factors that encourage the formation and growth of new firms and industries. Although these firms may be small at first, and their contribution to employment growth miniscule in comparison to large established employers, some of these firms individually, like Microsoft in Seattle, or as a cluster around the IMEC Institute in Leuven may change the socio-economic shape of a city or region. On the other hand, in some peripheral regions growth factors may be lacking and start-ups remain small or leave for a better environment. San Francisco and Linköping exemplify what can be accomplished by support measures to incentivize the foundation and growth of start-ups. In the past, the typical Swedish response to loss of a large employer, such as SAAB aircraft fighters, would be to ask Stockholm to find another large employer who could be relocated to the region. Whether from a simple lack of candidates or realization that an alternative strategy is possible, a significant shift in regional innovation may be identified in Sweden and elsewhere.

Knowledge-based regional innovation has become a virtually universal development strategy. It is an "implicit policy" in the US, arising through the synthesis of a variety of initiatives in contrast to Sweden where, in the early twenty-first century, it was an explicit objective of the VINNOVA agency. There is no federal agency directly responsible for integrating science into regional development in the US (Eisinger, 1997), although various national policies certainly have regional effects Indeed, translating the science-base of federally supported research at universities into economic activity has become the centerpiece of an immanent US regional development strategy. Science is becoming as important as the classic resources of land labor and capital to regional development. The changing resource base affects the constellations of actors, and their relative importance in the local political economy. With knowledge assuming increased significance as a factor of production, in both high technology and older manufacturing industries, knowledge-producing institutions such as universities assume a greater role in regional growth coalitions (Feldman, 2003; Etzkowitz and Klofsten, 2005).

Business-government relations, formerly the mainstay of local and regional economic development, increasingly include universities as significant actors (Abelson, 1986). While a broad international trend towards triple helix strategies may be identified in Brazil and Russia, peripheral regions such as Northeast UK continue to rely on dual helix business/government and university government formats. As university, industry and government interact more intensively, new organizational formats to achieve economic and social development are invented and linked to one another. These networks typically include liaison and transfer offices, inter-disciplinary centers, angel networks, venture capital, and firm-formation support structures such as incubators and quasi-political entities such as hi-tech councils. These developments reflect a transition from an industrial to a knowledge-based society, bringing with it a shift in emphasis from improving business climate and subsidizing firm retention and relocation to encouraging knowledge-based start-ups. In this chapter, we show the origins of an immanent regional development policy of firm formation from advanced research, its spread to Sweden, and across the US university and regional policy scene.

Regional innovation policy

Most of what can be called science-based regional policy is an accretion of efforts over decades that result in a distinctive format that is rarely recognized as a "policy." Since the results of these efforts were located in a particular region, they could be seen as components of a virtual science-based regional economic development policy. For example, building upon the substrate of a practically oriented university, Frederick Terman, Dean of Engineering at Stanford in the early post-war developed a strategy of capturing federal R&D funds to expand the engineering school and to support electronics research. The university also established a contract research institute to provide research support for Western companies and encouraged its graduates to form electronics companies nearby the university, playing a significant role in the early growth phase of Silicon Valley. These various initiatives might have remained interesting anomalies save for the development of an iterative process in which an innovative use of resources attracts imitators and becomes a significant flow of activity in other regions, as well. In recent decades, university towns such as Boulder, Colorado Ann Arbor, Michigan and Ithaca, New York have become the core of nascent hi-tech clusters.

Translating mostly federally funded academic research in the post-war into a stream of start-ups that then continues to produce descendants and expands beyond its academic sources may be viewed as an immanent innovation policy. Regional and local actors have become active players in stimulating science and technology-based innovation as a regional development strategy (Goldstein and Bergman, 1986; Markusen *et al.*, 1986; Eisinger, 1997). A science-based regional identity is created building upon and transcending an Italian or Danish industrial district in shoes or food products, or even a cluster developed from new business concepts such as New York City's Silicon Alley multimedia firms (Indergaard, 2004). Clusters may also be created from an advanced research stream as an assisted linear model, proceeding from research but guided and facilitated by policy initiatives in the guise of intellectual property regularization and academic development initiatives, the quintessential US model. Alternatively, they may arise from business opportunities, such as the Lagos computer market, where software application firms have arisen on the base of the original computer refurbishment and sales cluster. The next step in development, hopefully, will be to link to local universities developing expertise in computer science so that cluster and academia nay assist each other's development as in the Porto Digital project in Recife, Brazil.

Supporting academic research to improve existing industrial capabilities or generate new economic growth is part of many state and local economic development policies. These initiatives are often either an overlay or precursor to establishment of new universities (e.g. University of California at San Diego and Merced) or enhancement of existing campuses in special fields (e.g. University at Albany and University of Puerto Rico). Much as San Diego's biotechnology industry was sourced from the founding of a new science-based campus of the University of California; New York hopes to achieve the same result in nanotechnology by adding a new heavily supported research focus to a traditionally social science oriented university that supported the state government functions in fields such as criminal justice, public administration and education. The University of Puerto Rico buttresses a pharmaceutical industry acquired through tax incentives with a research center that meets current industry needs and hopefully advances the state of the art of pharmaceutical chemistry in the longer term. An Electronic Paper Institute is the key research strength of a new campus in Norrköping, Sweden, an old industrial town whose paper mills have declined but the wealth derived from them is now committed to a renewal strategy based on academic research in a related field. This contemporary Swedish effort at science-based regional economic revival reprises the early twentieth-century strategy of New England, also a declining industrial region at the time.

The major alternative strategy to a forward linear approach, of course, is to infuse existing industries with advanced technologies, whether through research or incremental innovation gained through manufacturing experience. Automotive clusters in Germany and heavy industry regions in China exemplify this approach. There is more than one path to innovation and too quick abandonment of traditional regional assets, as some say occurred in the northern UK during the 1970s and 1980s, likely exacerbate employment loss and economic decline. Nevertheless, it is a difficult decision whether to abandon a declining cluster, as with textiles in New England, and commit resources to a regional development strategy based on advanced academic research. Sometimes, a dual interactive strategy is warranted, as with the attempted link between Ann Arbor's emerging software cluster and the Detroit automotive industry. We suggest that an optimal strategy is more likely to emerge from a region's academic, business and governmental leadership, joining together to formulate a regional development strategy rather than leaving matters to evolve on their own. In the absence of regional organizing and strategy development efforts, especially in peripheral regions, dominant trends are likely to continue to persist to the benefit of leading clusters.

Science and regionalism

The location of research would seem to be irrelevant to the achievement of excellence. Nevertheless, clustering in research, as well as in manufacturing, creates positive benefits and thus has been encouraged both in research intensive and in aspiring regions. Space becomes differentiated and is shaped into valleys and corridors, emphasizing various objectives for a knowledge region. Regionalism has been an important factor shaping the development of the US starting from the divergent regional interests between the north and the south that had to be reconciled in order to establish the country. Science-based regional policies followed indirectly from these fundamental regional differences. The early nineteenth-century industrial growth in the Boston region provided a platform for the establishment of a technological university, MIT, according to a plan of a University of Virginia professor who moved to the Boston region to gain the support of area political and industrial leaders for the project. As a result, MIT received both private support and one-third of the state's "land grant," provided by the Morill Act to support the development of a new class of universities to provide a scientific base for agricultural development, the primary industry at the time. However, the A and M (agricultural and mechanical) rubric was broad enough to include the MIT project as well.

Indeed, encouraging the development of universities in each state with an emphasis on practical activities in contrast to the classical liberal arts colleges designed to train members of the clergy and traditional professions, can be said to be the nation's regional science policy. The origins of US science-based innovation policy and the format according to which it developed can be seen in the founding of the land grant university system. These schools combined research and education, in a cost effective format. These practically oriented universities focused on agriculture, the major industry of the time, in contrast to the traditional universities that prepared students for the classic professions Although the land grant model was generalized to all states by the Morill Act of 1862, the model had its origins in the Connecticut farmers lobbying their legislators to provide assistance to their efforts to apply science to agriculture and to educate their children to be better farmers (Rossiter, 1973). Several of those foundations, including the University of California, Berkeley, grew to become the leading universities of the country, focused on a broad range of arts and sciences. To this day, Berkeley has a School of Natural Resources and a county agent attached to it, responsible for liaison with area farmers, even as the major focus of agricultural R&D was moved to a new campus at Davis, the former site of Berkeley's experimental farm.

During the post-Civil War era, a southern counterpart to MIT, the Georgia Institute of Technology in Atlanta, was founded. In contrast to MIT, which was founded to infuse an existing industrial region with inputs form science and technology, the Georgia Tech project was the basis of a strategy to promote new industrial development in the South. A major component of the university was a contract research organization that provided research support to local industry and later attracted federal research grants. These two schools were examples of a broader movement to establish universities with a practical orientation in various parts of the country, e.g. Stanford University on the West Coast, with a focus on engineering, and New York University, with a focus on commerce, on the east coast. Nevertheless, rather than narrow technical or business institutes, these universities gained distinction as liberal arts research universities, encompassing a full range of disciplines. Academic breadth laid the basis for future interdisciplinary collaboration and the creation of new disciplines, such as computer science as a synthesis of elements of applied math, electrical engineering, psychology and philosophy.

The use of the university as a locus of research to further national goals was expanded greatly during the Second World War. The realization that basic research could lead to discontinuous innovation in weapons that could affect the course of the war led a group of leading academics to propose an R&D program to the federal government. Some of them had worked in military laboratories during the First World War where they felt that their work was impeded by having to work under military direction. Having originated the proposal to establish wartime advanced research program, academics were appointed to prominent positions in running the program. Not surprisingly, major wartime research projects such as the Rad Lab for radar at MIT were located on university campuses, although these were not major research centers at the time in comparison to corporate central laboratories at General

Electric or Bell Telephone. With virtually unlimited funding available, academics research experience was irrevocably transformed. Many of the basic scientists, working as engineers during wartime, experienced a double epiphany. On the one hand, they realized that they were deriving new ideas for basic research from their practical work, and on the other that they could accept governmental funds without loss of control over research direction, given that fellow scientists were in control of the disbursement of these funds. In the post-war, federal support of R&D by agencies such as NSF, NIH and DARPA continued the concentration of research funding in a relatively few regions.

Entrepreneurial academic model

Universities became the new engines of economic and social advancement. Science-based regional development in Boston (Route 128) and Silicon Valley arose from the seminal roles of MIT and Stanford in setting in motion a dynamic of science-based innovation in the late nineteenth and early twentieth centuries. For example, the venture capital firm was invented during the 1930s and 1940s as the project of a coalition of New England university–industry–government leaders (Etzkowitz, 2002). Their primary objective was not to make money for themselves, but to renew a declining industrial region. A new engine was required to power the regional economy on which their institutions ultimately relied for viability. To this end, they designed a strategy based upon the useful elements of the regional infrastructure left behind from its recent industrial past, in particular universities and financial institutions, to create a new innovation model. A firm-formation dynamic was set in motion by inventing the venture capital firm to create companies based on academic research projects, largely from MIT. Silicon Valley venture capitalists expanded upon this model, loosening its ties to academia by creating successive generations of new start-ups from previous firm formation efforts.

The land grant tradition of technology transfer to agriculture was reinterpreted to make it work for industry. Technology transfer organizations were invented to seek out industrially relevant research findings and move them to industry, via the mechanism of the patent system. Some academics, like Vannevar Bush, took further steps, participating in the formation of firms when existing companies were unable or unwilling to commercialize useful discoveries (Bush, 1945). Thus, the land grant model to support agricultural innovation evolved into an entrepreneurial academic model to stimulate industrial innovation (Etzkowitz, 2002). The model of local initiative translated into policy at the federal level that then spreads an initiative to other regions holds to this day. Thus, the Bayh Dole Act of 1980, granting universities control over intellectual property generated from federally funded research, was instituted at the initiative of a few universities that had early established technology transfer programs. The Bayh-Dole Act was intended to affect universities but, of course, has a differential regional effect since university research is unevenly distributed.

The Bayh-Dole act of 1980 formalized an incipient policy model of technology transfer form academia through licensing and start-ups, based on the expansion of federally funded research during the post-war. In other countries, the law was sometimes put in place before an underlying culture and local mechanism for technology transfer were in place, thereby not achieving immediate results. In the US, by contrast, the Act regularized a process that was already in place and encouraged other potential actors to get into the game. While disparities between regions will very likely persist; they will be between "have more" and "have less," rather than between regions that "have a lot" and those that "have none." On the one hand, regions such as Boston and the Bay area of California have an advantage in reaping the rewards from Bayh Dole by expanding their technology transfer activities. On the other hand, the Act implicitly encourages less research-intensive regions to build up the research capacities of their local universities. The objective is to be more competitive for federal research funds and to create jobs and new companies based on the intellectual property that will be generated from this research. Of course, the research-intensive regions also engage in capacity building efforts, such as the California Stem Cell initiative. The key to translate research strength into a source of regional development is a cluster of start-ups that then attract larger firms to access their innovations.

Venture capital is one element of a "start-up" model of economic development that can be substituted for by a variety of public and private initiatives such as angel networks and state and federal government programs (Etzkowitz *et al.*, 2000). Similarly, the business incubator synthesizes elements drawn from academia and business into a format to train organizations. Follow-on regions synthesize these exemplars into more or less coherent plans for science-based development. Thus, the venture model was transferred to northern California as firm formation activity expanded in the emerging semiconductor industry. A critical mass of university R&D with commercial potential, tapped by a technology transfer office, supported by incubator facilities, angel networks and the development of sources of venture capital, legal, accounting and the business development assistance, culminating in a science park, has become the de facto policy model for science-based regional development.

There has been a convergence toward a common entrepreneurial academic model combining elements of the research and land grant university traditions. Moreover, the scale and scope of university research, especially in the sciences and engineering, was transformed. The wartime experience taught academic researchers and universities how to manage large projects and significant amounts of money. A more traditional group of universities, both old and new foundations, had focused on pure research, from the late nineteenth century. Based on American interpretations of the German academic model, these universities moved closer to the land grant model as a result of their wartime experience. The land grant schools had themselves been moving to incorporate basic research elements, like genetics, into their more practically oriented research programs.

Universities, in collaboration with government and industrial actors, are the basis for creating interdisciplinary centers and new foci of research with intellectual and technological potential. Explicit transfer mechanisms and informal networks linking emerging entrepreneurs with previously successful mentors enhance a culture of entrepreneurship within academia. Emerging technology regions emulate these characteristics through programmatic intervention, even as organizational entrepreneurs, such as founders of angel networks in successful regions, franchise their models to other regions.

Sweden's innovation renovation

Three models of innovation and their relevance to Sweden have been set forth: the American model of new firm formation, the Irish model of foreign direct investment and the European model of large firms diversifying into new technologies. Although lack of foreign direct investment was considered a problem a few years ago, this has changed recently. However, foreign investment has focused on areas in which Sweden is already strong. It appears around "Swedish strongholds" and adds strength to what is already there rather than changing the productive structure. Significant firm formation activity was essentially a collaboration between two elements: researchers and elements of a large firm in a region, whether the firm still exists there or whether it has disappeared and left behind people and networks as in Uppsala, with the departure of a considerable portion of the R&D activities of Pharmacia to the US after its merger with Upjohn. Many firms started as collaborations between researchers from university and people from Pharmacia, leftover as a pool of skilled workforce available for new projects and ventures. Venture capital is willing to be part of such ventures because team building was available that they could have confidence in.

Three sources of hi-tech firm formation have been identified: start-ups that emerge from large firms versus those that arise from universities, as well as in between hybrids collaboration between employees of large firms and academics. Those that arise from companies typically have already identified markets in contrast to those from universities. People who leave large firms know customers already. Indeed, they have already built IT for their future customers. On the other hand, university based start-ups show faster growth. The most highly successful cases were said to where the incubator and science park were established in connection with a university, with long-standing links to a large firm. The effect of the large firm can take place even in its absence when the networks between academia and industry left behind after a large firm has relocated from the region become the basis for future firm formation.

An obstacle to innovation was firms staying with what they know, rather than developing new areas. A study of firm formation in the IT sector found that the major support structures were existing, large firms. Of 500 IT firms examined, most were working on mid-ware and deeper layers of technology rather than exploring technologies of system integration at higher level. The researchers divided firms into subsectors, visited web pages, examined products, etc. The work of these firms was complementary to what is explored by larger firms. The larger firms let other firms do things and maintain a relationship between them. There was a measurable effect of Ericsson. An emerging model is that instead of funding research on its own, Ericsson goes to the Stockholm School of Technology (KTH). People come from Ericsson to work on a project jointly with researchers from KTH. If successful, a firm is spun out.

Swedish venture capital, its availability, stage of development and the relationship between public and private venture capital were discussed. The shortage of seed capital was said to be a major obstacle to firm formation. Seed money has to be flexible, with the public sector entering when there is a downturn. Indeed, a key element in firm formation is public venture capital. A study found that it would not have been replaced by private venture capital if it had not been available for start-ups. 80 percent of start-up firms reported that there was no private alternative. Thus, it was felt that there was no competition between the private and public actors. However, the availability of venture capital in large enough amounts to properly support the later stages of firm-formation was questioned.

It has been suggested that Swedish venture capital focuses mainly on early stage investment due to insufficient capital to address the growth phases. In the first round of finance there is a focus on the idea. There was a difference of opinion over whether enough funding was available for new firms to build management and marketing teams and reach the market. Entrepreneurs underestimate the costs of getting sales. However, if the target identified is one in which the market is an existing firm then that may be the answer as to why there is less capital to reach the market as an independent firm. Perhaps that is not the objective of many firms being formed? They may be seeking sale to a large firm as an exit strategy rather than attempt a growth strategy as an independent firm. Even Skype, the most successful Nordic start-up, was successively sold to two American multinationals.

The issue of the national attitude towards entrepreneurship was reflected in the oft heard statement that, "You should have a real job. Why start a firm when you can be hired and get a real income?" Nevertheless, attitudes, at least among some students, appeared to be changing. It was said that students at the Stockholm School of Economics (SSE) students do not want to work for large firms. Recently, the entrepreneur has become a hero at SSE, previously a factory to produce people for large companies. The negative aspects of new firm-formation were also noted, such as the heavy paperwork burden on entrepreneurs. It inhibits them mentally from starting up when they hear about problems of filling out forms for tax authorities.

One of the reasons for low growth in the past several decades has been held to be the limited role of the university in innovation beyond supplying capable people. As a research organization, it should be a complement to the market, taking the role of doing the things not done by firms. Universities were found to be alike in the formation of a variety of institutional arrangements, e.g. holding companies, patenting and licensing, entrepreneurship courses, meeting places for role models and students. Nevertheless, there are tremendous differences among regions with respect to the existence of support structures. The concept of technopolis, concentrations of innovative activity, may be applied only to a few areas, such as Stockholm/Uppsala, Linköping and Gothenburg. One starting point for improvement is entrepreneurship training. A role for the regional colleges can be a new specialization in entrepreneurship competencies.

The science park has played a significant role in developing the large firm/university relationship. Larger firms felt they needed a recruitment base closer to universities. This was an important reason for Ericsson to locate units near Linköping and Lund. The science park location gave the opportunity for people at the university to work as consultants for the firms. The science park is also a vehicle for changing the industrial structure. Firms formed through the entrepreneurship programs at Linköping found a location at the Merdovi science park adjacent to the university. In time, this industrial formation of a couple of hundred firms directed toward new markets in electronics replaced the 5,000 jobs lost in constructing fighter aircraft.

Developing an institute sector as an interface between industry and academia was controversial. On the one hand, it was argued an expanded industrial institute sector was needed since R&D intensity in small- and medium-sized firms (SMEs) is lower than on neighboring countries such as Finland and Norway. Institutes could fill this gap. It was said that, in institutes, there is more market knowledge than in universities. "Academic actors are thinking in another way, not connected with people in companies. In institutes, there is a commercial approach to research. Institutes are used to time limits whereas universities are not." On the other hand, an expanded institute sector was rejected on the grounds that universities were flexible and could fill the role, while some of the existing institutes were not functioning that well. A synthesis of the two approaches, that large universities could have institutes, was also suggested.

An alternative view of the future is of an industrial structure with companies of different sizes, especially new mid-sized structures, not Ericsson or Nokia. One observer noted that Schumpeter was correct in that new structures promote better use of resources and enhances welfare through a high degree of new firm entrances. However, another questioned whether creative destruction is always a good thing since existing sales channels are lost. An industrialist asked, " Is it wonderful to see large firms go down and new ones emerge? Which small firm will grow to the size of Ericsson?" As these skeptical questions suggest, the Swedish model is a middle way in innovation policy, including a process of start-up in new industrial areas but weighted heavily in the direction of reinvigorating existing technological industries with potential for re-invention (Svensson *et al.*, 2011).

US science-based regional innovation

A seemingly self-generating industrial sphere of hi-tech business activity has its roots in academic and governmental initiatives. The components of a science-based regional innovation environment can be built on the base of resources put in place for other purposes and objectives. For example, the US "War on Cancer" may still be a long way form its stated objective, but it helped to create a biotechnology industry along the way. The following instances illustrate the various stages of this process, starting with: (1) the unintended consequences of federal actions undertaken for other purposes; (2) the use of state government political influence at the federal level to attract additional resources (national resources are drawn upon in regional development efforts but the initiative is local); and (3) followed by university initiatives to mobilize regional resources to encourage start-up activity. The unintended consequences of federal R&D policies and bottom-up technology transfer initiatives that were subsequently generalized by federal law in the Bayh-Dole Act exemplify the US policy model.

An "assisted linear model" of linked organizational innovations that develop inventions into businesses or in other ways perform the function of "innovation organizer," a catalyst for cluster formation, superseding a laissez-faire linear model in which innovation is expected to grow spontaneously from a well-funded science base. Science-based regional development proceeds from large-scale national expenditures for R&D that is supplemented at the state level by an estimated $3 billion per annum (Etzkowitz *et al.*, 2000). Not surprisingly, in less research-intensive areas, the focus of local effort is on measures to enhance R&D activity as a prerequisite hi-tech economic growth. Nevertheless, a strong regional science base by itself is not sufficient to achieve this objective. New York and Chicago, for example, have strong concentrations of biomedical research but little to speak of, as yet, in terms of economic growth from this foundation. How can these contradictions be explained and addressed?

The path-dependent elements in US regional innovation policy are an enhanced role for universities and an ambivalent role for national government. Nevertheless, knowledge-based regional development has been shaped by the unintended consequences of federal policies, as well as measures targeted at encouraging universities to participate in innovation. Federal research funds, largely confined to support of agricultural research before the Second World War, became available for a variety of civilian and military purposes on an ongoing basis after the war. Agricultural research funds were broadly distributed around the country, primarily to the land grant universities in each state, focused on local crops and agricultural conditions. Wartime research projects, on the other hand, gathered researchers from across the country to a very few sites on university campuses, such as MIT, or newly constructed research establishments, such as Los Alamos. Research centers, an unusual organizational format before the war, became the modal format in the post-war, linking research groups around a common goal, often inspired by the opportunity to attract a large multi-year grant or series of grants from a range of governmental and industrial sources.

The post-war expansion of federally funded research had officially unintended regional implications that were not entirely unexpected. Since the funds were almost entirely distributed on merit, rather than geographical criteria, it could be foreseen that they would largely end up at well-established research universities. The apparently neutral workings of the peer review process favored regions like Boston that already had concentrations of research capabilities, exemplified by the presence of Harvard and MIT, and were better able to capture these funds. The development of Stanford as a major research university during this period was in part based on an academic development strategy to access this new source of research funding. Other regions have since employed a variety of "catch up" strategies to capture a share of these funds in recent decades. The entry of additional universities and regions into the competition for federal research funds has made these funds harder to get despite growing budgets due to increased competition. It has also increased pressure for these funds to be expanded and explains much of the political support for initiatives to double the sizes of the budgets of the National Institutes of Health and the National Science Foundation.

Regions may be viewed as "thick" or "thin" depending upon the presence or absence of innovation support structures, whether informal or formal. A region that is rich in business development requisites will not have to develop as many explicit organizational mechanisms as one where these are lacking. Thus, whether it is important for a region to create new organizational mechanisms depends upon whether firm formation is already taking place, for example, supported by a network of angel investors, or requires a formal support structure, such as an incubator facility, to take off. Although there is no single set of interventions appropriate to all regions, some common elements, that foster hi-tech regional development, can be identified such as the presence of an entrepreneurial university with the means to assist technology transfer and firm formation.

Efforts to develop R&D capacity are the necessary base of science-based regional economic policy but it is not sufficient. The development of methods to translate R&D capacity into economic growth and development is required. The director of the Technology Transfer Office at the University of West Virginia took the initiative in organizing local resources to establish venture capital funds, a missing element to promote science-based development in that area. Between the effort to seek out potential technologies for development from research at the university and public/private co-operation to provide venture capital, a hi-tech firm-formation process was initiated.

The effort drew upon national policies that had an indirect effect, in encouraging a local hi-tech conurbation. As part of the "War on Poverty" during the Kennedy/Johnson administrations, Appalachia, a long-depressed region, received some attention. The major initiative was road construction to link the area to other parts of the country in order to link it to the national economy, a necessary if not sufficient condition for the success of the local initiative discussed above.

The great US experiment with direct regional policy took place as part of the response to the great depression of the 1930s. These polices were based on the problems and potential of natural features such as watersheds. Large-scale initiatives to promote conservation and growth simultaneously were envisioned. The Tennessee Valley Authority, cross cutting state boundaries, was expected to be a model for other regions. Flood control, electrification and other objectives were achieved by focusing on construction of dams. Nevertheless, regional development policy on such a scale was never repeated, given the onset of the Second World War and strong political opposition to large-scale central government initiatives. However, unintended effects of this flood control project became the basis of regional R&D capabilities.

The electrical generation capacities created by the Tennessee Valley Authority became a military asset during the Second World War when it became the location of choice for the Manhattan Project effort to make purified uranium for the atomic bomb project. The R&D facilities created for the project were kept in place during the post-war as the Oak Ridge National Laboratory. A region with virtually no R&D capabilities became a major R&D center, at least in a specialized area, as an unintended consequence of the previous effort to tame a watershed and create electricity production facilities. The subsequent interaction between the Oak Ridge National Laboratories and the University of Tennessee helped build up that university's research capabilities, preparing the way for it to become a source of science-based innovation for its region. An R&D conurbation superseded a watershed as the focus of regional development.

Concentration of national research resources at a relatively few leading US universities is no longer acceptable to other regions now that the role of academic research in creating new firms and jobs has been widely recognized (Dougherty and Etzkowitz, 1996). Thus, in response to criticism, federal R&D programs have taken regional disparities into account to a limited extent. For example, the Experimental Program to Stimulate Co-operative Research (EPSCoR) is a set aside program that is in effect for 13 states and the Commonwealth of Puerto Rico, which have been designated as less research-intensive areas. A sum is provided that is competed for within the state, according to the same criteria used nationally. Not surprisingly, the leading research institutions in each EPSCoR state are the greatest recipient of funds. NSF runs workshops to teach researchers in EPSCoR states how to improve their chances to compete in NSF competitions for research centers. There is also a de facto national "program" that directly targets areas with little or no research resources – the "pork barrel legislation": members bills that fund research centers at local and major national universities.

Conclusion: An assisted linear model

Science-based regional development can no longer be considered as uniquely characteristic of a very few regions (Dorfman, 1983). What is new is the spread of innovation policy to virtually all regions, irrespective of whether they are research or industrially intensive. The role of university and government is apparent in the early stages of science-based regional development. As regions become successful, hi-tech development appears to be a business driven process, as new firms spin off of old ones. However, in economic downturns, mature regions revert to strategies that characterized their earlier development but are typically later forgotten even by the initiating organization. Typically, this involves reviving or organizing a regional organization representing academic, industrial, governmental and other actors to brainstorm and implement new foci for knowledge-based innovation (cf. Etzkowitz and Ranga, this volume).

Through its innovation agency, VINNOVA, Sweden incentivized regions to bring together triple helix collaborations through a grants competition for regional development. In the US, lacking explicitly formulated policies at the national level, regions draw upon each other's successful projects as models for replication. A variety of regions have demonstrated the capability to induce hi-tech growth, even as some efforts fail and others are restructured. Most initiatives focus upon some combination of background conditions, e.g. presence of R&D resources (Casas *et al.*, 2000) through development of universities and research institutes; facilitating factors, e.g. availability of highly skilled human resources (Florida, 2002) and a conducive legal and regulatory environment; and specific financial and organizational mechanisms such as venture capital firms, science parks (Qunitas *et al.*, 1991) incubators and liaison and transfer offices. A patchwork of innovation mechanisms that gradually become linked may be constructed to create a regional innovation system, translating inventions and market needs into economic activity via firm-formation.

The conditions for hi-tech economic growth are not spontaneous creations; rather, they can be identified and put in place by explicit measures, by various actors, governmental and non-governmental. In regions that are just emerging as high-technology centers, the university TTO often plays an active, organizing role in creating the conditions to encourage firm-formation in co-operation with other actors. However, in highly developed regions, angels and venture capitalists have become less involved with universities as a dynamic of firm-formation from previous start-ups takes hold.

Nevertheless, in periods of economic downturn, there is a return to collaboration with other regional actors, including universities, in order to identify the next technological paradigm for start-up development (Bathelt, 2001). This development occurred concomitantly with the transition from industrial to knowledge-based society, heightening the significance of research for its economic as well as its knowledge potential.

As old technological paradigms are exhausted, even the most successful regions go back to the first phases of the model and focus on creating new sources of research in order to insure a flow of inventions into the firm formation process. Silicon Valley is ultimately dependent upon the flow of new inventions into the firm-formation pipeline. Thus, there is a periodic return to developing new research fields, with government funds in universities, as the source of science-based regional development. Biotechnology, nanotechnology, stem cells and clean-tech are the current foci of attention. At the regional level, universities attempt to develop unique variations on these themes in order to differentiate themselves from competitors in the search for research advantage. Translational research and incubating virtual firms in academic labs bridge the gap between fundamental investigation and realizing the economic potential of new knowledge.

The linear model reconfigured as an "assisted linear model" is the implicit framework of an emergent US regional policy for knowledge-based economic development, focused on generation of start-ups with growth potential and encouraging their local clustering.

The significant support for S and T in the Obama Administration's stimulus package recognized the significant role that the knowledge base plays in US economic development. Nevertheless, the automotive and financial industries were supported on a much larger scale. To seriously address the underlying causes of the great recession, support for emerging knowledge-based industries will have to be scaled up, with commitments made in a longer time frame. Sweden's VINNOVA Innovation Agency initiative, created in response to the 1992 economic crisis, provides a model that should be considered for transfer to the US in order to more effectively target support for regional innovation.

Notes

1. The discussion of Sweden draws upon presentations and comments at a Workshop on Innovation Policy, sponsored by the Centre for Business and Policy Studies (SNS) at the Villla Soderas, 2002.

References

Abelson, P. (1986). "Evolving State–University–Industry Relations." *Science* 240 (4850): 265.

Bathelt, H. (2001). "Regional Competence and Economic Recovery: Divergent Growth Paths in Boston's High Technology Economy." *Entrepreneurship and Regional Development* 13 (4) October: 287–314.

Bush, V. (1945). *Science: The Endless Frontier.* Washington, DC: US Government Printing Office.

Casas, R., De Gortari, R. and Santos, M. (2000). "The Building of Knowledge Spaces in Mexico: A Regional Approach to Networking." *Research Policy* 29 (2): 225–241.

Dorfman, N. (1983). "Route 128: The Development of a Regional High Technology Economy." *Research Policy* 12: 299–316.

Dougherty, K. and Etzkowitz, H. (1996). "The States and Science: The Politics of High Technology Industrial Policy at the State Level." Report to the National Science Foundation Sociology and Regional Science Programs.

Eisinger, P. (1997). *The Rise of the Entrepreneurial State.* Madison, WI: University of Wisconsin Press.

Etzkowitz, H. (2002). *MIT and The Rise of Entrepreneurial Science.* London: Routledge.

Etzkowitz, H. and Klofsten, M. (2005). "The Innovating Region: Towards a theory of knowledge based regional development." *R&D Management* 35 (3): 243–255.

Etzkowitz, H., Gulbrandsen, M. and Levitt, J. (2000). *Public Venture Capital.* New York: Harcourt.

Feldman, M. (2003). "Entrepreneurship and American Research Universities: Evolution in Technology Transfer," in Hart, D.M. (ed.), *The Emergence of Entrepreneurship Policy: Governance, Start-ups, and Growth in the US Knowledge Economy*, Cambridge: Cambridge University Press, 92–112.

Florida, R. (2002). *The Rise of the Creative Class.* New York: Basic Books.

Goldstein, H. and Bergman, E. (1986). "Institutional Arrangements for State and Local Industrial Policy." *Journal of the American Planning Association*, 52: 265–276.

Indergaard, M. (2004). *Silicon Alley: The Rise and Fall of a New Media District.* London: Routledge.

Markusen, A., Hall, P. and Glasmeier, A. (1986). *High Tech America.* Boston, MA: Allen and Unwin.

Quintas, P, Wield, D. and Massey, D. (1991). *High Tech Fantasies: Science Parks in Society, Science and Space.* London: Routledge.

Rossiter, M. (1973). *Justus Liebig and Emergence of Agricultural Science in America* New Haven, CT: Yale University Press.

Svensson, P., Klofsten, M. and Etzkowitz, H. (2011). "A Knowledge-Based Strategy for Renewing a Declining Industrial City: The Norrköping Way." *European Planning Studies*. London: Routledge

14 Lessons on regional innovation governance in open economies

Staffan Laestadius and Annika Rickne

Introduction

Classical schoolbook assumptions of "closed economies" have little relevance in today's analyses on how to govern regional economic development. Small countries – such as Sweden and Belgium – are more open than closed, as are regions in large countries such as the US. This is what the book is about: how to govern, or even create, dynamic "hot spots" in a slippery space. In this final chapter, we intend to summarize and to draw general conclusions from the different approaches and empirical illustrations of the chapters of the book. In short, we identify that the book has contributions in four different – although highly related – areas.

Implementing a concept-driven innovation policy

Innovation policy in Sweden – as in many other places – rests on the analytical approaches of innovation systems and triple helix. As pinpointed by Laestadius and Rickne, such frameworks build on underlying assumptions and are more or less efficient in capturing innovation dynamics (Chapter 2). This concept-based character of the Swedish policy actors' reality and of the innovation policy (as performed by VINNOVA) is analyzed in several chapters (cf. Chapters 2, 8, 9). Such reliance on analytical frameworks is not only the case in the formation of VINNOVA – which is explicitly constructed to perform innovation systems policy – but also in the detailed formulation of policies and in the allocation of funding. Differing from most research councils, this government body is proactively conditioning much of its funding according to a rhetoric based on innovation systems theory in general and triple helix approaches, in particular. In fact, this was a way for the Swedish policy system to grapple with the new innovation paradigm, as pointed out by Persson in Chapter 8.

In practice, significant work is performed to translate the general concepts to scripts and simple rules useful for applications, funding and performance evaluation (cf. Chapter 9). The VINNVÄXT program is a good illustration to this translation logic. As pointed out by Laestadius and Rickne (Chapter 2), adapting an innovation system approach and focusing on system changes (with a functionality perspective) might have enabled Swedish policy actors to efficaciously grasp global dynamics. In addition, the interactions between theory, policy and industrial practice has explored the usefulness of frameworks.

However, there are also some potential problems integral to the translation logic. The studies in this book demonstrate that sometimes policy in its implementation becomes interventionistic and detailed rather than general. In addition, there seems to be an underlying assumption that it is possible for a policy actor to "pick the winners" (cf. Chapter 9). This may, on the side of applicants, lead to a tendency to opportunistic behavior as regional applicants organize themselves to live up to the rhetoric of (regional) innovation systems connections and engagement of exactly the "expected" triple helix partners in each dimension. This turns the world upside down: from being tools for understanding innovation process the concepts become rules for the game.

If interpreted within a traditional administrative context, this may contribute to a two-dimensional lock in: regional as well as cognitive. This is also what seems to have happened in several of the cases studied in this book. Actors have been directed primarily by the policy scripts, instead of by the factual needs of, for example, producers or markets. This was surely not the policy-makers' intention, and has not been beneficial for reaching the goals.

Nevertheless, there is an indication in our material that the VINNVÄXT program has indeed contributed to overall and significant success stories. There are several good and positive achievements, some of which are mentioned in this book. Thus, even if the linkage between theory, policy scripts, actor interpretation and behavior has not always been straightforward, Swedish policy initiatives have been influential in a positive way in many instances. This mirrors well the findings in the international studies – Chapter 12 by Carlsson and Chapter 13 by Etzkowitz – where public policy has shown its instrumental character in several successful regions.

The complexity of proximity

An important underlying dimension of such innovation governance is that the classical meaning of the concept "proximity" has changed. As shown in several of the chapters, cognition activities related to innovative processes can, to a large extent, take place with little or no necessary physical proximity. And so they do. Company strategies for knowledge production, diffusion and use, for example, involve distribution over space, in that various parts of the knowledge and assets needed for innovation and production come from partners located far outside of the region.

As shown in the contributions by Laage-Hellman *et al.* (Chapter 5) and Gustavsson *et al.* (Chapter 6), firms that are innovative and large enough to be of importance for a regional policy agenda are often also active on a global scale. They do not only interact with firms in other regions and countries, but may also take part in many "triple helix"-type structures worldwide, and be integral to the government–industry–academia web of interaction. We provide in this book no answer on the question of whether different forms of knowledge can be ranked as regards their need for physical proximity. In fact, it may be argued that such a ranking is impossible and a non-static phenomenon. Our chapters show that Swedish actors exchange all forms of knowledge (tacit/codified, analytic/symbolic/synthetic, etc.) with partners and sources, not independently of location but in global patterns. For example, in the case of biomedicine, there are certain types of knowledge exchanges that are – presently, at least – regional in character, and others that are national or international (Chapter 5). Such patterns are, of course, industry, spatially and time dependent. As revealed in these chapters, firms develop intra-firm innovation processes, as well as open innovation processes on a global scale, utilizing the achievements in modern communication technology and management capabilities. That is not the same as postulating that physical proximity is of no importance – we will return to that below – but more a conclusion that significant aspects of direct human contact can be compensated for by industry or industrial actors. As we learn to communicate globally, and as we learn to manage cultural differences, we may overcome yet more constraints of physical proximity.

Naturally, the RISs transform when the operations of firms become increasingly global. This is far from new from a Swedish perspective: the Swedish economy belongs to the most open economies in the world, and there is a large set of international firms operating from Sweden. This creates experiences from openness, as well as challenges as regards policy formulations – something that is in focus in many of the chapters. The question here is how such configurations can be handled by actors in the innovation system. To have a chance to govern it, it is important to develop a general understanding on what the competitive advantage of the regional innovation system is for each industrial sector and whether there is a change over time. Another way to put this is to say that one should understand which types of knowledge need physical proximity and which do not, and in which sub-categories of actors in the value chain the region might be most relevant. By this, one may see what is needed to build or strengthen in the regional context.

The concept of a region

Highly related to the proximity discussion, our study reveals the problems with the "region" concept. Normally understood as a piece of territory, implicitly or explicitly including a more or less complete set of actors located in, or with activities related to, a certain geographical space, the "region" concept has, at least for industrial purposes, lost some of its former relevance.

It may be argued that less need for physical proximity on a daily basis enlarges the functional regions. Physical interaction on a non-daily basis – which is normal in many innovative activities – can take place over large territories, spanning over several political borders. This has implications not only for the organization of innovative work and processes, but also for the provision of human resources. In fact, "modern functional industrial regions" may well get a character of networks of cognitive and physical flows between non-physically connected hubs, distributed across formal regional and national borders. These networks may include industrial as well as academic actors. Several of the chapters in the book approach these questions from various angles.

Laage-Hellman *et al.* (Chapter 5) show that biomedical innovation processes have a strong networking character, that they, to a large extent, are open and that they take place on a national and international level rather than on a regional level as traditionally defined. In fact, local networking is more or less absent in some of the industrial segments analyzed. This is also the case with the globalized robotics industry, as shown in Chapter 6. Intra-firm R&D processes, as well as academic co-operation, far exceed the locale for the robotics firms located in the Mälardalen region in Sweden. The longitudinal analysis by Teräs and Ylinenpää in Chapter 4 shows not only similar tendencies of inter-regional networking, but, interestingly enough, also that local/regional co-operation has declined relatively in importance.

Obviously, regions of various sizes cannot be easily compared. While success examples around the globe are relevant, many regions are, of course, very large in absolute size and in volume of resources as compared to a country such as Sweden or its regions. For comparative analyses, the whole of small countries such as Sweden or of Finland may be relevant units of analyses, rather than regions such as Gothenburg, Luleå or Oulu.

In addition, there is another translation problem in common use of regional analogies. For example, Silicon Valley – being a role model for much of modern regional analysis and policy – should probably not be understood as a science park or a valley primarily, but as a cognitive and cultural context, although with a specific focal point. Indeed, there are many explanations to such regional dynamics, including the phase of the development cycle, where, for example, the Silicon Valley success coincided with the take off phase of the ICT cycle. Such patterns cannot easily be repeated in new areas. As discussed by Etzkowitz and Ranga (Chapter 3), mechanisms for spurring regional development may vary substantially between settings, and identifying such variances and/or creating appropriate conditions is partly the responsibility of regional policy-making.

Partly, the issue of what a "region" really means may also be looked upon as related to position. Indeed, regions have various positions in terms of regional hierarchy in the nation, and this defines their attractiveness and access to resources, where the ones at the top have larger possibilities. Clearly, position is a relative indicator: perhaps a smaller RIS on the top has larger opportunities than a large RIS that is lower ranked. The "starting value" is thus lesser for lower-ranked regions, so these conditions may require larger input and more persistency to reach the same level as higher-ranked regions.

This is also what can be learned from Chapter 7 by Andersson and Johansson: they confirm that it necessitates unyielding efforts to change a region's position, but with such persistency, one can indeed change. However, even when sustained over long periods, policy efforts might not be successful, as highlighted by Carlsson (Chapter 12). As discussed in Chapter 4, although similar in several dimensions, Oulu and Luleå have different positions in their national contexts. Oulu – although small – was and became a core node in the Finnish ICT economy, while Luleå has always been a marginal player in the Swedish system. This national embeddedness seems to have conditioned their development paths to an extent that makes it difficult for analysts to discriminate from local influences.

Our understanding of regions may, however, also depart from the perspective of human resources. Irrespective of the level of networking in the innovative process, the hubs of activity have to be located somewhere, and firms that compete on qualified human resources have to adapt to where and how to get access to qualified staff. Building high-level education and R&D units are slow processes compared to the growth rate of rapidly transforming industries, which creates incentives for relocation of innovative units to regions with excess supply of human resources. As shown, human resource availability was obviously a strong reason for the biomed cluster in Gothenburg (Chapter 5) and the ICT clusters in Oulu and Luleå (Chapter 4). Once established, these clusters may also develop path-dependent processes as regards development of competence and expertise, thus creating strong arguments for further localization; and this with no necessary increase in local/regional inter-firm networking. Rapidly growing markets may, of course, strengthen these arguments, as shown in Chapter 8. Also, long-term investments in qualified human resources may, *ceteris paribus*, be a strong means for regional policy.

On path-dependent processes and national coordination

Strong regional innovation systems or clusters manage to continuously renew and invigorate themselves (see Etzkowitz, Chapter 13). At the same time, the development and formation of innovative regions/clusters is strongly path-dependent, which is revealed in many chapters in this book totally in line with conventional wisdom. A core topic as regards this issue is the role of policy in path creation. While Chapter 4 by Teräs and Ylinenpää rather indicates a hesitant role for policy – pointing to the strong long-term paths of regional development (low-populated regions lose their population to the already highly populated ones), Chapter 12 by Carlsson indicates that triggering events – also initiated by policy – may initiate new processes and/or change existing trends.

Not only the theoretical discussions by Andersson and Johansson (Chapter 7), but also the empirical analyses (e.g. Chapters 5, 6, 10) indicate a limited role for innovation policy to create new regional innovation systems or to fundamentally change existing tendencies. Initial conditions must be favorable. Such conditions – industrial strength and specialized research expertise – was exploited and developed in the case of ProcessIT Innovations, balancing risks of regional lock-in effects, as discussed by Johansson and Ylinenpää (Chapter 10). Carlsson resolves that, while clusters may be initiated and reinforced by policy arrangements, the evidence from successful clusters around the globe also exhibits that spontaneous, self-sustaining development is common (Chapter 12).

Such empirical illustrations, coupled with the findings by Andersson and Johansson on the role of path dependency and persistency, lead us to question if it is viable to try to build entirely new regional clusters, which is done in Sweden in initiatives such as "Printed Electronics Arena," which aims to spur development as based on research environments, or "Peak Innovation," focusing on sports equipment and tourism in the Northern part of the country. While both these regions start with some specific strong endowments, several of the components in a regional innovation system are missing, and there is only a modest industrial activity. While innovation and entrepreneurial activity is unexpected in character, and these initiatives may very well prove highly successful, it is nevertheless important to know what levels of engagement are required over long time periods, and what output can be expected.

The strength of path dependency in innovation policy is also shown in Chapter 11 by Malin Lindberg, in how gender is mainstreamed in Swedish regional innovation policy. The mainstreaming process takes place on the surface – like recruiting more women to the (traditionally male dominated) steel industry to solve the emerging problems of qualified human resources recruitment in many industrial fields – while achievements on more fundamental levels still are weak.

The empirical studies in the book indicate that enlarging the "unit of policy" to at least national level, and looking upon the regional actors as hubs in a larger network, may contribute to a more realistic view on what can be created, transformed and achieved. Regional policy-making across established administrative borders will introduce more complexity in the policy-making – assuming that existing borders prevail. Based on the analyses in this book, it is easy to argue that one implication of globalization is that small countries should look over their regional borders. Under all circumstances, regions in small, open economies need to be tightly connected to a functioning whole on the national level. Regional strength, in the case of Sweden, is primarily dependent on the strength of the national innovation system: what is good for the nation is good for the region. When it comes to international competitiveness of clusters, Sweden is too small to be governed primarily from a (traditional) regional perspective. Regions within Sweden should, at most, be regarded as hubs in a national network, linked to and interdependent of other regions of various sizes and locations globally. A national hub-based network may very well involve nodes outside of Sweden.

The factual situation in Sweden has been the other way around: policy has, to some extent, promoted individual regions at the expense of the national level. For example, within life science, there are three regional initiatives in the VINNVÄXT program, financed by government but not coordinated to contribute to national goals. Also, the Swedish Research Council has promoted stem cell related activities, but has not demanded harmonization between regions. The need for coordination does not, however, mean "picking the winners" or detailed control of industrial dynamics. Instead, we perceive one role of policy to facilitate the growth of as many competitive hubs as possible, but also linking these into national networks with strong global ties.

Thus, small countries must have an industrial and innovation policy based on functional hub-based networks, rather than a strict singular regional approach. However, while in the cluster initiative RobotValley, for example, the idea from VINNOVA has been to build a national hub-based network, this is not easily achieved (Chapter 6). Instead, the policy reality, with local triple helix-based incentives and regional growth goals, and the requirements of creating job opportunities within the region, leads to sub-optimization on the regional level. It may, in some cases, even prove difficult not to sub-optimize within the region. The balance thus entails promoting regions on their own accords and not guided by strict scripts, but at the same time not losing national focus.

Importantly, these governance structures seldom involve government actors only, and can hardly be seen as "top down." Instead, governance arrangements often come about as cumulative results of interaction between various stakeholders (firms, academia, NGOs, policy organizations, etc.). This is vividly illustrated by Johansson and Ylinenpää (Chapter 10), stressing the complexity of systems and how their development relies on learning and experiences gained by the constituting actors. Mutual trust and substantial time periods are required to build regional innovation systems. In accordance, Carlsson relates how various actors contribute to fulfilling the functions in an innovation system (Chapter 12). Also, multi-actor governance may involve other classifications of actors than as done above, and gender may be one important dimension, as discussed by Lindberg (Chapter 11). Accordingly, evolution cannot be orchestrated or planned to any large degree. Visualizing a multidimensional model of innovation, Etzkowitz and Ranga point to the importance of an intertwining of what they call "spheres and spaces" in order to reach full potential for such evolution paths (Chapter 3).

Index

Page numbers in *italics* refer to tables or figures